Roger H. Hildebrand

INTERSTELLAR MATTER

INTERSTELLAR MATTER

Proceedings of the Second Haystack Observatory Meeting

A Symposium on Interstellar Molecules and Grains in Honor of Alan H. Barrett

Cambridge, Massachusetts, USA

10–12 June 1987

Edited by

James M. Moran

Paul T. P. Ho

Harvard–Smithsonian Center for Astrophysics

Cambridge, Massachusetts, USA

GORDON AND BREACH SCIENCE PUBLISHERS

New York London Paris Montreux Tokyo Melbourne

Gordon and Breach Science Publishers

Post Office Box 786
Cooper Station
New York, New York 10276
United States of America

3-14-9, Okubo
Shinjuku-ku, Tokyo
Japan

Post Office Box 197
London WC2E 9PX
England

Private Bag 8
Camberwell, Victoria 3124
Australia

58, rue Lhomond
75005 Paris
France

Library of Congress Cataloging-in-Publication Data

Haystack Observatory Meeting (2nd : 1987 : Cambridge, Mass.)
Interstellar matter : proceedings of the Second Haystack Observatory Meeting : a symposium on interstellar molecules and grains in honor of Alan H. Barrett, Cambridge, Massachusetts, USA, 10-12 June 1987 / edited by James M. Moran, Paul T.P. Ho.
p. cm.
Includes indexes.
ISBN 0-677-21910-5
1. Interstellar matter—Congresses. 2. Cloud physics—Congresses. 3. Cosmochemistry—Congresses. 4. Molecules—Congresses. I. Barrett, Alan H. (Alan Hildreth) II. Moran, James M. III. Ho, Paul T. P., 1951– IV. Haystack Observatory. V. Title.
QB495.H39 1987 523.1′135—dc 19 88-4336

Alan H. Barrett

Dedication

This volume is dedicated to Alan H. Barrett: scientist, teacher, and mentor to a generation of students at M.I.T. Barrett was raised in Springfield, Massachusetts, and spent two years in the Navy after graduating from high school in 1944. He attended Purdue University and obtained his Ph.D. degree under the guidance of Charles Townes at Columbia University in 1956 with a thesis on microwave spectroscopy. He recognized the bright future in the fledgling but rapidly growing field of radio astronomy, and began his search for interstellar OH as a postdoctoral fellow at the Naval Research Laboratory. He then moved to the University of Michigan, and finally settled at M.I.T. in 1961 as associate professor of electrical engineering. Within a few years, he established research programs in remote sensing of the atmosphere with microwave radiometers, which eventually led to whole earth surveillance programs from satellites; and in radio astrophysical spectroscopy of molecules, a field he created. He earned tenure, moved to the physics department, and raised a family. Many of his achievements are described in detail in this volume. Most notable among them are the discovery of limb darkening of Venus, which proved that the planet had a hot surface and warm atmosphere, and the development of a successful model to explain the greenhouse effect in the atmosphere; the discovery of the radio lines of interstellar OH; the development of VLBI for studying masers; and the discovery of OH masers in late type stars.

During his 25 years at M.I.T., Barrett invested a great deal of his energy in his teaching. From the list of his 19 Ph.D. students and their thesis titles on the following pages, one can see the breadth of Barrett's interests. He was in many regards an ideal mentor, critically involved in the research of his students, allowing them room to grow and formulate their own ideas. In the classroom, he was an inspiring teacher – clear and perceptive in his presentations. In the laboratory and at the telescope, he led by his example and efforts. He infected his students with enthusiasm and trained them in his meticulous style. His unique blend and balance of challenge, exhortation, and trust extracted the best from each of his students. In his retirement, we as his students marvel that he is once again a step ahead of us, with a twinkle in his eye. We wish him Godspeed.

Ph.D. Students of Alan Barrett

David H. Staelin *The Microwave Spectrum of Venus*	1965	(EE)
William B. Lenoir *Remote Sounding of the Upper Atmosphere by Microwave Measurements*	1965	(EE)
Ronald J. Allen *Observations of Several Discrete Radio Sources at 3.64 and 1.94 Centimeters*	1967	(P)
Norman E. Gaut *Studies of Atmospheric Water Vapor by Means of Passive Microwave Techniques*	1967	(M)
Alan E. E. Rogers *Emission and Absorption of Microwave Radiation by Interstellar OH*	1967	(EE)
James M. Moran *Interferometric Observations of OH Emission*	1968	(EE)
John A. Ball* *Observations of OH Radio Emission Sources*	1969	(A)
Thomas H. Wilheit *Studies of Microwave Emission and Absorption by Atmospheric Oxygen*	1970	(P)
William J. Wilson *OH Radio Emission Associated with Infrared Stars*	1970	(EE)
Philip R. Schwartz *Radio Frequency Spectral Lines Associated with Infrared Objects*	1971	(P)

*Thesis supervisors were Barrett, A. E. Lilley, and M. L. Meeks

Philip C. Myers 1972 (P)
Microwave Observations of Molecular Spectral Lines in Galactic Clouds

Kai S. Lam 1976 (P)
Application of Pressure Broadening Theory to Atmospheric Microwave Absorption

Robert N. Martin 1976 (P)
Microwave Spectral Lines in Galactic Dust Globules

Paul T. P. Ho 1977 (P)
Studies of the Ammonia Molecule in the Interstellar Medium

Matthew H. Schneps 1979 (P)
Stellar Winds in the Interstellar Medium: Elephant Trunk Globules and Bubble Nebulae

Richard B. Buxton 1981 (P)
A Microwave Study of Regions of Small Scale Star Formation

John T. Armstrong 1983 (P)
Molecular Clouds in the Region of the Galactic Center

James M. Jackson 1985 (P)
Radio Observations of Starburst Galaxies

David C. Murphy 1985 (P)
A Comparative Study of Regions of High and Low Mass Star Formation

---

EE = Electrical Engineering Department, M.I.T.
P = Physics Department, M.I.T.
M = Meteorology Department, M.I.T.
A = Astronomy Department, Harvard University

Contents

Preface

The papers in this volume were presented at the Second Haystack Observatory Conference entitled "Interstellar Matter," which was held at M.I.T. in Cambridge, Massachusetts, on June 10–12, 1987 in honor of Alan H. Barrett, Professor of Physics at M.I.T., who retired this year. One hundred and eleven scientists from seven countries attended the scientific sessions. The conference opened with addresses by Kenneth A. Smith, Associate Provost and Vice President for Research at M.I.T., and Joseph E. Salah, Director of Haystack Observatory. The scientific sessions on June 10 and 11 consisted of 13 invited papers and 42 poster papers on the topics: Molecules as Probes of Cloud Chemistry; Molecules as Probes of Cloud Physics; Grains and Large Molecules; and New Techniques for Studying Interstellar Matter. The last day of the conference consisted of 14 talks on Barrett's career as a scientist, colleague, and teacher, and a discussion of the ramifications and implications of his work. Many of his associates from M.I.T. and colleagues and friends were in attendance, along with Ginny Barrett, his wife, and their two children, Rick and Bonnie, as well as Nan and Garry Hough, long-time friends of the Barretts.

The conference included a reception on June 9, and a visit to Haystack Observatory and a cookout on June 10. On June 11, a banquet was held, after which Charles Townes gave a talk on Barrett's career. The pictures from this talk, which are reprinted in this volume, were provided by Ginny Barrett and by the M.I.T. Lincoln Laboratory. A reception was held at the close of the conference. The frontispiece of Alan Barrett and the other candid photographs shown here were taken at that event by Barry Heatherington of the M.I.T. News Office.

We thank Joe Salah for his enthusiastic support and assistance with all aspects of the meeting. Philip Myers, with the help of the organizing committee, coordinated the scientific and reminiscence programs and related activities. He also organized a book of mementos presented to Al at the conference. Kathleen McCue, Al's long-time secretary, and Kathleen Walsh of Haystack Observatory made most of the local arrangements for the conference. We thank Charles Townes for thoughtful insights on Al's life and career. Finally, we thank Carolann Barrett and Cristina Brigham, whose hard work and attention to detail greatly aided in the preparation of the conference proceedings.

J. M. Moran
P. T. P. Ho

Organizing Committee

Bernard Burke
George Clark
Aubrey Haschick
Paul Ho
Kathleen McCue
James Moran
Philip Myers (chair)
Joseph Salah
Philip Schwartz
Lewis Snyder
David Staelin
Patrick Thaddeus
Kathleen Walsh

Foreword

It is with great pleasure that we dedicate these proceedings of the Second Haystack Observatory Meeting on *Interstellar Matter* to our colleague Alan H. Barrett, on the occasion of his retirement. The meeting was held at the Massachusetts Institute of Technology in Cambridge, Massachusetts, on June 10–12, 1987.

I wish to express my appreciation to the scientific and local organizing committees, and in particular to Phil Myers, Paul Ho, Jim Moran, Kathleen Walsh, and Kathy McCue for their dedicated efforts in insuring the success of this meeting.

I also wish to acknowledge the support provided for this meeting by the National Science Foundation, the Northeast Radio Observatory Corporation, the MIT Department of Physics, and the Associate Provost and Vice President for Research at MIT.

Joseph E. Salah

Haystack Observatory

List of Participants

Jonathan Allen, MIT, Cambridge, MA
Ronald J. Allen, University of Illinois, Urbana, IL
J. Thomas Armstrong, I. Physikalisches Inst., Cologne, FRG
Willem A. Baan, Arecibo Observatory, Arecibo, PR
John A. Ball, Haystack Observatory, Westford, MA
John Bally, AT&T Bell Labs, Holmdel, NJ
Thomas M. Bania, Boston University, Boston, MA
Alan H. Barrett, MIT, Cambridge, MA
Sara C. Beck, Northeastern University, Boston, MA
Cecilia Benton, Wellesley College, Wellesley, MA
Victoria E. Buch, Center for Astrophysics, Cambridge, MA
Bernard F. Burke, MIT, Cambridge, MA
Sylvie Cabrit, University of Massachusetts, Amherst, MA
Kenneth L. Carr, M/A-Com, Inc., Burlington, MA
Sean Casey, Yerkes Observatory, Williams Bay, WI
Wen Ping Chen, State University of New York, Stony Brook, NY
Albert Cohen, Electronic Space Systems Corp., Concord, MA
Brian E. Corey, Haystack Observatory, Westford, MA
Alexander Dalgarno, Center for Astrophysics, Cambridge, MA
Thomas M. Dame, Center for Astrophysics, Cambridge, MA
Hélène Dickel, University of Illinois, Urbana, IL
Dennis Downes, IRAM, Grenoble, France
John W. Dreher, MIT, Cambridge, MA
Johanna Eisenberger, Electronic Space Systems Corp., Concord, MA
Moshe Elitzur, Goddard Space Flight Center, Greenbelt, MD
Bruce G. Elmegreen, IBM, Yorktown Heights, NY
Gregory Engargiola, University of Chicago, Chicago, IL
Vladimir Escalante, Center for Astrophysics, Cambridge, MA
Margaret A. Frerking, Jet Propulsion Laboratoy, Pasadena, CA
Jerome I. Friedman, MIT, Cambridge, MA
Ronald S. Friedman, Center for Astrophysics, Cambridge, MA
Guido Garay, Universidad de Chile, Santiago, Chile
J. Antonio Garcia-Barreto, UNAM, Ensenada, Mexico
Alfred E. Glassgold, New York University, New York, NY
Paul F. Goldsmith, University of Massachusetts, Amherst, MA
Dinos P. Gonatas, University of Chicago, Chicago, IL
Alyssa A. Goodman, Center for Astrophysics, Cambridge, MA
Carl A. Gottlieb, Harvard University, Cambridge, MA
Roland Gredel, Center for Astrophysics, Cambridge, MA

Lincoln J. Greenhill, Center for Astrophysics, Cambridge, MA
Tor Hagfors, Cornell University, Ithaca, NY
Aubrey D. Haschick, Haystack Observatory, Westford, MA
Michael G. Hauser, Goddard Space Flight Center, Greenbelt, MD
Eric Herbst, Duke University, Durham, NC
Jacqueline N. Hewitt, Haystack Observatory, Westford, MA
Mark H. Heyer, Carnegie Institution of Washington, Washington, DC
Roger H. Hildebrand, University of Chicago, Chicago, IL
Richard E. Hills, Cavendish Laboratory, Cambridge, UK
Paul T. P. Ho, Center for Astrophysics, Cambridge, MA
David J. Hollenbach, NASA Ames, Moffett Field, CA
William E. Howard III, Naval Space Command, Dahlgren, VA
Richard P. Ingalls, Haystack Observatory, Westford, MA
William M. Irvine, University of Massachusetts, Amherst, MA
James M. Jackson, University of California, Berkeley, CA
Eric R. Keto, Center for Astrophysics, Cambridge, MA
Yong Ha Kim, State University of New York, Stony Brook, NY
Edwin F. Ladd, Center for Astrophysics, Cambridge, MA
William D. Langer, Princeton University, Princeton, NJ
Dariusz C. Lis, University of Massachusetts, Amherst, MA
Irene Little-Marenin, Hanscom Air Force Base, Bedford, MA
Robert N. Martin, Steward Observatory, Tucson, AZ
Robert D. Mathieu, Center for Astrophysics, Cambridge, MA
John S. Mathis, University of Wisconsin, Madison, WI
Cornell H. Mayer, Naval Research Lab, Washington, DC
Gordon C. McIntosh, University of Massachusetts, Amherst, MA
M. Littleton Meeks, Meeks Associates, Lincoln, MA
Gary J. Melnick, Center for Astrophysics, Cambridge, MA
James M. Moran, Center for Astrophysics, Cambridge, MA
Philip C. Myers, Center for Astrophysics, Cambridge, MA
Nathan Netzer, Princeton University, Princeton, NJ
David Neufeld, Center for Astrophysics, Cambridge, MA
Thomas A. Pauls, Naval Research Lab, Washington, DC
Thomas G. Phillips, Caltech, Pasadena, CA
Stephen R. Platt, University of Chicago, Chicago, IL
C. Read Predmore, University of Massachusetts, Amherst, MA
Mark J. Reid, Center for Astrophysics, Cambridge, MA
Kurt W. Riegel, National Science Foundation, Washington, DC
Suzanne M. Rodday, Brown University, Providence, RI
Alexander Rudolph, University of California, Berkeley, CA
Norman L. Sadowsky, Faulkner Hospital, Jamaica Plain, MA

Joseph E. Salah, Haystack Observatory, Westford, MA
Rudolph E. Schild, Center for Astrophysics, Cambridge, MA
Matthew H. Schneps, Center for Astrophysics, Cambridge, MA
Philip R. Schwartz, Naval Research Lab, Washington, DC
Frank H. Shu, University of California, Berkeley, CA
Ronald L. Snell, University of Massachusetts, Amherst, MA
Lewis E. Snyder, University of Illinois, Urbana, IL
Philip Solomon, State University of New York, Stony Brook, NY
J. Gregory Stacy, Center for Astrophysics, Cambridge, MA
David H. Staelin, MIT, Cambridge, MA
Daryl A. Swade, University of Massachusetts, Amherst, MA
John Szczepanski, MIT, Cambridge, MA
Toshiaki Takano, Toyokawa Observatory, Toyokawa, Japan
Jan A. Tauber, University of Massachusetts, Amherst, MA
Patrick Thaddeus, Center for Astrophysics, Cambridge, MA
Charles H. Townes, University of California, Berkeley, CA
Barry E. Turner, NRAO, Charlottesville, VA
Ewine van Dishoeck, Center for Astrophysics, Cambridge, MA
Frederick M. Walter, University of Colorado, Boulder, CO
William D. Watson, University of Illinois, Urbana, IL
Anthony J. Weitenbeck, St. Cloud State University, St. Cloud, MN
William J. Welch, University of California, Berkeley, CA
Jerome E. Wiesner, MIT, Cambridge, MA
William J. Wilson, Jet Propulsion Laboratory, Pasadena, CA
Gisbert F. Winnewisser, I. Physikalisches Inst., Cologne, FRG
Adolf N. Witt, University of Toledo, Toledo, OH
H. Alwyn Wootten, NRAO, Charlottesville, VA
Stanley H. Zisk, Haystack Observatory, Westford, MA
Xing-Wu Zheng, Center for Astrophysics, Cambridge, MA
Henry J. Zimmermann, MIT, Cambridge, MA
Lucy M. Ziurys, University of Massachusetts, Amherst, MA

(1) X.W. Zheng, (2) W.A. Baan, (3) M.H. Heyer, (4) S. Cabrit, (5) F.H. Shu, (6) M.G. Hauser, (7) P.F. Goldsmith, (8) C.R. Predmore, (9) W.M. Irvine, (10) D.J. Hollenbach, (11) P. Thaddeus, (12) M.J. Reid, (13) S. Casey, (14) J.E. Salah, (15) S.R. Platt, (16) K. Riegel, (17) T.A. Pauls, (18) T. Takano, (19) T.M. Dame, (20) B.G. Elmegreen, (21) W.D. Watson, (22) P.T.P. Ho, (23) G. Engargiola, (24) M.A. Frerking, (25) R. Gredel, (26) W.E. Howard, (27) M. Elitzur, (28) E. Herbst, (29) S.C. Beck, (30) J. Bally, (31) J.G. Stacy, (32) W.D. Langer, (33) D.P. Gonatas, (34) A.H. Barrett, (35) A.D. Haschick, (36) P.R. Schwartz, (37) J.S. Mathis, (38) H. Dickel, (39) R.H. Hildebrand, (40) C.H. Mayer, (41) R.N. Martin, (42) W.J. Welch, (43) J.A. Tauber, (44) J.A. Ball, (45) L.J. Greenhill, (46) A. Rudolph, (47) N. Netzer, (48) R.S. Friedman, (49) G.J. Melnick, (50) P. Solomon, (51) B.F. Burke, (52) A.N. Witt, (53) D.A. Swade, (54) T. Hagfors, (55) E.F. Ladd, (56) J. Szczepanski, (57) W.J. Wilson, (58) A. Dalgarno, (59) K. McCue, (60) C.H. Townes, (61) B.E. Turner, (62) J.M. Jackson, (63) A.A. Goodman, (64) Y.H. Kim, (65) G. Hough, (66) R.P. Ingalls, (67) A. Cohen, (68) G.C. McIntosh, (69) D.C. Lis, (70) T.G. Phillips, (71) W.P. Chen, (72) R.E. Hills, (73) J.T. Armstrong, (74) G.F. Winnewisser, (75) L.E. Snyder, (76) I. Little-Marenin, (77) P.C. Myers

I. Molecules as Probes of Cloud Chemistry

THE CHEMISTRY OF DIFFUSE CLOUDS

A. DALGARNO
Harvard-Smithsonian Center for Astrophysics, Cambridge, Massachusetts, USA

Abstract The chemistry of diffuse interstellar clouds is summarized and the observational data on the abundances of simple diatomic molecules are discussed with an emphasis on the differences between the measured column densities and the results of theoretical models. Modifications arising from shock-induced formation mechanisms and from a possible significant component of large molecules are considered.

INTRODUCTION

Diffuse interstellar clouds are concentrations of gas and grains extensive enough to absorb detectable amounts of optical radiation from the stars that lie behind them but not so opaque that no radiation is transmitted. Diffuse clouds with measurable column densities of molecules have visual extinctions A_V up to the order of unity corresponding to column densities of hydrogen up to the order 10^{21} cm^{-2}. The column densities are determined from the measured absorption strengths. Their derivation may be complicated by the existence of more than one cloud in the line of sight and by velocity gradients within the clouds.

The chemical composition depends on the interstellar radiation field which varies with depth into the cloud as it is progressively absorbed by the gaseous constituents and by the grains. In order to compare with observation it is necessary to construct models which incorporate the variations of the physical parameters within the clouds.

Information about the density and temperature of the gas and about the intensity of the radiation field to which it is subjected may be obtained from analyses of the measured fine-structure populations of atomic oxygen and carbon and the measured rotational populations of molecular hydrogen and carbon and carbon monoxide. Because it is possible by such considerations to specify with some accuracy the physical conditions in which the molecules are created and destroyed, diffuse clouds provide a critical test of the interstellar chemistry that with appropriate modifications applies to the dense clouds where the complex organic species are found.

The molecules that have been detected in diffuse molecular clouds are the diatomic species H_2, HD, OH, CH, CH^+, CO, C_2, CN and possibly SiO and NO. Upper limits have been obtained for NH and HCl and possibly H_2O and H_2O^+.

CHEMISTRY

Molecular hydrogen is formed by recombination on the surface of grains. Reactions in the gas phase appear to be necessary to reproduce the abundances of the other molecules detected in diffuse clouds. It is convenient to divide the chemistry into an oxygen chemistry, a carbon chemistry and a nitrogen chemistry, though they are not independent and must be considered together in constructing models of the clouds.

Neutral oxygen atoms do not react with molecular hydrogen at the low temperatures characteristic of diffuse clouds and the oxygen chemistry is initiated by cosmic ray ionization to produce H^+ and H_2^+. The H^+ ions undergo charge transfer

$$H^+ + O \rightarrow H + O^+ \qquad (1)$$

to produce O^+. In a diffuse cold cloud, the oxygen atoms reside in the lowest fine-structure level and reaction (1) has an energy deficit equivalent to 232 K. The O^+ ions enter a sequence of abstraction reactions leading to H_3O^+, which then undergoes dissociative recombination

$$H_3O^+ + e \rightarrow H_2O + H \tag{2a}$$

$$\rightarrow OH + H_2 \quad . \tag{2b}$$

The H_2O is destroyed by photodissociation

$$H_2O + h\nu \rightarrow OH + H \tag{3}$$

and by reactions with positive ions. The reaction with C^+,

$$C^+ + H_2O \rightarrow HCO^+ + H \ , \tag{4}$$

leads to CO through

$$HCO^+ + e \rightarrow H + CO \quad . \tag{5}$$

The OH is destroyed by photodissociation

$$OH + h\nu \rightarrow O + H \tag{6}$$

and by reactions with O,

$$O + OH \rightarrow O_2 + H \tag{7}$$

and C^+,

$$C^+ + OH \rightarrow CO + H^+ \tag{8}$$

$$\rightarrow CO^+ + H \quad . \tag{9}$$

If reaction (9) takes place, CO^+ is converted to CO by

$$CO^+ + H_2 \rightarrow HCO^+ + H \tag{10}$$

followed by reaction (5).

The H_2^+ ions react with H_2 to form H_3^+,

$$H_2^+ + H_2 \rightarrow H_3^+ + H \quad . \tag{11}$$

Proton transfer

$$H_3^+ + O \rightarrow OH^+ + H_2 \tag{12}$$

produces OH^+ which reacts further with H_2 to yield again H_3O^+.

The oxygen chemistry reflects the cosmic ray ionization and the abundance of OH provides a quantitative measure of the ionizing flux.

The molecule HD is made, like H_2, on the surface of grains but in addition it is created by the chemical reactions

$$H^+ + D \rightarrow H + D^+ \quad (13)$$

$$D^+ + H_2 \rightarrow HD + H^+ \ . \quad (14)$$

The abundance of HD is also a measure of the cosmic ray flux ζ, but it depends in addition on the cosmic [D]/[H] ratio. The ratio of HD/OH is independent of ζ and proportional to [D]/[H]. HD is removed by photodissociation

$$HD + h\nu \rightarrow H + D \quad (15)$$

and by reactions with H_3^+,

$$H_3^+ + HD \rightarrow H_2D^+ + H_2 \ . \quad (16)$$

In diffuse clouds, the carbon atoms are ionized by the interstellar radiation field and the carbon chemistry is believed to begin with the radiative association of C^+ and H_2,

$$C^+ + H_2 \rightarrow CH_2^+ + h\nu \ , \quad (17)$$

to form CH_2^+. Then either

$$CH_2^+ + e \rightarrow CH + H \quad (18)$$

occurs or

$$CH_2^+ + H_2 \rightarrow CH_3^+ + H \quad (19)$$

followed by

$$CH_3^+ + e \rightarrow CH_2 + H \quad (20)$$

$$\rightarrow CH + H_2 \quad (21)$$

$$CH_2 + h\nu \rightarrow CH + H \ . \quad (22)$$

Photodissociation of CH_3^+

$$CH_3^+ + h\nu \rightarrow \text{products} \quad (23)$$

may contribute to the destruction of CH_3^+. If the electron density is low, CH_3^+ may undergo radiative association

$$CH_3^+ + H_2 \rightarrow CH_5^+ + h\nu \tag{24}$$

to form CH_5^+ which then forms CH_2 through

$$CH_5^+ + e \rightarrow CH_2 + H_2 + H \tag{25}$$

and some methane through

$$CH_5^+ + CO \rightarrow CH_4 + HCO^+ \quad . \tag{26}$$

The molecule CH^+ is formed by photoionization

$$CH + h\nu \rightarrow CH^+ + e \tag{27}$$

and possibly photodissociation

$$CH_3^+ + h\nu \rightarrow CH^+ + H_2 \ , \tag{28}$$

and removed by

$$CH^+ + H \rightarrow C^+ + H_2 \tag{29}$$

$$CH^+ + H_2 \rightarrow CH_2^+ + H \tag{30}$$

$$CH^+ + e \rightarrow C + H \tag{31}$$

$$CH^+ + h\nu \rightarrow C^+ + H \tag{32}$$

$$C + H^+ \quad . \tag{33}$$

An upper limit of 1×10^{-15} $cm^3 s^{-1}$ for the rate coefficient of the initiating reaction has been measured at 10K.[1] If the rate coefficient is chosen so that theoretical models match the measured abundances of CH, a value of 5×10^{-16} cm^3 s^{-1} is obtained.[2]

Carbon monoxide CO is formed by reactions of atomic oxygen with CH, CH^+, CH_2^+ and CH_3^+ and destroyed by photodissociation

$$CO + h\nu \rightarrow C + O \quad . \tag{34}$$

Molecular carbon C_2 is formed by

$$CH + C^+ \rightarrow C_2^+ + H \tag{35}$$

$$C_2^+ + C \rightarrow C_2 + C^+ \tag{36}$$

$$C_2^+ + H_2 \rightarrow C_2H^+ + H \tag{37}$$

$$C_2H^+ + e \rightarrow C_2 + H \tag{38}$$

and destroyed by photodissociation

$$C_2 + h\nu \to C + C \quad . \tag{39}$$

The nitrogen chemistry is uncertain. The proton transfer reaction analogous to (12) is endothermic. The unusual H_2^+ transfer reaction

$$N + H_3^+ \to NH_2^+ + H \quad , \tag{40}$$

which is exothermic, is often invoked but it is likely that it has a substantial activation energy.[3]

Cosmic rays ionize nitrogen atoms but it has been determined recently that

$$N^+ + H_2 \to NH^+ + H \tag{41}$$

is endothermic.[1,4-6] The reaction may nevertheless be a major removal mechanism for the N^+ ions, the main competitive reactions being

$$N^+ + CO \to CO^+ + N \tag{42}$$

$$\to NO^+ + C \quad . \tag{43}$$

If (41) does proceed, the sequence runs through to NH_3. The interstellar molecule CN can be formed by reactions of C^+ with NH, NH_2 and NH_3 to produce HCN^+ and H_2CN^+, followed by dissociative recombination and photodissociation. Additional sources of CN are

$$C_2 + N \to CN + C \tag{44}$$

$$CH + N \to CN + H \tag{45}$$

$$C_2^+ + N \to CN + C^+ \quad . \tag{46}$$

Nitric oxide may be produced by

$$OH + N \to NO + H \tag{47}$$

$$NH + O \to NO + H \tag{48}$$

$$N^+ + O_2 \to NO + O^+ \tag{49}$$

and

$$N^+ + CO \to NO + C^+ \quad . \tag{50}$$

It is destroyed by photodissociation

$$NO + h\nu \rightarrow N + O \tag{51}$$

and reactions with N,

$$NO + N \rightarrow N_2 + O \tag{52}$$

with O

$$NO + O \rightarrow N + O_2 \tag{53}$$

and with C^+

$$NO + C^+ \rightarrow NO^+ + C \quad . \tag{54}$$

Hydrogen chloride is formed from Cl^+, produced by the photoionization of Cl, by the sequence

$$CL^+ + H_2 \rightarrow HCl^+ + H \tag{55}$$

$$HCl^+ + H_2 \rightarrow H_2Cl^+ + H \tag{56}$$

$$H_2Cl^+ + e \rightarrow HCl + H \quad . \tag{57}$$

It is removed by photodissociation and photoionization and by reaction with C^+.

Silicon monoxide SiO is formed and destroyed by essentially the same processes as CO with Si^+ in place of C^+.

CLOUD MODELS

The most elaborate models of diffuse molecular clouds are those of van Dishoeck and Black.[2] In Table 1, I select one of their models and compare the results with the observations towards ζ Ophiuchi.

The agreement for OH is brought about by choosing the cosmic ray flux to be 1.3×10^{-16} s^{-1}. However, the chemistry adopted by van Dishoeck and Black[2] needs to be modified. In contrast to the conventional assumption that the branching ratio of reactions (2a) and (2b) is unity, Bates[7] has argued that dissociative recombination of H_3O^+ leads only to H_2O. The effect will be small because H_2O is rapidly photodissociated to produce OH. More serious may be the choice of 1×10^{-9} cm^3 s^{-1} for the rate coefficient of the sum of reactions (8) and (9). Recent studies of reactions with dipolar molecules suggest values near to 1×10^{-8} cm^3 s^{-1} at 30K.[8,9] The rate coefficient of reaction (7) at low temperatures may be underestimated also.[10] Any deficiency in OH could be remedied by increasing the cosmic ray flux with a consequent reduction in [D]/[H].

The upper limits on H_2O and H_2O^+ are too large to place useful constraints on the chemistry.

The agreement between the model and the measured abundance of HD is brought about by the choice of [D]/[H]= 1.5×10^{-5}. The inferred value of [D]/[H] is consistent with measurements of D/H near the Sun[11] and with values derived from the DCO^+/HCO^+ ratio in dense clouds.[12]

The agreement for CH is obtained by postulating a rate coefficient of 5×10^{-16} cm^3 s^{-1} for reaction (17). The same value works for all four clouds studied by van Dishoeck and Black[2] and is consistent with the experimental upper limit.[1]

The model is successful for C_2 whose existence at the correct abundance was predicted before the molecule was detected.[13] There is a discrepancy for CN but there are still too many uncertainties in the nitrogen chemistry for the discrepancy to be meaningful. The theoretical prediction for NH is very uncertain but it is in any case consistent with the measured upper limit.

The models fail for CH^+ and CO. The failure for CH^+ is of long standing and can probably be remedied only by identifying a new source of CH^+. Elitzur and Watson[15] suggested that the source is shock-heated gas in which the endothermic reaction

$$C^+ + H_2 \rightarrow CH^+ + H \quad . \qquad (58)$$

occurs. In a hydrodynamic shock, it is difficult to avoid a simultaneous overproduction of OH through the neutral reaction[15-17]

$$O + H_2 \rightarrow O + H \quad . \qquad (59)$$

Constraints are also provided by the measured rotational populations of H_2. The difficulties can be overcome by postulating a magnetohydrodynamic shock in which the systematic velocity difference between the ions and the neutral molecules contributes to the kinetic energy of 0.4eV needed for the reaction (57) to occur.[18-20]

By adopting a shock velocity of $9 kms^{-1}$, a pre-shock magnetic field of 4.5μG and a preshock H_2/H ratio of 3, Draine[19] was able to reproduce the CH^+ abundance towards ζ Ophiuchi and in conjunction with the quiescent cloud results of van Dishoeck and Black[2] to obtain rotational populations of H_2 in reasonable agreement with observations. The inferred cosmic ray flux is reduced by the contribution to OH formation in the shocked gas. A value of the ionization rate of 7×10^{-16} s^{-1} gives acceptable agreement for all the clouds considered by van Dishoeck and Black[2] who obtain also a common value of $(1.5 \pm 0.5) \times 10^{-5}$ for [D]/[H].

There exists a general correlation between CH and highly excited rotational levels of H_2[21] which provide further support for the view that CH^+ is an indicator of shocked gas. There should occur velocity differences between molecules produced in the shocked gas and in the cool diffuse cloud. A study of many clouds by Federman[22] indicates that in only half the clouds is there a positive indication of velocity differences. The resolution of the measurements may be inadequate. Recent millimeter observations towards ζ Ophiuchi in emission in the 1-0 and 2-1 rotational lines reveal a much more complex velocity structure than that indicated by the optical absorption line studies.[23, 24]

There remain problems in explaining the observed line widths[23,24] and Langer, Glassgold and Wilson[24] have proposed that some CH^+ may be produced by cloud-cloud collisions in which the translation motion provides the 0.4eV needed to drive reaction (58).

An alternative to shocked gas has been developed for the Pleiades by White[25], who constructed a model of warm gas heated by photodissociation by a nearby early type star.

The various models do not take explicit account of the reactions of C^+ with vibrationally excited hydrogen[26]

$$C^{+} + H_2(v \geq 2) \rightarrow CH^{+} + H \quad . \qquad (60)$$

Laboratory studies[27] have confirmed that reaction (60) is rapid at low temperatures.

The shock hypothesis does not remove the discrepancy for CO. It had been conjectured that an accurate treatment of the photodissociation of CO would resolve it. The work of Letzelter et al.[28] has greatly clarified the physics of CO. It appears to lead to an enhanced destruction rate[29,30] and so enlarge the discrepancy. Shielding of CO by absorption in the broad lines of H and H_2 reduces the rate but probably not enough.[29,30] An increase in the rate coefficients of reactions (8) and (9) helps but also probably not enough, a major source of OH being the sequence initiated by reaction (12). Unless the adopted temperature and density distributions are seriously in error, an additional source of CO appears to be required.

It has been suggested that polycyclic aromatic hydrocarbons exist in the diffuse interstellar gas with an abundance ratio relative to hydrogen between 10^{-7} and 10^{-6}.[31,32] In the diffuse molecular clouds, the PAH molecules will be distributed between PAH^{+}, PAH and PAH^{-}.[33,34] The possibility arises that reactions of atomic oxygen with any of these forms is a source of CO. However, the PAH molecules are removed by such reactions and cannot be a permanent source. Indeed, Duley and Williams[35] have argued that reactions with atomic oxygen and atomic hydrogen destroy PAH molecules so rapidly that they cannot survive in the abundances that have been advocated.

If somehow large molecules are present in the diffuse molecular clouds, they will modify the ratios of ionized to neutral constituents such as C^{+}/C, Mg^{+}/Mg and Fe^{+}/Fe through the reactions[34]

$$I^{+} + LM \rightarrow I + LM^{+} \qquad (61)$$

$$I^{+} + LM^{-} \rightarrow I + LM \quad . \qquad (62)$$

Table 2 reproduces the observational data on C^+/C for four clouds and compares them with the predictions of the models of van Dishoeck and Black.[2] The discrepancies could be ascribed to observational uncertainties but they may be removed by postulating an LM abundance ratio of about 2×10^{-7}. Detailed calculations are in progress[36] which should at least place an interesting upper limit on the LM abundances.

Recently two more molecules have been added to the list of those seen in diffuse clouds. Using IUE data, Pwa and Pottasch[37] claim to detect SiO and NO towards ζ Ophiuchi with column densities of 1×10^{13} cm^{-2} and 2.6×10^{13} cm^{-2} respectively. There are major uncertainties in the reaction scheme and the rate coefficients for the formation and destruction of SiO.[38] Within these uncertainties, 1×10^{13} cm^{-2} is large but not impossible. For NO, the main sources are reactions (47)-(5) and the major sinks are (51)-(54). Photodissociation in the normal unattenuated radiation field proceeds at a rate of 4×10^{-10} s^{-1}.[39] Even with the most optimistic assumptions about the chemical reactions, it is difficult to obtain a column density in excess of 10^{11} cm^{-2}.[40] Shocked gas is unlikely to produce much NO, so that if the IUE identification of NO is correct, gas phase chemistry is inadequate.

ACKNOWLEDGMENT

This work was supported by the National Science Foundation, under Grant AST-86-17675.

TABLE I Calculated and observed molecular abundances seen towards ζ Ophiuchi.*

	OH	H_2O	H_2O^+
Observed	4.8±0.5 (13)	<2.2 (13)	<1.5 (13)
Model	4.5 (13)	5.5 (11)	1.4 (11)
	HD	CH	CH^+
Observed	2.1±0.2 (14)	2.5±0.1 (13)	2.9±0.1 (13)
Model	2.9 (14)	2.1 (13)	3.2 (10)
	C_2	CO	NH
Observed	1.5±0.2 (13)	2.0±0.3 (15)	<7.5 (12)
Model	1.7 (13)	1.3 (14)	3.8 (11)
	CN	HCl	
Observed	2.5±0.1 (12)	<4.5 (11)	
	1.2 (12)	3.0 (11)	

*Column densities in units of cm^{-2}.

TABLE II Observed and model abundances of C and C^+ towards four stars*

	ζ Per		ζ Oph	
	Observed	Model	Observed	Model
C	3.3±0.4 (15)	3.3 (15)	3.2±0.6 (15)	3.1 (15)
C^+	3.0+1.0 (17)	6.7 (17)	0.93±0.45 (17)	4.5 (17)
	χ Oph		o Per	
	Observed	Model	Observed	Model
C	1-10 (15)	3.4 (15)	3.3-6.4 (15)	3.6 (15)
C^+	0.3-4 (17)	8.7 (17)	0.1-2.0 (17)	6.1 (17)

*From reference 2.

REFERENCES

1. J. A. Luine and G. H. Dunn, Ap. J. Lett., 299, L67 (1985).
2. E. F. van Dishoeck and J. H. Black, Ap. J. Suppl., 62, 109 (1986).
3. E. Herbst, D. J. DeFrees and A. D. McLean, Ap. J. in press (1987).
4. N. G. Adams and D. Smith, Chem. Phys. Lett. 117, 67 (1985).
5. J. B. Marquette, B. R. Rowe, G. Dupeyrat and E. Roueff, Astr. Ap., 147, 115 (1985).
6. K. M. Irvin and P. B. Armentrout, J. Chem. Phys., 86, 2659 (1987).
7. D. R. Bates, Ap. J. Lett., 306, L45 (1986).
8. J. B. Marquette, B. R. Rowe, G. Dupeyrat, G. Poissant and C. Rebrion, Chem. Phys. Lett., 122, 431 (1985).
9. D. C. Clary, D. Smith and N. G. Adams, Chem. Phys. Lett., 119, 320 (1985).
10. D. C. Clary and H.-J. Werner, Chem. Phys. Lett., 112, 346 (1984).
11. A. Vidal-Madjar, A. and Gry, C., Astr. Ap., 138, 285 (1984).
12. A. Dalgarno and S. Lepp, Ap. J. Lett., 287, L47 (1984).
13. J. H. Black and A. Dalgarno, Ap. J. Suppl., 34, 405 (1977).
14. A. Dalgarno, in Atomic Processes and Applications, edited by P. G. Burke and B. L. Moiseiwitsch (North Holland, Amsterdam,

1976), Chap. 5, pp. 110-132.
15. M. Elitzur and W. D. Watson, Ap. J., 236, 172 (1980).
16. J. B. A. Mitchell and G. D. Watt, Ast. Ap., 151, 121 (1985).
17. M. Graff and A. Dalgarno, Ap. J., 317, 432 (1987).
18. B. T. Draine and N. Katz, Ap. J., 310, 392 (1986).
19. B. T. Draine, Ap.J., 310, 408 (1986).
20. G. Pineau des Forets, D. R. Flower, T. W. Hartquist and A. Dalgarno, M.N.R.A.S., 220, 801 (1986).
21. D. L. Lambert and A. C. Danks, Ap. J., 303, 401 (1986).
22. S. R. Federman, in Astrochemistry, edited by M. S. Vardya and S. P. Tarafdar (Reidel, Dordrecht, 1987), pp. 123-132.
23. R. M. Crutcher and S. R. Federman, Ap. J. Lett., 316, L71 (1987).
24. W. D. Langer, A. E. Glassgold and R. W. Wilson, Ap. J., 332, 450, (1987).
25. R. E. White, Ap. J., 284, 695 (1984).
26. T. P. Stecher and D. A. Williams, Ap. J. Lett., 177, L141 (1974); M.N.R.A.S., 168, 51P (1974).
27. M. E. Jones, S. F. Barlow, G. B. Ellison and E. F. Ferguson, Chem. Phys. Lett., 130, 218 (1986).
28. C. Letzelter, M. Eidelsberg, F. Rostas, J. Breton and B. Thieblemont, Chem. Phys. 114, 273 (1987).
29. G. A. Mamon, A. E. Glassgold and P. J. Huggins, in preparation (1987).
30. E. F. van Dishoeck and J. H. Black, in preparation (1987).
31. A. Leger and J. L. Puget, Astr. Ap. 137, L5 (1984).
32. M. K. Crawford, A. G. E. M. Tielens and L. J. Allamandola, Ap. J. Lett. 293, L45 (1985).
33. A. Omont, Astr. Ap. 164, 159 (1986).
34. S. Lepp and A. Dalgarno, Ap. J. in press (1987).
35. W. W. Duley and D. A. Williams, M.N.R.A.S., 219, 859 (1986).
36. S. Lepp, A. Dalgarno, E. F. van Dishoeck and J. H. Black, Ap. J. in press (1987).
37. T. H. Pwa and S. R. Pottasch, Astr. Ap. 164, 116 (1986).
38. J. Turner and A. Dalgarno, Ap. J. 213, 386 (1977).
39. E. F. van Dishoeck in Astrochemistry, edited by M. S. Vardya and S. P. Tarafdar (Reidel, Dordrecht, 1987).
40. S. P. Tarafdar and A. Dalgarno, in preparation (1987).

NEWLY DETECTED MOLECULES IN DENSE INTERSTELLAR CLOUDS

WILLIAM M. IRVINE[*], L.W. AVERY[+], P. FRIBERG[#], H.E. MATTHEWS[+] AND L.M. ZIURYS[*]

[*]FCRAO, University of Massachusetts, Amherst, MA 01003, USA

[+]Herzberg Inst. Astrophysics, NRC, Ottawa, Canada, K1A OR6

[#]Onsala Space Observatory, S-43900 Onsala, Sweden

Abstract The last year or so has seen the identification of several new interstellar molecules, including C_2S, C_3S, C_5H, C_6H, and (probably) HC_2CHO in the cold, dark cloud TMC-1; and the discovery of the first interstellar phosphorous-containing molecule, PN, in the Orion "plateau" source. Further interesting results include the observations of $^{13}C_3H_2$ and C_3HD, and the first detection of HCOOH (formic acid) in a cold cloud.

INTRODUCTION

The chemical complexity in dense interstellar clouds has literally increased by orders of magnitude since the pioneering studies of Weinreb and Barrett, with some 70 molecular species now known in such regions. Moreover, a rapidly growing chemical inventory has been discovered in the cold, dark clouds (Table 1), the current interest in which owes much to the work of Alan Barrett and his colleagues[1]. In the present article we discuss the most recent such detections, with particular emphasis on the cold clouds, and including some of our own unpublished data.

SULFUR-CONTAINING CARBON CHAINS

In 1984 Suzuki et al. reported the detection of an unusual unidentified line in the cold cloud TMC-1.[37] Although it was very strong in that source (T_R=4.2 K), it was present only rather weakly in

TABLE I Interstellar Molecules Found in Cold, Dark Clouds[36,38]

SIMPLE HYDRIDES, OXIDES, SULFIDES, AND RELATED MOLECULES:			
H_2	CO	CS	
NH_3	SO_2	CC	
		OCS	
NITRILES, ACETYLENE DERIVATIVES, AND RELATED MOLECULES:			
HCN	$HC\equiv C-CN$	$H_3C-C\equiv C-CN$	$H_2C=CH-CN$
H_3C-CN	$H(C\equiv C)_2-CN$	$H_3C-C\equiv CH$	HNC
$C\equiv C-CO$	$H(C\equiv C)_3-CN$	$H_3C-(C\equiv C)_2-H$	$HN=C=O$
$C\equiv C-CS$	$H(C\equiv C)_4-CN$	$H_3C-(C\equiv C)_2-CN$?	
$HC\equiv CCHO$?	$H(C\equiv C)_5-CN$		
ALDEHYDES, ALCOHOLS, ETHERS, KETONES, AND RELATED MOLECULES:			
$H_2C=O$	H_3COH	$HO-CH=O$?	
$H_2C=S$		$H_2C=C=O$?	
$H_3C-CH=O$			
CYCLIC MOLECULES:		IONS:	
C_3H_2		CH^+	HCS^+
C_3H		HN_2^+	$HCNH^+$?
		HCO^+	SO^+ ?
RADICALS:			
CH	C_2H	CN	SO
OH	C_3H	C_3N	C_2S
	C_4H		
	C_5H		
	C_6H		

Sgr B2 and was undetectable in other standard sources such as Orion A or IRC+10216. These properties were similar to those of a line at 81.5 GHz observed in TMC-1 during the Onsala spectral scan[2] by two of us, M. Guelin, and Aa. Hjalmarson. The observed properties suggested that both lines might be low-energy transitions of the same molecular species, which was likely to be carbon-rich, given the known properties of TMC-1. In an effort to provide more evidence, we have carried out a search in TMC-1 for additional emission features at frequencies corresponding to known, but weak, unidentified lines in Sgr B2. Two such lines were detected, and are illustrated in Figure 1 together with higher resolution data for the previously known lines. The detection of additional, related lines at lower frequencies by Kaifu et al.[3] and laboratory microwave spectroscopy by Saito et al.[4] have now identified the carrier of all of these features as the $^3\Sigma$ radical C_2S. The distribution of C_2S is illustrated by the spatial-velocity diagram in Figure 2, showing a cut along the TMC-1 ridge. The C_2S emission peaks in the vicinity of the maximum intensity observed for other carbon-chains, such as the cyanopolyynes. Our data suggest a rotational temperature of about 5 K, at the lower end of the range given by Saito et al.[4] Unlike better known sulfur-containing organic molecules such as CS and HNCS, whose abundance relative to their oxygen analogs is approximately in the cosmic sulfur-to-oxygen ratio, the abundance of C_2S in TMC-1 is more than an order of magnitude greater than the upper limit obtained for the C_2O abundance[5], and thus is almost three orders of magnitude greater than the cosmic S/O ratio.

Another three strong unidentifed lines in TMC-1 were observed in the same spectral scan by Kaifu et al.[3] which led ultimately to the identification of C_2S. Subsequently, the same laboratory spectroscopic procedure used to study C_2S resulted in the identification of the carrier of the new lines as the linear molecule C_3S.[6] Taken together, these results establish the existence of a new series of sulfur-containing carbon-chain molecules with the general formula C_nS in the dark clouds in Taurus. As with C_2S, C_3S is vastly more abundant than its oxygen analog, C_3O, which itself had only recently been identified in the interstellar

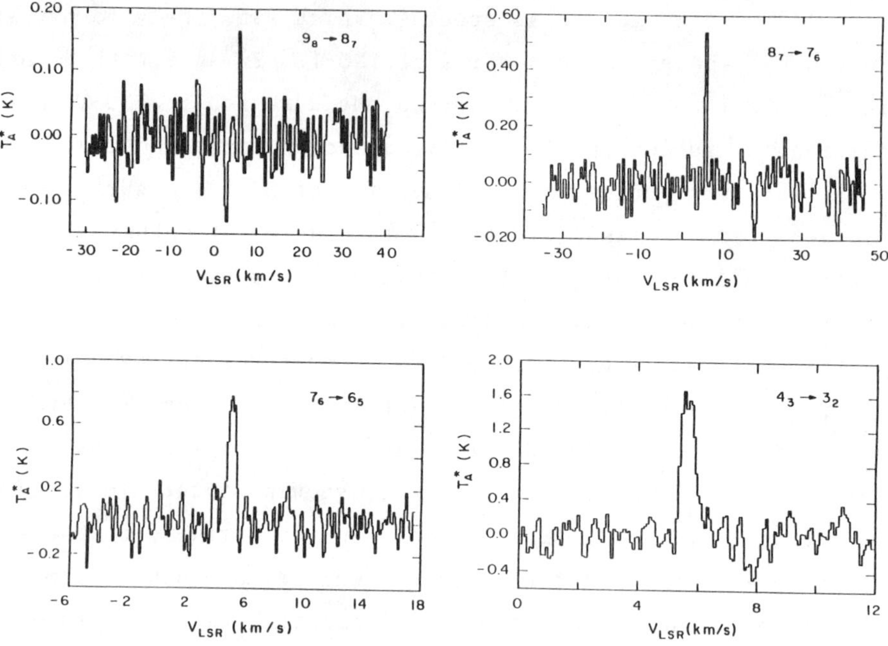

FIGURE 1 Four transitions of the newly identified radical C_2S, observed at the cyanopolyyne peak in the cold cloud TMC-1 with the FCRAO 14 m antenna. Transitions labeled with quantum numbers J_N. Velocity scales have not been corrected to the accurate frequencies determined by Saito et al.[4] Spectral resolution is 100 kHz (9_8-8_7 and 8_7-7_6), 25 kHz (7_6-6_5), and 12.5 kHz (4_3-3_2).

medium[7]. E. Herbst (private communication) has suggested that the fundamental reason for these apparently anomalously high abundances may be related to the failure of the S^+ ion to react with molecular hydrogen, leaving it free to react with acetylene and higher-order carbon chains to insert sulfur in such molecules. In contrast, O^+ reacts with H_2 as part of a scheme which leads ultimately to OH and H_2O.

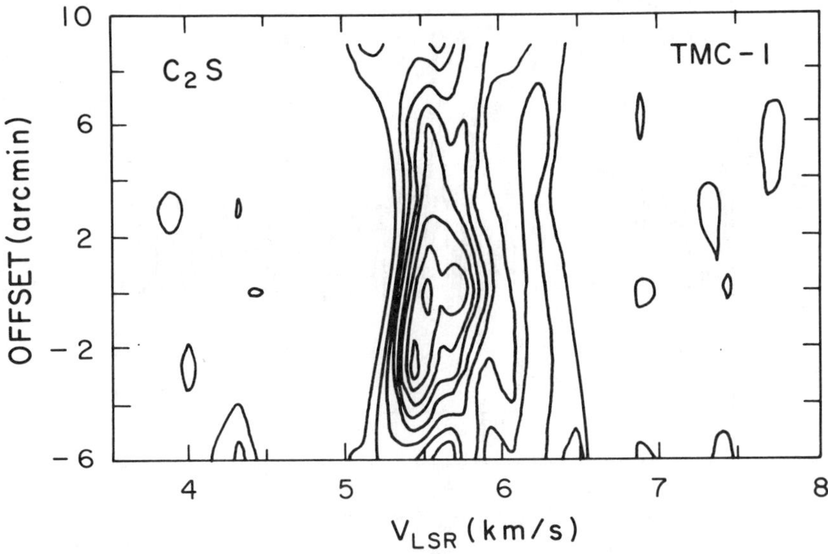

FIGURE 2 Spatial-velocity diagram along the TMC-1 ridge showing emission from the 4_3-3_2 transition of C_2S, as observed at FCRAO. The (0,0) position is the cyanopolyyne peak, RA = $4^h38^m38^s$, DEC = $25^o35'45''$ (1950). Velocity scale should be incremented by 0.3 km/s to correspond with the measured frequency of 45379.046 MHz.[4]

NEW OXYGEN-CONTAINING ORGANIC MOLECULES

The identification and abundance determination for tricarbon monoxide in TMC-1 illustrated the importance of symbiotic interaction between quantum chemistry, laboratory molecular spectroscopy, and radio astronomical observations[8]. The principal scheme suggested for the production of C_3O leads through the $H_3C_3O^+$ ion via dissociative electron recombination:

$$H_3C_3O^+ + e \rightarrow C_3O + H_2 + H \quad . \tag{1}$$

Although there is no laboratory data for the branching ratios of such recombination, it seems likely that an additional product of

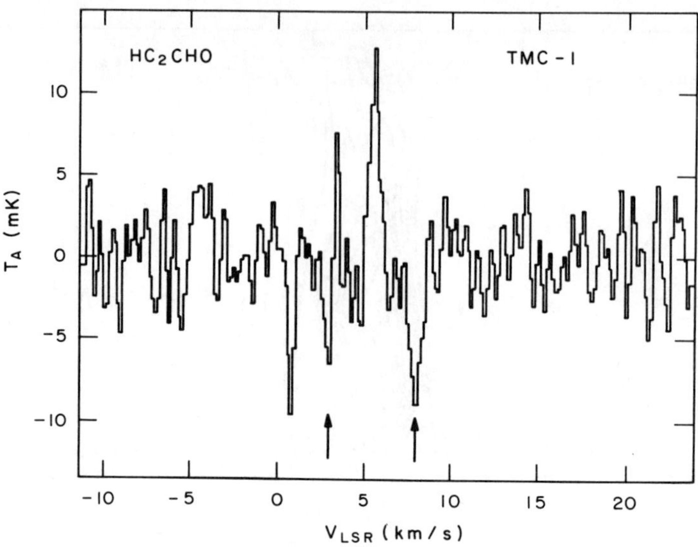

FIGURE 3 Line assigned to the 2(0,2)-1(0,1) transition of propynal (HC_2CHO), observed in TMC-1 with the NRAO 43 m telescope. Spectral resolution is 20 kHz, arrows indicate frequency-switching artifacts, position is (0,0) location in Figure 2, and frequency is from Winnewisser.[9]

reaction (1) would be H_2C_3O + H. The former molecule has more than one isomeric form, but one stable species is certainly propynal (HC≡CCHO). Propynal had been suggested as a possible interstellar species by Winnewisser as far back as 1973[9], and an upper limit on its abundance in TMC-1 had been obtained at Onsala by Askne[10]. As part of a search for new oxygen-containing molecules in cold clouds, we have recently detected a line in TMC-1 which we assign to the 2(0,2)-1(0,1) transition of HC_2CHO (Figure 3). Although with the high sensitivity of current radio astronomical receivers one must be aware of the possibility of chance coincidences, we feel quite confident that the assignment is correct. For the standard TMC-1 velocity of 5.8 km s^{-1}, the observed line

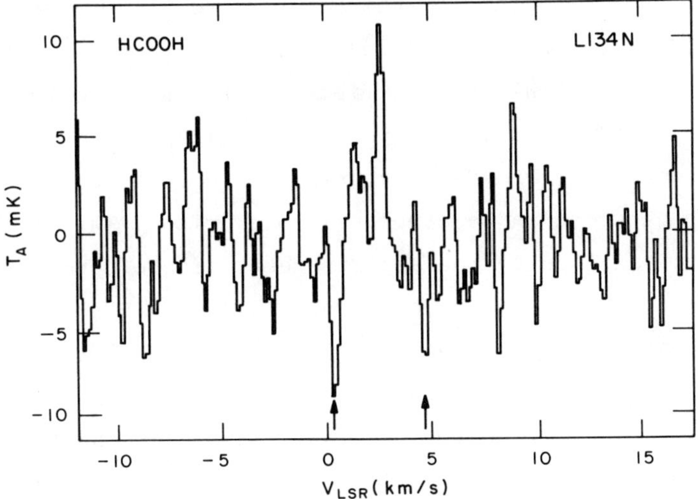

FIGURE 4 Line assigned to the 1(0,1)-0(0,0) transition of formic acid[16] (HCOOH), observed in L134N with the NRAO 43 m telescope. Position is RA = $15^h51^m30^s$, DEC = -2 43'30" (1950); observing procedure as in Figure 3.

agrees with the measured frequency within the laboratory uncertainity[9], or 5 parts in 10^7. To verify our assignment, we have sought the 2(1,1)-1(1,0) line and found a weak (2-3σ) feature at the corresponding frequency. This putative feature is only about 1/3 as strong as would be predicted on the basis of intrinsic line strengths alone. This weakness can be understood, however, as a result of the structure of HC_2CHO. Because of the presence of components of the electric dipole moment along both the a and b molecular axes, b-type transitions will tend to drain population from energy levels with $K_p \neq 0$ into the $K_p = 0$ stack, in a manner similar[11] to that for HNCO.

In any case, assuming the line in Figure 3 is due to propynal, that it is optically thin, and that the rotational temperature in TMC-1 is between 5 and 10 K, a column density $N(HC_2CHO)$ =

$5 \pm 3 \times 10^{12}$ cm^{-2} is deduced, which may be compared with the value of $1.4 \pm 0.4 \times 10^{12}$ cm^{-2} found for C_3O.

Formic acid (HCOOH) was the first organic acid to be observed in the interstellar medium, in Sgr B2. Curiously enough, until the recent detection of weak features in the spectrum of Orion A,[12] it had been detected in no other molecular cloud. As part of the same project that led to the apparent detection of propynal, we sought formic acid in the cold clouds TMC-1 and L134N. We have detected it at a low level only in L134N (Figure 4). The only published route for HCOOH production within ion-molecule chemistry of which we are aware involves radiative association of HCO^+ and H_2O, followed by dissociative electron recombination. The predicted formic acid abundance[13] is almost two orders of magnitude larger than the observed value in L134N. We have recently been informed by E. Herbst (private communication), however, that this reaction is probably much slower then the published estimate, and that a more likely route produces the precursor ion $CH_3O_2^+$ from the ion-molecule reaction of CH_4 and O_2^+. The latter species may provide a clue as to why formic acid was detected in L134N and not TMC-1. Note that SO and SO_2 also appear to be more abundant in L134N, while most carbon-rich species are considerably more abundant in TMC-1.[14] This suggests that at least portions of TMC-1 may have a significantly larger carbon/oxygen ratio then does L134N. The possibility of gradients in this ratio within L134N has recently been discussed by Swade[15].

NEW HYDROCARBON RADICALS

The series of hydrocarbon radicals C_nH has recently had two new interstellar detections: C_5H and C_6H. The entire series C_nH(n=1,...6), and also the cyclic species C_3H_2, share some interesting properties. They were all observed astronomically before their rotational spectra were measured in the laboratory, and all (except CH and C_2H) were first seen in the envelope of the evolved carbon star IRC+10216, and then subsequently observed in TMC-1. Only C_3H_2 has a $^1\Sigma$ (closed shell) ground electronic state; both C_2H and C_4H are $^2\Sigma$, while CH, C_3H, and C_5H are $^2\Pi_{1/2}$. Surprisingly, C_6H differs from the pattern for the other radicals, in that its ground state is $^2\Pi_{3/2}$ making it similar to the OH radical whose radio frequency spectrum has been described by Barrett[17].

C_5H was first identified from a series of doublets in the spectrum of IRC+10216 observed with the new IRAM 30 meter telescope; the hyperfine structure within these lambda doublets could not be resolved in this broad-line source. Subsequently, observations of TMC-1 with the 100 meter telescope at Effelsberg were successful in detecting hyperfine structure at 21 GHz, and the identification has been confirmed by laboratory spectroscopy[18,19].

Most recently, the C_6H radical has been independently identified in TMC-1 with the Nobeyama Radio Telescope,[20] and by astronomers at IRAM and Effelsberg, who observed first a series of doublets in IRC+10216, and subsequently confirmed the identification by resolving the hyperfine structure in TMC-1.[21,22]

As pointed out by both groups, C_3H and C_5H are some two orders of magnitude less abundant than C_2H and C_4H, while C_6H is also much less abundant then would be predicted by extrapolating the C_2H-C_4H abundances to longer chains as can be done empirically for the cyanopolyynes ($HC_{2n}CN$). Suzuki et al.[20] speculate that the low abundances may reflect the higher reactivity of radicals in the $^2\Pi$ state.

Cyclopropenylidene (C_3H_2) is the first hydrocarbon ring molecule discovered in the interstellar medium[23]. It is very widely distributed in interstellar clouds and is quite abundant in the cold Taurus clouds[24]. Among the most interesting recent results have been the detections of both the 13-carbon and the deuterated isotopic species, resulting in better estimates of the

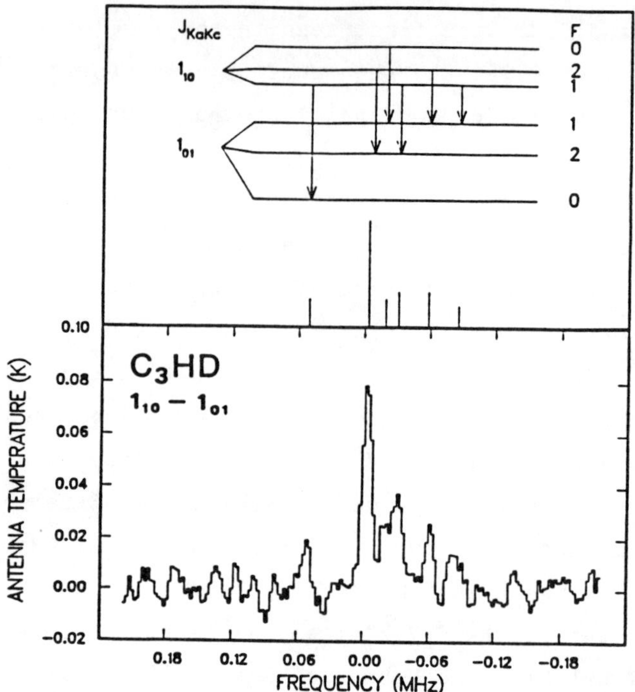

FIGURE 5 (above) Energy level diagram and relative intensities of the six hyperfine components of the 1(1,0)-1(0,1) transition of C_3HD. (below) Spectrum of the dark cloud L1498 obtained with the NRAO 43 m antenna.[30]

abundance of C_3H_2 and the deuterium fractionation [25-28]. It appears that the deuterium fractionation is among the highest ever observed for an interstellar molecular species[29]. In addition, observations of the very cold, quiescent clouds L1498 and TMC-1 have for the first time allowed the hyperfine structure due to the deuterium nucleus to be resolved, as is shown in Figure 5.

A NEW ELEMENT IN THE CHEMISTRY OF DENSE CLOUDS: PHOSPHORUS

In spite of the observation that the element phosphorus is almost undepleted in diffuse clouds, searches for phosphorus-containing

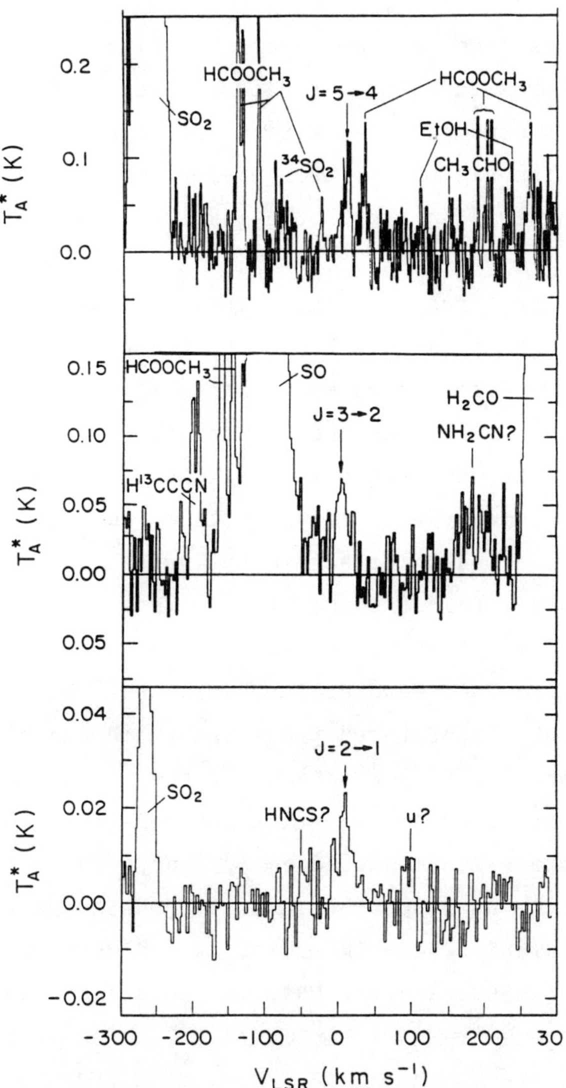

FIGURE 6 Three transitions of phosphorous nitride (J=2-1, J-3-2, and J=5-4) observed toward the Orion molecular cloud with three different FCRAO receivers.[32]

molecules in dense clouds have heretofore been unsuccessful. As evidence increases that some interstellar molecular material survived the formation of the solar nebula and was incorporated into primitive objects such as comets and meteorites[31], the chemical state of phosphorus in molecular clouds becomes of increasing interest. Phosphorus plays a critical role in terrestrial biochemistry, and the history of this element in the solar nebula and subsequently in planetary formation is fundamental to the origin of life.

The first phosphorus-containing interstellar molecule, phosphorous nitride (PN), has now been detected by Ziurys[32] and by Turner and Bally[33], following up on hints in the Owens Valley survey of the one millimeter spectrum of the Orion molecular cloud.[12] Figure 6 shows three transitions of PN observed by Ziurys; the emission appears to arise from the inner edge of the Orion "plateau", a region of hot, dense gas which is outflowing from the embedded infrared source IRc2. This location for PN suggests that, like other molecular species containing elements from the second row of the periodic table, the chemistry of PN may involve "high temperature" processes which can overcome energy barriers that prevent such reactions in the colder, quiescent clouds. The corresponding situation for some silicon and sulfur-containing species has been studied by Ziurys and Friberg[34,35].

This research was supported in part by NASA grant NAGW-436 and NSF grant INT-8512284. The FCRAO is supported in part by NSF grant AST-8512903, and the NRAO is operated by AUI under contract with the NSF.

REFERENCES

1. P. T. P. Ho, R. N. Martin, P. C. Myers, and A. H. Barrett, Ap.J.(Lett.), 215, L29 (1977).
2. L. E. B. Johansson, C. Andersson, J. Ellder, P. Friberg, Aa. Hjalmarson, B. Hogland, W. M. Irvine, H. Olofsson, and G. Rydbeck, Ast. Ap., 130, 227 (1984).
3. N. Kaifu, H. Suzuki, M. Ohishi, T. Miyaji, S. Ishikawa, T. Kasuga, M. Morimoto, and S. Saito, Ap.J.(Lett.), 317, L111 (1987).
4. S. Saito, K. Kawaguchi, S. Yamamoto, M. Ohishi, H. Suzuki, and N. Kaifu, Ap.J.(Lett.), 317, L115 (1987).

5. W. M. Irvine, in Astrochemistry, ed. M. S. Vardya and S. P. Tarafdar (D. Reidel, Dordrecht, 1987), pp. 245-252.
6. S. Yamamoto, S. Saito, K. Kawaguchi, N. Kaifu, H. Suzuki, and M. Ohishi, Ap.J.(Lett.), 317, L119 (1987).
7. H. E. Matthews, W. M. Irvine, P. Friberg, R. D. Brown, and P.D. Godfrey, Nature, 310, 125 (1984).
8. R. D. Brown, P. D. Godfrey, D. M. Cragg, E. H. N. Rice, W. M. Irvine, P. Friberg, H. Suzuki, M. Ohishi, N. Kaifu, and M. Morimoto, Ap.J., 297, 302 (1985).
9. G. Winnewisser, J. Mol. Spec., 46, 16 (1973).
10. W. M. Irvine and Aa. Hjalmarson, in Cosmochemistry and the Origin of Life, ed. C. Ponnamperuma (D. Reidel, Dordrecht, 1983), pp. 113-142, Table 6.
11. W. H. Hocking, M. C. L. Gerry, and G. Winnewisser, Ap.J. (Lett.), 187, L89 (1974).
12. E. C. Sutton, G. A. Blake, C. R. Masson, and T. G. Phillips, Ap.J.(Suppl.), 58, 341 (1985).
13. E. Herbst and C. M. Leung, M.N.R.A.S., 222, 689 (1986).
14. W. M. Irvine, F. P. Schloerb, Aa. Hjalmarson, and E. Herbst, in Protostars and Planets II, ed. D. C. Black and M. S. Matthews (U. Arizona, Tucson, 1985), pp. 579-620.
15. D. A. Swade, Ph.D. Thesis, Univ. Massachusetts (1987).
16. E. Willemot, D. Dangoisse, N. Monnanteuil, and J. Bellet, J. Phys. Chem. Ref. Data, 9, 59 (1980).
17. A. H. Barrett, IEEE Trans. Ant. Prop., 12, 822 (1964).
18. J. Cernicharo, M. Guelin, and C. M. Walmsley, Ast. Ap., 172, L5 (1986).
19. C. A. Gottlieb, E. W. Gottlieb, and P. Thaddeus, Ast. Ap., 164, L5 (1986).
20. H. Suzuki, M. Ohishi, N. Kaifu, S. Ishikawa, and T. Kasuga, Pub. Ast. Soc. Japan, 38, 911 (1986).
21. M. Guelin, J. Cernicharo, C. Kahane, J. Gomez-Gonzalez, and C. M. Walmsley, Ast. Ap., 175, L5 (1987).
22. J. Cernicharo, M. Guelin, K. M. Menten, and C. M. Walmsley, Ast. Ap., 181, L1 (1987).
23. P. Thaddeus, J. M. Vrtilek, and C. A. Gottlieb, Ap.J.(Lett.), 299, L63 (1985).
24. H. E. Matthews and W. M. Irvine, Ap.J.(Lett.), 298, L61 (1985).
25. S. C. Madden, W. M. Irvine, and H. E. Matthews, Ap.J.(Lett.), 311, L27 (1986).
26. J. Gomez-Gonzalez, M. Guelin, J. Cernicharo, C. Kahane, and M. Bogey, Ast. Ap., 168, L11 (1986).
27. M. Gerin, H. A. Wootten, F. Combes, F. Boulanger, W. L. Peters, T. B. H. Kuiper, P. J. Encrenaz, and M. Bogey, Ast. Ap., 173, L1 (1987).
28. M. B. Bell, P. A. Feldman, H. E. Matthews, and L. W. Avery, Ap.J.(Lett.), 311, L89 (1986).

29. M. B. Bell, L. W. Avery, H. E. Matthews, P. A. Feldman, J. K. G. Watson, S. C. Madden, and W. M. Irvine, Ap.J., in press (1987).
30. M. B. Bell, J. K. G. Watson, P. A. Feldman, H. E. Matthews, S. C. Madden, and W. M. Irvine, Chem. Phys. Lett., 136, 588 (1987).
31. E.g., J. F. Kerridge and S. Chang, in Protostars and Planets II, ed. D. C. Black and M. S. Matthews (U. Arizona Press, Tucson, 1985), pp. 738-754.
32. L. M. Ziurys, Ap.J.(Lett.), 321, L81 (1987).
33. B. E. Turner and J. Bally, Ap.J.(Lett.), 321, L75 (1987).
34. L. M. Ziurys, Ap.J., in press (1987).
35. L. M. Ziurys and P. Friberg, Ap.J.(Lett.), 314, L49 (1987).
36. W. M. Irvine and R. F. Knacke, in Origin and Evolution of Planetary and Satellite Atmospheres, ed. S. Atreya, J. Pollack, and M. Matthews (Univ. Arizona, Tucson), in press.
37. H. Suzuki, N. Kaifu, T. Miyaji, M. Morimoto, M. Ohishi, and S. Saito, Ap.J., 282, 197 (1984).
38. S. Yamamoto, S. Saito, M. Ohishi, H. Suzuki, S. I. Ishikawa, N. Kaifu, and A. Murakami, Ap.J.(Lett.), 322, L55 (1987).

THE CHEMISTRY OF SHOCKS AND OUTFLOWS

WM. J. WELCH
Radio Astronomy Laboratory, University of California, Berkeley, CA

Abstract The effects of winds from young stellar objects on the properties of the surrounding gas and dust have been studied in most detail in the Orion A source. The composition shifts from being what one would expect in a mainly carbon rich environment in the large cloud to being sulfur and silicon enhanced and apparently oxygen rich close to the outflow. The observational evidence concerning the composition and physical distributions near the outflow is briefly summarized. Maps of SO in a few other regions suggest that the Orion A results are probably typical.

INTRODUCTION

Since the first discovery of OH in the interstellar medium by Professor Barrett and his colleagues,[1] some 70 additional molecules have been detected in many transitions. On the large scale the chemical compositions of molecular clouds show fairly uniform abundances, and ion–molecule chemical schemes explain the observed abundances of most of the species reasonably well.[2] Near sites of active star formation, where there are energetic outflows from the protostars, there are striking changes in the chemistry. The recent construction of larger single millimeter antennas and interferometers has enabled the greater angular resolution necessary to study these more compact regions. The following will briefly summarize some aspects of this recent work.

LINE SURVEYS OF ACTIVE REGIONS

The detections of new molecules have been reported by a number of active observers (cf. Irvine et al.[3] for a review). Of particular value are the extensive line surveys of both the SGRB2 and Orion A regions. Cummins et al.[4] surveyed SGRB2 over 70–150 GHz with the 7m BTL dish, detecting over 300 lines of 21 molecules. Johansson et al.[5] scanned SGRB2 over the spectral region 70–93 GHz with the Onsala 20m telescope and observed 170 molecular lines. Most recently, there was a survey of Orion A with one of the OVRO 10m telescopes in the range 208–263 GHz which detected over 800 lines.[6,7]

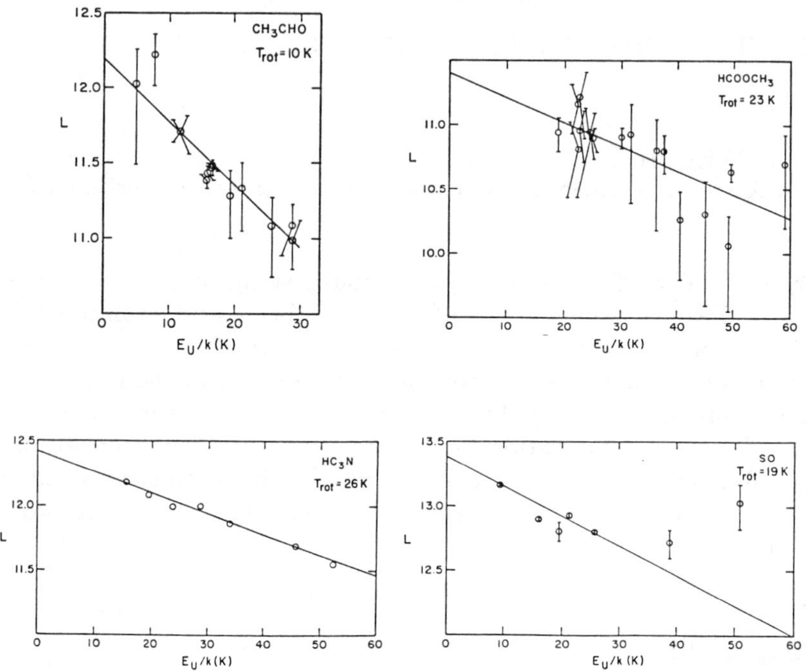

FIGURE 1 Rotation diagrams for molecules in SGRB2.[4]

In many cases, several lines of the same molecule are observed in these surveys, so that rotational analyses can be performed. That is, plots of integrated line strength as functions of rotational energy level can be constructed, and these often are straight lines, showing the levels to be in LTE and the lines to be largely optically thin. These analyses permit accurate determination of rotational temperatures and column depths. Figure 1 is a composite of four such plots taken from the study of Cummins et al..[4] The temperature of HC_3N is particularly well determined, for example. Note that the data for SO show a component of gas at 19K and a substantial amount at very much higher temperatures. This suggests that SO, in contrast to the HC_3N, for example, has a significant abundance close to the warm outflow sources. Note also that for many of these molecules to be in LTE due to collisions requires that the density of the dominant species, H_2, must be greater than $10^6\,cm^{-3}$. Since the average densities cannot be this large, the regions must be very clumpy.

The OVRO study of Orion A has the special advantage of good angular resolution, about 30″, in a nearby region, so that the emission from the material close to the outflow is especially well represented. The other advantage of this survey is that at 1.3 mm wavelength the lines are especially strong and the spectrum is very rich.

Figure 2 shows a multiline map of the SGRB2 region made with the Nobeyama 45m telescope.[8] With the angular resolution of this large instrument, about 15″, enhanced emission from several species is clearly evident toward two regions of current star formation, locations which show compact HII regions, masers, outflows, and IR point sources. The sulfur compounds SO and OCS are prominent as is vibrationally excited HC_3N. This is an example of a general circumstance: In the near neighborhood of hot spots of star formation, the chemistry appears to shift from being weakly ionized carbon rich to neutral oxygen rich, there are strong enhancements in S and Si compounds, and some species show significant vibrational excitation.

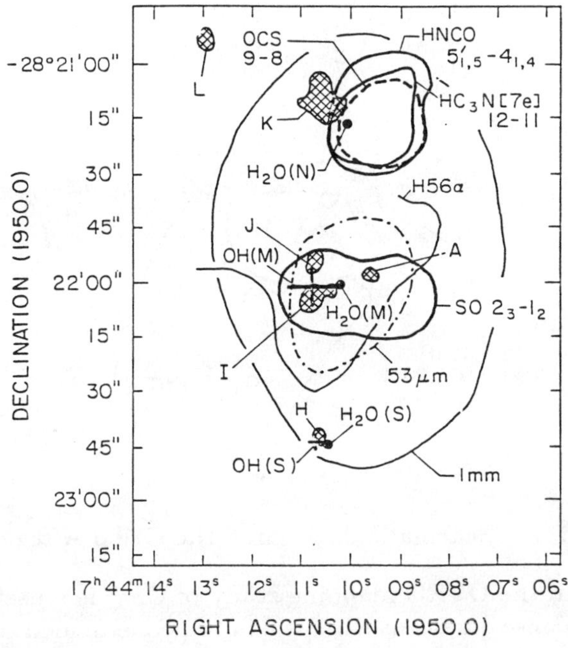

FIGURE 2 Maps of different molecular lines in SGRB2.[8]

THE ORION A STAR FORMING REGION

This is the nearest and best studied region of massive star formation. Early evidence for outflows in Orion A comes from CO spectra showing a broad shoulder of emission from a source less than one minute in diameter, suggesting wind velocities greater than 100 km/sec.[9] Genzel et al.[10] traced the outflow in a proper motion study of water masers. There are a number of point IR sources, of which three are self-luminous: BN, IRC2, and IRC9.[11] Beckwith et al.[12] detected emission from shock excited H_2. Watson et al.[13] have observed CO at temperatures greater than 700K, and Werner et al.[14] detected intense emission from the 63μ fine structure line of OI. The outflow from SiO shows IRC2 to be the current principal source of winds in the region.[15]

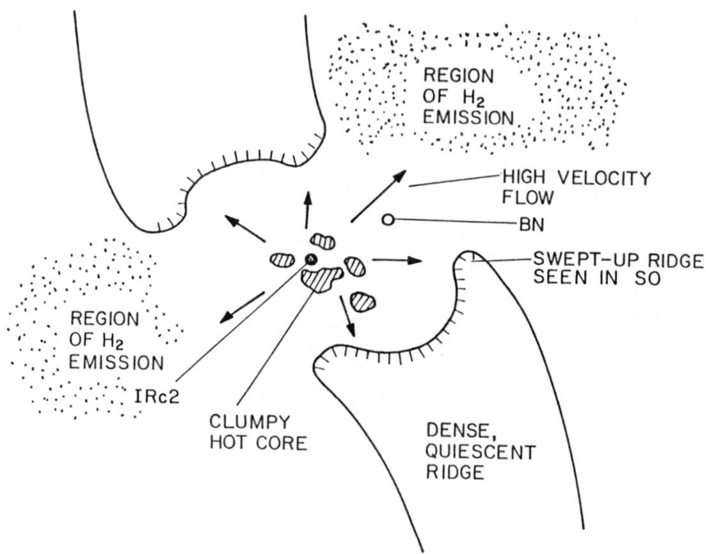

FIGURE 3 Schematic diagram of the Orion A region.[16]

Figure 3 from the OVRO mapping study of CS[16] is a useful schematic of the region. The dense quiescent ridge is a flattened extended structure with composition like that of the cloud as a whole. It's line width is $\simeq$ 4km/sec and molecular hydrogen density 10^{-3}–10^{-4} cm^{-3}, and its molecular rotational temperatures are 50–60K for large dipole moment species. Near IRC2, the winds sweep up material in all directions, driving shocks into the ridge at 15–20 km/sec, and expanding at speeds in excess of 100 km/sec in directions perpendicular to the ridge where emission from (high velocity) shock excited molecular hydrogen is detected. The "hot core"[17] consists of one or more clumps of material close to IRC2 which is evidently too dense to be accelerated by the wind. There are SiO masers in the atmosphere of IRC2.[18] Water masers are found there, in the swept up ridge, and in the higher velocity flow normal to the ridge.[10]

The "plateau source" is a term applied to the gas in the high velocity wings seen in the spectra of many molecules, although it corresponds to different distributions in the different molecules. This is shown by the top three and lower left panels of Figure 4[19,20]. All four panels are maps at a radial velocity of 19 km/sec, well removed from the large scale cloud or "hot core" velocities. The thermal SiO(1–0) is strongly peaked on IRC2. The HCN(1–0) and $SO(2_2{-}1_1)$ are similar to each other and distinct from the SiO. Most different is the ion HCO^+(1–0), which is rather extended. Its distribution is like that of the 2μ H_2 emission and is probably produced in the same shocks. The HCN and SO show a high velocity dispersion, >35 km/sec, and apparently outline the "swept–up ridge", the boundary of the cavity cleared by the winds from IRC2.

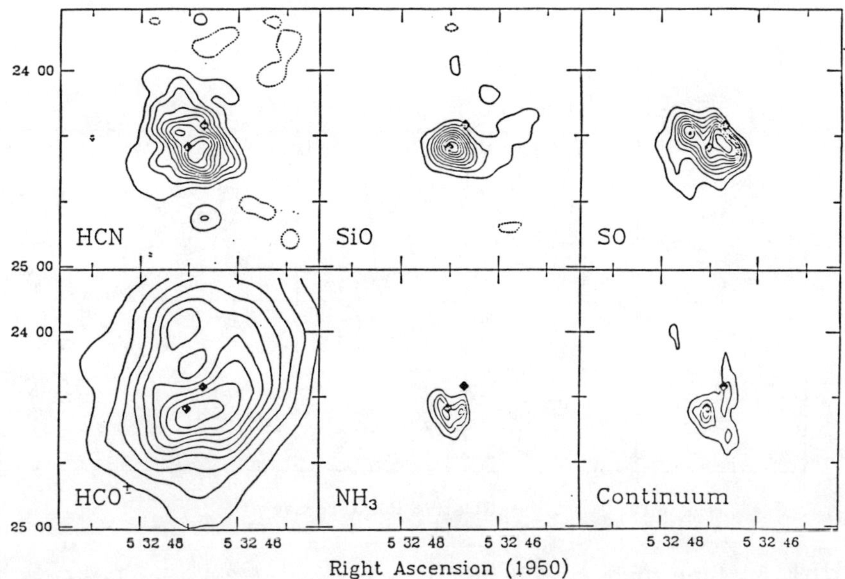

FIGURE 4 Orion A maps of 3mm continuum, integrated ammonia, and other lines at 19 km/sec (VLSR).[19,20]

The "hot core" is hot, with molecular rotational temperatures greater than 200K[7], because of its proximity to IRC2 . Gases show line widths of 15–20 km/sec centered at 7 km/sec. This region is evident in the lower central panel of Figure 4, which is a map of the integrated $NH_3(3,3)$ line.[21] The peak emission is a few seconds south of IRC2, but the material appears to surround IRC2. NH_3 is very enriched here,[22] probably because the high temperature has evaporated it from the grains. It is also very dense, n> 10^7 cm^{-3}, with a high column density of gas and dust. This is apparent from the three millimeter continuum map in the lower right panel of Figure 4.[23,24] Figure 5 shows maps of a number of molecules both at the "hot core" velocity of 7.4km/sec, and at a higher velocity.[25] Whereas the SO_2 is a broad line like SO, outlining the interior and walls of the wind blown cavity, molecules like CH_3CH_2CN are found mainly in the "hot core". This highly saturated molecule appears to be the product of high temperature and high density neutral chemistry in the "hot core" material. Note that the absence of a substantial peak in the ion HCO^+ toward the "hot core" implies that ion chemistry is probably unimportant in this region. HDO is very enriched and also found mostly in this region as shown in Figure 5.[26] It is suggested that, as in the case of the NH_3, it is the result of grain sublimation at high temperature.

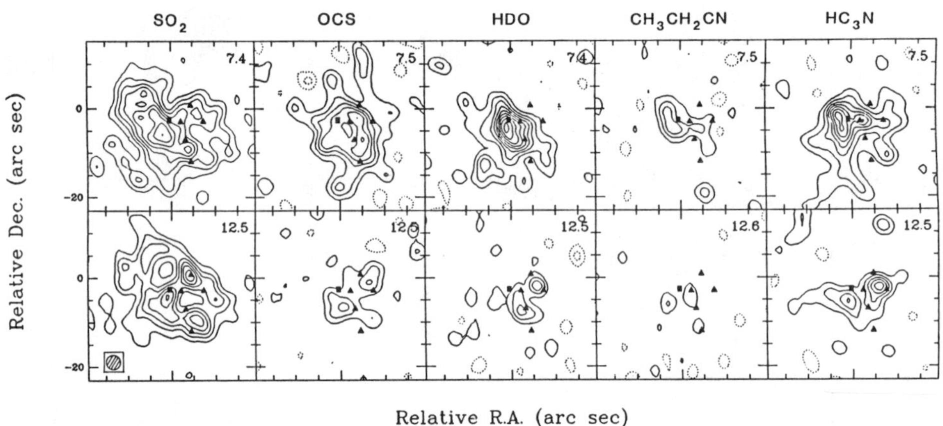

FIGURE 5. Line maps at 7.4 and 12.5 km/sec at 3.5″ resolution.[26] The square marks the position of IRc2, and the triangles mark the positions of other prominent infrared sources.

The compact broad line molecules like SO, SO_2, and SiO show regions where gas is being swept up at high speeds. Molecules containing S and Si are very enriched, especially the oxygen containing varieties. There is little evidence of CN and C_3H_2 here. Mundy et al.[16] find the map of high velocity CS near IRC2 similar to that of SO; however, only 10–20% of the single dish flux toward IRC2 in CS is in this compact distribution, whereas more than 80% of the fluxes from SO and SO_2 are here in the compact maps. Note also in Figure 4 that the ion HCO^+ shows no strong peak toward IRC2, suggesting that the region around that source where the SO, etc, are found is mostly neutral. The substantial enrichment of SO and SO_2 relative to CS is expected for warm neutral regions where $O/C>1$.[27] These circumstances suggest the following speculations for the swept–up gas.[7] The mantles are evaporated by the winds, releasing oxygen from water and methanol, so that the gas becomes oxygen rich. In addition, some of the grain cores are destroyed releasing sulfur and silicon, which then appear in the enhanced abundances of the oxides of these refractory species.

Using largely the spectroscopic signatures of these different regions, Blake et al.[7] were able to separate out the relative abundances of a number of different species. Figure 6 shows their derived relative abundances of these molecules in the "hot core", "plateau" (what we have been calling the "compact swept–up gas"), and the ridge.

ABUNDANCES NEAR OTHER OUTFLOWS

Only the Orion outflow has been studied in any great detail in many different species. In fact, the striking enhancement of SO near outflows has been detected elsewhere. Note the map of SGRB2 of Figure 1.[8] Jackson et al.[28] have observed $SO(2_2\text{–}1_1)$ in W49, W51, G34.3+0.2, and SGRB2. In each case, all of the SO flux is concentrated to regions where there is other evidence of outflows, water masers, CO flows, etc, exactly as in Orion. Thus, the detailed Orion studies probably do reveal the typical chemical effects of early stellar winds in other regions.

SUMMARY

In the Orion A source, material in different locations exists in strikingly different physical and chemical states, apparently as a result of the interaction of the stellar wind with the surrounding gas and dust. The highest flow velocities are observed at the greatest distance in emission from H_2 and HCO^+

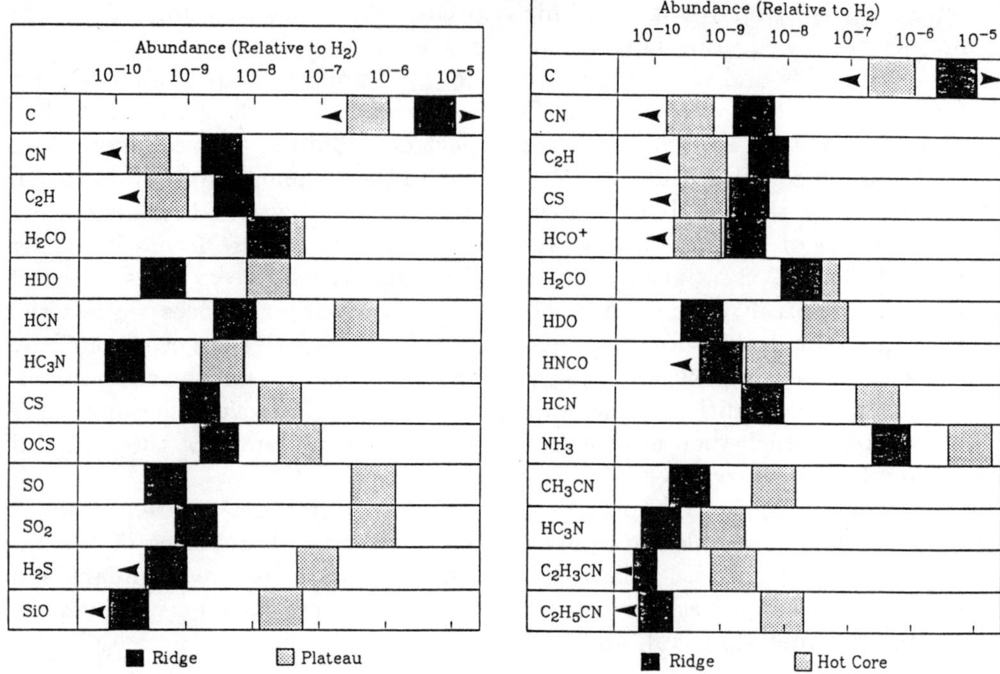

FIGURE 6 Abundances of "hot core" and "plateau" molecules.[7]

which appear to be excited by MHD shocks.[29] The "ridge" reveals a composition like that of the large scale cloud, carbon rich with an abundance of CN, C_3H_2, and C. The "hot core" is material close to IRC2 which is too dense to be pushed by the wind. It is substantially heated by radiation from the star and shows evidence of gases evaporated from grains and neutral chemistry at high density and temperature. The high velocity swept–up gas close to the star, perhaps ablated from the "hot core" clumps, appears oxygen rich with a large enhancement of molecules containing S and Si, possibly due to the destruction of grains and grain mantles. The observation of highly concentrated high velocity width SO clouds in several other regions suggests that the Orion A source is typical of other outflow regions.

REFERENCES

1. S. Weinreb, A. H. Barrett, M. L. Meeks, and J. C. Henry, Nature, 200, 829(1963).
2. E. Herbst and W. Klemperer, Astrophys. J., 185, 505(1973).
3. W. M. Irvine, P. F. Goldsmith, and A. Hjalmarson, in Interstellar Processes, edited by D. Hollenbach and H. A. Thronson, (Reidel, Dordrecht, 1987), p. 561.
4. S. E. Cummins, R. A. Linke, and P. Thaddeus, Astrophys. J., 60, 819(1986).
5. L. E. B. Johansson, C. Andersson, J. Ellder, P. Friberg, A. Hjalmarson, B. Hoglund, W. M. Irvine, W. M. Olofsson, and G. Rydbeck, Ast. Ap., 130, 227(1984).
6. E. C. Sutton, G. A. Blake, C. R. Masson, and T. G. Phillips, Astrophys. J. Suppl., 58, 341(1985).
7. G. A. Blake, E. C. Sutton, C. R. Masson, and T. G. Phillips, Astrophys. J.,315, 621(1987).
8. P. F. Goldsmith, R. L. Snell, T. Hasagawa, N. Ukita, Astrophys. J., submitted.
9. B. Zuckerman, T. Kuiper, and E. Rodriguez–Kuiper, Astrophys. J., 209, L137(1976).
10. R. Genzel, M. Reid, J. Moran, and D. Downes, Astrophys. J., 244, 884(1981).
11. C. G. Wynn–Williams, R. Genzel, E. E. Becklin, and D. Downes, Astrophys. J., 281, 172(1984).
12. S. Beckwith, S. E. Persson, G. Neugebauer, and E. E. Becklin, Astrophys. J., 223, 464(1978).
13. D. M. Watson, J. W. V. Storey, C. H. Townes, E. E. Haller, and W. L. Hansen, Astrophys. J., 239, L129(1980).
14. M. Werner, M. K. Crawford, R. Genzel, D. J. Hollenbach, C. H. Townes, and Dan M. Watson, Astrophysical J., 282, L81(1984).
15. M. C. H. Wright, R. L. Plambeck, S. N. Vogel, P. T. P. Ho, and W. J. Welch, Astrophys. J., 267, L41(1983).
16. Lee G. Mundy, N. Z. Scoville, L. G. Baath, C. R. Masson, and D. P. Woody, Astrophys. J., 304, L51(1986).
17. M. Morris, P. Palmer, and B. Zuckerman, Astrophys. J., 237, 1(1980).
18. M. C. H. Wright and R. L. Plambeck, Astrophys. J., 267, L115(1983).
19. R. L. Plambeck, M. C. H. Wright, W. J. Welch, J. H. Bieging, B. Baud, P. T. P. Ho, S. Vogel, Astrophys. J., 259, 617(1983).
20. S. N. Vogel, M. C. H. Wright, R. L. Plambeck, and W. J. Welch, Astrophys. J., 283, 655(1984).

21. R. Genzel, D. Downes, P. T. P. Ho, and J. Bieging, Astrophys. J., 259, L103(1982).
22. C. M. Walmsley, W. Hermsen, C. Henkel, R. Mauersberger, and T. L. Wilson, Astron. Astrophys., 172, 311(1987).
23. C. R. Masson, M. J. Claussen, K. Y. Lo, A. T. Moffet , T. G. Phillips, A. I. Sargent, S. L. Scott, and N. Z. Scoville, Astrophys. J., 295, L47(1985).
24. M. C. H. Wright and S. N. Vogel, Astrophys. J., 297, L11(1985).
25. R. L. Plambeck and M. C. H. Wright, private communication.
26. R. L. Plambeck and M. C. H. Wright, Astrophys. J., 317, L101(1987).
27. Sheo S. Prasad and Wesley T. Huntress, Astrophys. J., 260, 590(1982).
28. J. Jackson, J. Carlstrom, S. N. Vogel, P. Carral–Davila, A. Rudolph, and W.J. Welch, in preparation.
29. D. F. Chernoff, D. J. Hollenbach, and C. F. Mckee, Astrophys. J., 259, L97(1982).

II. Molecules as Probes of Cloud Physics

HEATING AND COOLING OF MOLECULAR CLOUDS AND THEIR SURFACES

DAVID HOLLENBACH
NASA Ames Research Center, Moffett Field, California, U.S.A.

Abstract Some of the basic mechanisms of heating molecular clouds and their surfaces are reviewed, including cosmic rays, ambipolar diffusion, radiation, and shock waves. Cooling by fine structure transitions of atoms and ions as well as by vibrational and rotational transitions of interstellar molecules is discussed. A theoretical application of these processes is made to the observations of warm (T$\sim$200 K) CO and vibrationally excited H_2.

I. Introduction

The emission from molecular clouds is rendered observable to the extent that the clouds are heated above the 3K blackbody temperature of the universe. The clouds then cool radiatively and the resultant emerging photons, often in the radio and infrared portion of the electromagnetic spectrum, can be detected above the microwave background. Generally, clouds are held at an equilibrium temperature by a balance of heating and cooling processes. The notable exception is heating by strong shock waves, where the gas is impulsively heated and subsequently cooled. In this brief discussion of the heating and cooling of molecular clouds and their surfaces, we shall first review the major heating and cooling mechanisms and then shall apply this knowledge to understanding the thermal balance and emission spectrum of very dense molecular cores illuminated by intense ultraviolet fluxes.

II. Gas Heating

Order of magnitude estimates can be made of the total heat input rate H to the gaseous component of molecular clouds in order to provide an estimate of the relative importance of various heating mechanisms. In the following analysis of H, we show that heating by ambipolar diffusion and cosmic rays is generally negligible compared to the total gaseous heat input rate caused by the stellar radiant fluxes incident upon the clouds.

In order to simplify the estimates of H, we use the empirical relation that on average $n_4R_{pc}{\sim}0.2$, where $n_4= n(H_2)/10^4cm^{-3}$ is the average molecular hydrogen density in a cloud and R_{pc} is the cloud radius in parsecs.[1] Note that this does not imply that *within* a given cloud the density decreases from center to surface, but rather that larger clouds have lower mean gas densities. This relation holds for a large statistical sample of clouds; individual clouds show large deviations from this average.

A. Cosmic Rays

Cosmic ray heating is a well known heat source which can maintain the interiors of opaque molecular clouds at 10 K.[2] Cosmic ray heating of molecular clouds is dominated by subrelativistic protons which ionize the molecular hydrogen. The energetic electrons which are produced lose energy by secondary ionizations, by excitation of molecular hydrogen, and by heating the gas. Multipling the heating rate per molecule[2] by the number of molecules in the cloud, we obtain:

$$H_{cr} \sim 0.1 n_4 R_{pc}^3 \ \ L_{\odot}$$

$$H_{cr} \sim 0.02 R_{pc}^2 \ \ L_{\odot},$$

where the latter equation is derived using the empirical relation between n_4 and R_{pc}.

B. Ambipolar Diffusion

Ambipolar diffusion can be a significant source of heating in clouds in which the gravitationally contracting neutrals slip past the magnetically-supported ionic component.[3] The randomization of the ordered neutral flow velocity by the collisions with the ions heats the neutrals. The cloud heating rate is given by:

$$H_{ad} \sim (\rho_n \rho_i \gamma v_d^2) \frac{4}{3} \pi R^3,$$

where γ=3.5x10^{13} $cm^3g^{-1}s^{-1}$ is a drag coefficient, ρ_n and ρ_i are the average mass densities of neutrals and ions, and v_d is the average drift velocity of the neutrals through the ions. Assuming a magnetically supported cloud and using the expression for v_d from Lisano and Shu[3], we obtain:

$$H_{ad} \sim 0.2 \left(\frac{x_i}{10^{-7}}\right) \left(\frac{B}{100\mu G}\right)^2 R_{pc}^2 \ L_{\odot},$$

where x_i is the average fractional ionization and B is the mean magnetic field in the cloud. These latter two parameters are quite uncertain in molecular clouds, but the normalization is chosen to approximately give factors of unity within the parentheses.

C. Gas-Grain Interaction

In contrast to the other heating mechanisms in this review, the gas-grain heating relies on the dust grains being warmer than the gas (presumably because of radiant heating whose source is nearby or embedded stars). Leung[4] has performed detailed calculations of the thermal balance equations which provide grain temperatures through clouds with embedded sources. Grain temperatures averaged through the cloud are typically 15-50 K. The cooler gas molecules colliding with the warm grains are, on average, heated in each encounter, with a mean energy transfer per collision of $2\alpha k(T_{gr}-T)$, where α is the thermal accommodation coefficient, T_{gr} is the grain temperature, and T is the gas temperature. Burke and Hollenbach[5] performed detailed calculations of α, and typically it is of order unity in molecular clouds. Assuming $T \sim 10$ K and $T_{gr}-T \sim 10$ K, we obtain:

$$H_{gr} \sim 0.04 R_{pc} \ L_{\odot}.$$

This coupling of the gas to the radiant stellar energy is weak; grains radiate most of the energy in the far infrared and transfer little energy to the gas. This can be seen by calculating the far infrared continuum luminosity of a cloud with 20 K dust grains:

$$L_{FIR} \sim 2 \times 10^3 Q_{-3} R_{pc}^2 \ L_{\odot},$$

where $Q_{-3} = Q/10^{-3}$ is the average far infrared absorption efficiency factor for 20 K interstellar grains.

The gas can be heated by warmer grains via a more indirect route. Atoms or molecules can absorb the far infrared radiation from the grains and subsequent collisional deexcitations by H_2 can transfer this internal energy into random translational motion (heat). Several authors have treated gas heating by far infrared excitation of rotational states of H_2O and the fine structure states of atomic oxygen.[6,7] Although these mechanisms can dominate the direct gas-grain collisional heating, they are typically never more than a factor of a few larger than H_{gr}.

D. X Rays

X rays can penetrate into molecular clouds and directly heat the gas by the photoionization of gaseous elements. In the primary ionization much of the X ray photon energy typically goes into the kinetic energy of the ejected electron and this energetic electron undergoes numerous encounters with other gas molecules, producing secondary ionizations, excitation and heating. Lepp and McCray[8] have studied the X ray heating of molecular clouds and show that typically ~0.1 of the absorbed X ray energy is deposited as heat.

Stars like the sun radiate too little X ray luminosity to appeciably heat molecular cloud gas as a whole. However, the accretion of molecular gas by neutron stars passing through the clouds can be a much more significant source of heating. Using the space density of neutron stars ($\sim 10^{-3}$ pc^{-3}) and the accretion luminosity expression of Lepp and McCray[8], we obtain:

$$H_X \sim 8 \times 10^{-2} v_6^{-3} R_{pc}^2 \ L_{\odot},$$

where $v_6 = v/10$ km s^{-1} is the velocity of the neutron star through the cloud.

E. Grain Photoelectric Heating

Although Lyman continuum photons are absorbed in HII regions and in a thin ($\Delta A_v \lesssim 0.1$) transition zone on the surfaces of neutral clouds, FUV($6eV \lesssim h\nu \lesssim 13.6eV$) photons penetrate into the neutral material to depths $\Delta A_v \sim$ several. Since average clouds have $n_4 R_{pc} \sim 0.2$, the column density from surface to center corresponds to $\Delta A_v \sim 6$. Therefore, FUV photons can play a role in the heating and chemistry of a significant portion of the cloud. FUV photons heat the gas when they are absorbed by grains and cause the photoelectric ejection of (hot) electrons into the gas. Grain photoelectric heating, like gas-grain collisions, is inefficient; typically $\gtrsim 96\%$ of the FUV photon energy is absorbed by the grain and radiated as far infrared continuum while $\lesssim 4\%$ is deposited by the electrons as heat to the gas.[7] Unlike gas-grain collisions, however, photoelectric heating can raise the gas temperature above the grain temperature.

The low efficiency of photoelectric heating can be understood by noting that FUV photons, absorbed by typical grain materials, photoelectrically eject electrons only 10% of the time and that the work functions of neutral materials are of order 6eV.[9] Thus, it requires the absorption of ten 10eV photons to deposit 4eV of heat to the gas. This is an upper limit on the heating since the photoelectric mechanism leaves the grains positively charged, and the ejected electrons generally must overcome the Coulomb potential as well as the work function of the grain.[7,9,10] Recently, numerous observations of the cooling of photoelectrically-heated molecular cloud surfaces ("photodissociation regions" or "PDRs") have confirmed that the gas cooling luminosity is $\lesssim$4% of the far infrared continuum luminosity of the grains.[7]

The heating of molecular cloud surfaces by grain photoelectric heating can be written:

$$H_{pe} \sim 0.5 G_o R_{pc}^2 \; L_\odot,$$

where G_o=FUV flux/1.6x10^{-3} erg $cm^{-2}s^{-1}$ is the ratio of the FUV flux to the average interstellar FUV flux in the solar neighborhood. An O8V star 0.1 pc from a cloud gives $G_o \sim 10^5$.

F. Shock Heating

The shock heating of molecular gas has been studied and reviewed extensively.[11–14] The heat input rate to the postshock column is given by $\sim 0.5\rho_o v_s^3$, where ρ_o is the mass density of the preshock gas and v_s is the shock velocity. The total shock heating H_s in a cloud therefore requires the knowledge of v_s and the total shock surface area. These parameters vary depending on the nature of the shock driving source, making even a crude estimate of H_s difficult. Direct observations of shock luminosities in various molecular clouds illustrate this point. In IC443 a molecular cloud is impacted by a supernova blast wave and the H_2 cooling behind the resultant shocks reveals a shock heat input rate of $\gtrsim$2000 $L_\odot$ [15], which far exceeds other possible heating sources. A similar situation is observed in the molecular cloud associated with a strong bipolar outflow source in DR21.[16] In Orion, on the other hand, very strong shocks are driven by the outflow from IRc2 in BN-KL, but the ~200 $L_\odot$ shock heating[17] is dwarfed by the ~1000 $L_\odot$ photoelectric heating of the gas at the cloud surface[7]. Comparisons of the CII(158μm) luminosities (from photoelectrically heated regions) to the H_2(2μm) or high J CO luminosities (from shock heated regions) in external galaxies reveals that photoelectric heating generally dominates shock heating when a galaxy-averaged ensemble of molecular clouds is taken.[18] This can be understood by noting that while the photoelectric heating is of order $10^{-2}L_*$, where L_* is the stellar luminosity of a galaxy, the shock heating must be less than the total mechanical luminosities of the stellar winds and supernovae, which is likely to be much less than $10^{-2}L_*$.[19]

As a crude estimate of H_s, we calculate the shock heating by calculating the embedded stellar luminosity L of a cloud and by assuming that the shock heating rate due to stellar winds is given by fL, where the fraction f is less than 10^{-2}. Hence

$$H_s \sim 4f_{-2}\left(\frac{L}{M}\right) f_* R_{pc}^2\ L_\odot,$$

where $f_{-2}=f/10^{-2}$, (L/M) is the luminosity to mass ratio of the stars in solar units, and f_* is the fraction of the cloud gas mass in stars. Although shock heating may not dominate the total heat input to the cloud, we note that the shock wave heating is localized to small regions in the cloud so that the heating results in high gas temperatures and the emission of characteristically high excitation transitions such as the vibrational 2μm spectrum of H_2.

G. H_2 Vibrational Heating

The collisional deexcitation of vibrationally excited molecular hydrogen transfers internal energy to heat. Newly-formed molecular hydrogen is ejected from grains with about 4eV of vibrational energy[20,21], and H_2 can be given ~2eV of vibrational energy in a process initiated by the absorption of an FUV photon ("UV pumping"). In the UV pumping mechanism the FUV photon electronically excites the H_2 which then radiatively decays to vibrationally excited states of the ground electronic state (~0.9 of the time) or to the vibrational continuum (i.e., dissociates ~0.1 of the time).[22] At low densities the vibrationally excited H_2 radiatively decays to the ground state, little heating is accomplished, and a fluorescent spectrum is produced with a low ratio (~2) of v=1-0/v=2-1 emission. The critical density to convert the UV pumped vibrational energy to heat is given by[12]:

$$n_{cr}^{H} = 6\times10^5 T^{-0.5} e^{(400/T)^2}\ cm^{-3},$$
$$n_{cr}^{H_2} = 7\times10^5 T^{-0.5} e^{+[18100/(T+1200)]}\ cm^{-3},$$

where the H and H_2 refer to collisions with hydrogen atoms or hydrogen molecules respectively. We note that the critical densities are quite temperature sensitive, and increase greatly at low gas temperatures. The H_2 vibrational heating can be the dominant heating mechanism under the special conditions which exist, for example, in the dense, hot, dissociated gas behind a shock or in the dense, hot, FUV-illuminated gas at cloud surfaces. We shall give an example of the latter case in Section IV.

III. Gas Cooling Mechanisms

Molecular clouds cool by the collisional excitation of molecules and atoms followed by the radiative emission of this energy from the cloud. In the opaque molecular interiors of clouds, the gas temperatures are generally ~10-30 K and the cooling is dominated by rotational excitation of the abundant CO molecules, whose low-lying states lie 5-30 K above ground.[23] Not all of the millimeter and submillimeter photons emitted by the CO molecule escape and cool the cloud, however, because the optical depths τ_{ul} in the transitions are large. For example, the optical depth in the 2.6mm CO J=1-0 line is given by:

$$\tau_{10} \sim \left(\frac{N_{CO}}{3\times 10^{15}\ cm^{-2}}\right)\left(\frac{10K}{T}\right)^2\left(\frac{1\ \mathrm{km\ s^{-1}}}{\Delta v}\right),$$

where N_{CO} is the total column of CO assuming LTE and Δv is the Gaussian velocity dispersion. Since a column 10^{21} cm^{-2} of H_2, corresponding to $\Delta A_v \sim 1$, contains $\sim N_{CO} \sim 10^{17}$ cm^{-2}, the optical depths in the CO lines will be rather large ($\gtrsim$ 100) in the cloud interiors. Goldsmith and Langer[23] provide cooling curves for the dominant species in molecular cloud interiors for optically thick conditions.

In the photodissociated surfaces of molecular clouds, the gas is primarily H, He, O, and C^+, in order of abundance, and the temperatures are generally 30-3000 K. H and He have no low-lying levels which can effectively cool the gas, and the fine structure transitions OI(63μm) and CII(158μm) generally dominate the cooling.[7] The optical depths in these transitions are typically of order unity so that the optically thin limit to the cooling is a reasonable approximation. Somewhat further into the cloud, the gas becomes molecular and the warm surface layer of molecules can also be treated in the optically thin limit.

Shock waves inside molecular clouds can also accelerate, heat, and dissociate the gas. Cooling in these shocks can often be approximated by the optically thin limit as well.

Figures 1 and 2 present optically thin cooling curves for molecular and atomic gas of density n=10^3 cm^{-3} and 10^7 cm^{-3}, respectively.[7,12] The curves are calculated using the recent total deexcitation rate coefficients for CO[24]

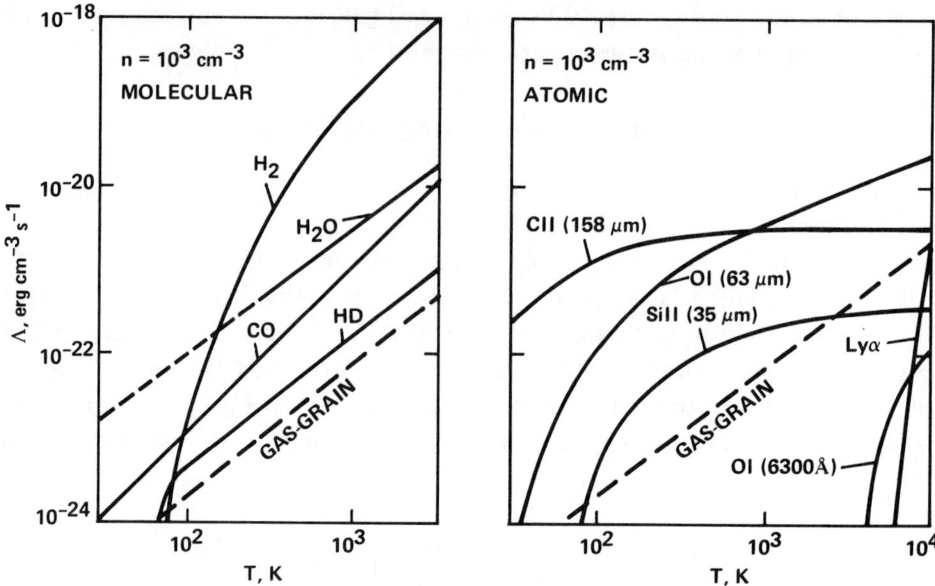

FIGURE 1: Optically thin cooling curves for gas of hydrogen nucleus density $n=10^3$ cm^{-3}. H_2O cooling below 300 K is an upper limit (see text for abundances of coolants).

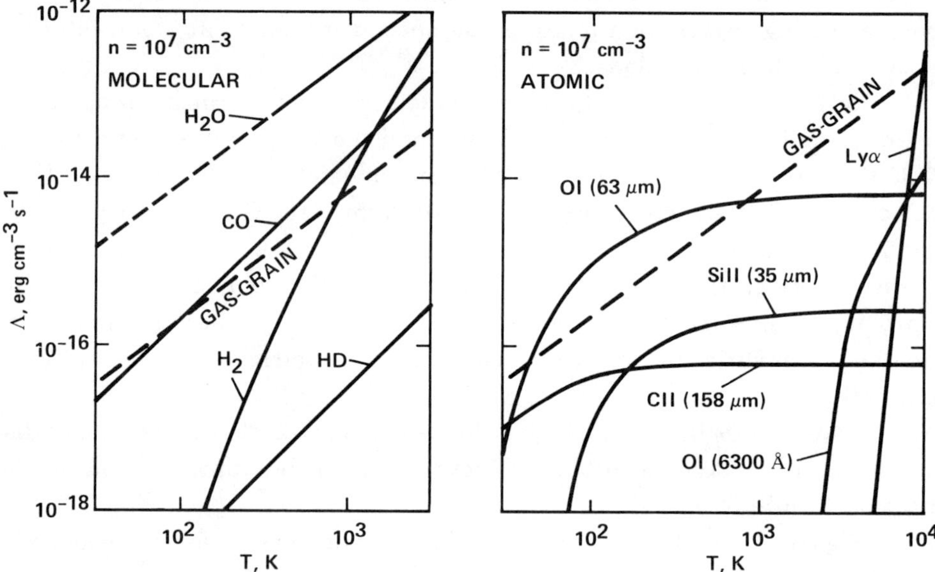

FIGURE 2: Same as Figure 1, but with $n=10^7$ cm^{-3}.

and H_2O.[25] Figures 1 and 2 utilize the following fractional abundances relative to hydrogen nuclei: (i) molecular gas, x(H_2)=0.5, x(CO)=3x10^{-4}, x(H_2O)= 3x10^{-4}, and x(HD)=10^{-5}; (ii) atomic gas, x(C^+)=3x10^{-4}, x(O)=6.7x10^{-4}, and x(Si^+)=3x10^{-5}. These abundances may be appropriate to the warm conditions often present in shocks and PDRs, where little material has frozen on the grains to form icy mantles. For $T \gtrsim 300$ K, all oxygen not in CO will be converted to H_2O in molecular gas with low FUV intensities[14]; however, at lower temperatures the gas may have substantially lower H_2O abundances. The H_2O cooling below 300 K is marked in dashed line to flag the possible overestimate of cooling in this regime. Gas-grain cooling is marked by a dashed line, assuming that $T_{gr} \lesssim T$.

In the atomic gas CII(158μm) dominates the cooling under low density and temperature conditions, while OI(63μm) dominates at higher densities and temperatures because of its higher critical density and excitation energy. At $T \gtrsim 4000$ K OI(6300A) emission plays a role in cooling and gas-grain cooling can dominate for extremely high densities ($n \gtrsim 10^7$ cm^{-3}).

In optically thin molecular gas warm CO dominates the cooling at low temperatures and H_2 dominates for $T \gtrsim 100$ K. Although H_2 is more abundant than CO, its lowest lying rotational level is $\Delta E/k = 512$ K above ground, a situation which diminishes its role as a low temperature coolant. H_2O can also be an important coolant if its abundance is high. In high density molecular gas, H_2O and CO are relatively more important compared to H_2 because of their higher critical densities. H_2O is often the dominant coolant in dense, warm, postshock molecular gas.[26]

IV. Application: Dense Cores Illuminated by Nearby OB Stars

A. Introduction to Photodissociation Region Model

In the following application we shall review recent research into the cooling of molecular clouds heated by external FUV fluxes, a heating mechanism shown to be of prime importance in Section II. This application uses a photodissociation region ("PDR") model[7] which calculates the chemistry and thermal balance of a cloud exposed to an FUV flux as a function of depth from the neutral surface. The main free parameters of the model are G_o and the cloud density n. Typically, the models are run to a depth $A_v \sim$5-10, since photodissociation is still important to these depths. Therefore, PDR regions include a large fraction of a typical cloud; in this respect, it is often misleading to call PDRs the "surfaces" of molecular clouds. The model includes heating by the photoelectric mechanism, UV H_2 pumping, gas-grain collisions, FIR excitation of O and H_2O, and cosmic rays; ambipolar diffusion, shock waves and X-ray heating are assumed negligible.

Previous work[7,27,28] has shown that most PDRs are characterized by several different layers. The surface layer into $A_v \sim 1$ is characterized by atomic chemistry and dominated by CII(158μm) and OI(63μm) cooling; heating is by the grain photoelectric mechanism, the gas is warmer than the grains, and the luminosity in the CII(158μm)+OI(63μm) lines is of order $\sim 10^{-2}$ to $10^{-3} L_{FIR}$, where L_{FIR} is the far infrared continuum. SiII(35μm), FeII(26,35μm), and CI(radio recombination, 9850A) emission also originates in this layer. A transition region extends from $A_v \sim 1$-3 where grain and self shielding allow atomic hydrogen to transform into H_2 and C^+ into CO; UV pumped H_2 (2μm) and the low J ^{12}CO emission largely originates in this region. The atomic carbon peaks in this transition zone and maintains a high fractional abundance relative to hydrogen over an extended region into the cloud. The CI(370,609μm) emission comes from this extended region, and the integrated line intensities are relatively insensitive to n and G_o. Tielens and Hollenbach[28] interpret the observed high abundances of atomic carbon[29] in molecular clouds as arising from the photodissociated gas which lies within $A_v \sim 5$ of the cloud surface.

B. Star Forming Cores in PDRs

^{12}CO 7-6 and 14-13 emission has recently been observed in M17.[30] These observations show that the temperature of the CO gas is of order 200 K, the linewidths are narrow ($\Delta v_{FWHM} \lesssim$ 5km s^{-1}), and the emission is associated with the PDR of M17. The possible shock origin of the warm CO seems unlikely on energetic grounds, utilizing the observed narrow linewidths.[30] None of the previous theoretical models[7,31], which covered the parameter space ($n=10^{2-5}$ cm^{-3}, G_o=1-10^6), produced significant quantities of warm CO in PDRs. Consequently, the origin of this warm CO was, at first, perplexing from a theoretical standpoint.

We have recently made higher density ($n=10^{6-7}$ cm^{-3}) PDR models in which warm, $T \gtrsim 200$ K CO is produced.[32] These models are motivated by the observations of dense cores ($n=10^{6-7}$ cm^{-3}, size$\lesssim$0.1 pc) in the M17 HII and PDR regions.[33-37] These $\lesssim 10\ M_\odot$ cores are likely to be gravitationally bound, or their high pressures would quickly ($\sim 10^{4-5}$ years) expand them.

On the other hand, they are probably supported against free-fall gravitational collapse, or they would form stars too quickly (10^{4-5} years).

The surfaces of these dense cores are exposed to the FUV flux from the O stars which power the M17 HII region, leading to $G_o \sim 10^4$. This flux is sufficient to warm the molecular CO gas to $T \gtrsim 100$ K if the density of the gas is sufficiently high to allow H_2 and CO self shielding to bring the C^+/CO transition region to $A_v \lesssim 1$ from the surface where grain photoelectric heating is at its maximum. $G_o/n \lesssim 10^{-3}$-10^{-2} cm^3 is required to satisfy this condition[32]; thus, $n \sim 10^{6-7}$ cm^{-3} is necessary.

Figure 3 shows the resultant chemical abundances and temperature

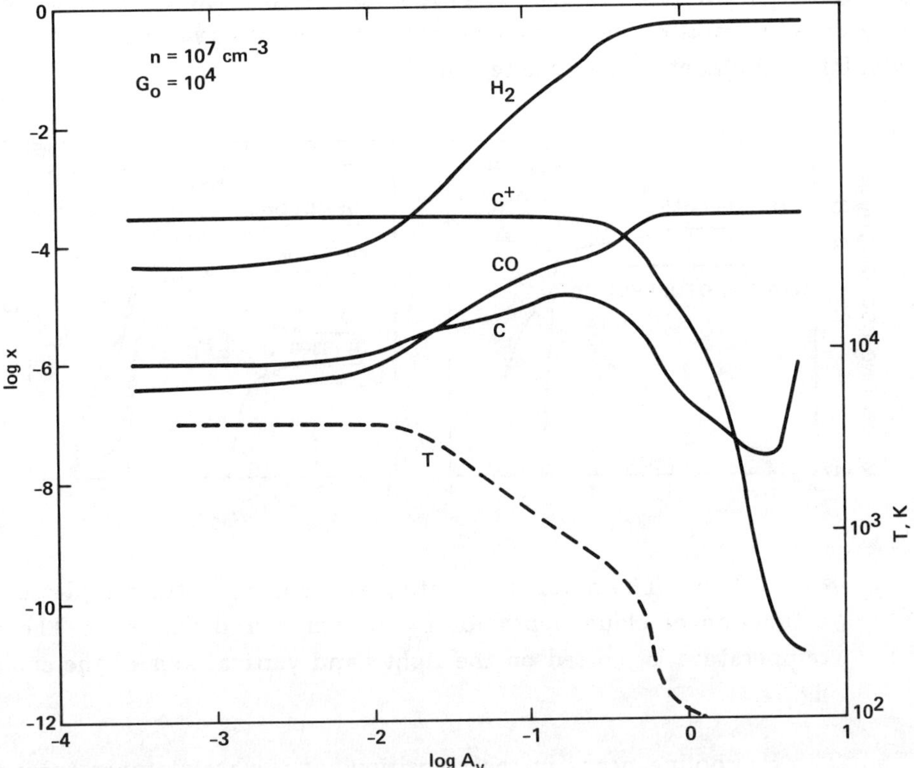

FIGURE 3: The fractional abundance x of various chemical species is plotted versus the depth A_v into a cloud of density $n=10^7$ cm^{-3} illuminated by $G_o=10^4$. The temperature (dashed line) is plotted on the right hand vertical axis. Note the large abundance of warm molecular gas.

structure of a PDR model for n=10^7 cm^{-3} and G_o=10^4.[32] One sees the various characteristic layers of a PDR (the atomic surface region, the molecular transition zone and the extended atomic carbon region), but for these low values of G_o/n, the zones are moved closer to the cloud surface than in the more typical cases with higher ratios of G_o/n.

Figure 4a shows the dominant heating sources as a function of depth into these cores, and Figure 4b shows the dominant coolants. At these high densities and FUV fluxes the UV pumped H_2 heating dominates in the atomic and transition zones, while the grain photoelectric heating dominates at somewhat higher depths. The atomic surface reaches ~4000 K, where grain-gas cooling dominates and OI(6300Å) production is significant. This emission from dense neutral clumps may contribute to the observed OI(6300Å) flux from HII and planetary nebulae regions.[38,39]

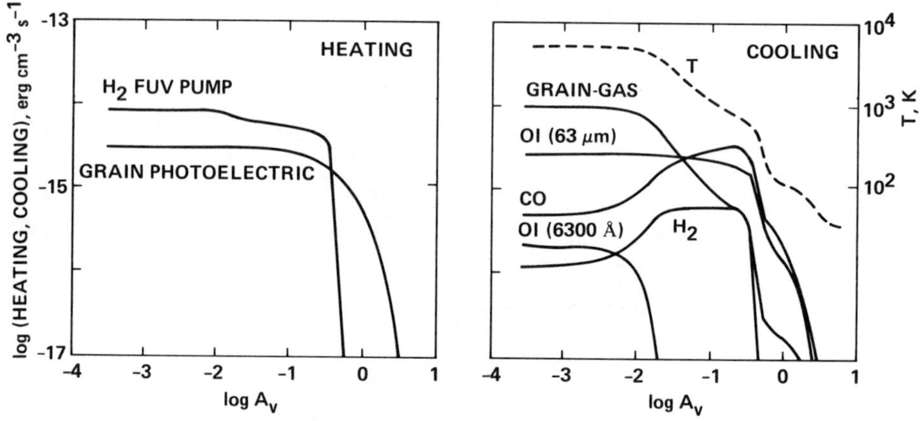

FIGURE 4: The dominant heating and cooling rates are plotted as a function of cloud depth for n=10^7 cm^{-3} and G_o=10^4. The gas temperature is plotted on the right hand vertical axis of the cooling figure.

The H_2 cooling peaks in the transition zone, where temperatures are of order 1000 K. At these temperatures the H_2 v=1-0 transitions are important but the v=2-1 transitions originate too high above the ground state ($\Delta E/k \sim 12,000$ K) to be significant. Thus, the H_2 2μm emission spectrum will resemble that of shocks which have high ratios (~10) of 1-0/2-1[14] rather than the low ratios (~2) of low density, UV-pumped H_2 vibrational cascades.[22] Although the *heating* is by UV pumping, the high gas density brings the H_2

vibrational levels into LTE. Thus, observers should use caution in identifying shock activity by the 1-0/2-1 ratio seen towards planetary nebulae, HII regions, or the Galactic center which may have dense gas illuminated by large FUV fluxes. Figure 5 plots this ratio and the absolute intensity of the 1-0S(1) line as a function of n for $G_o=10^4$ in a series of PDR models.[32]

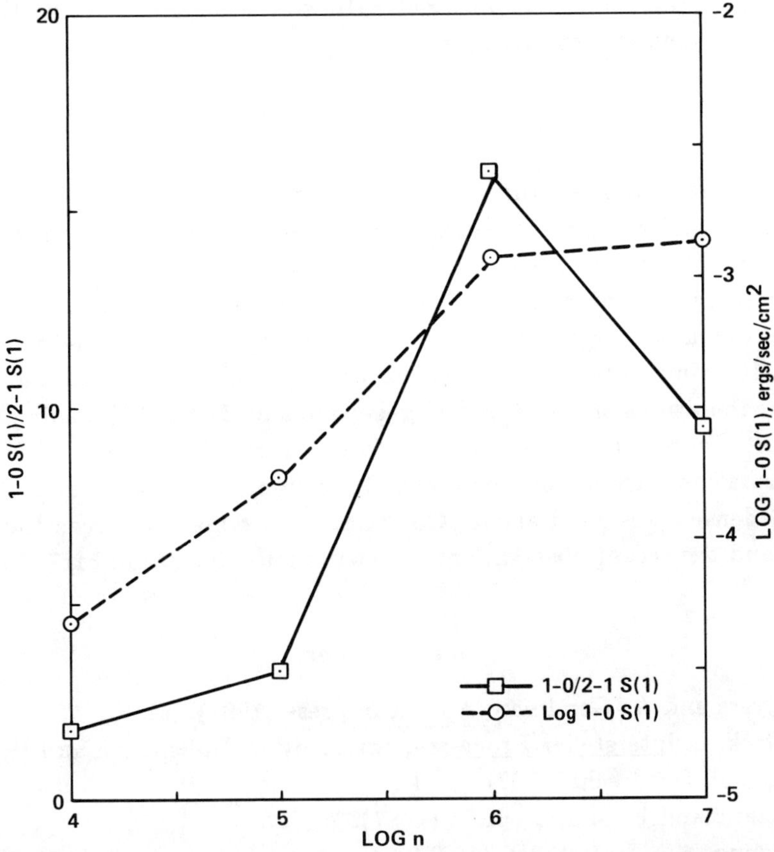

FIGURE 5: The ratio of the H_2 1-0/2-1 S(1) intensities (solid line, boxes) and the emergent 1-0 S(1) flux (dashed line, circles) are plotted for four theoretical models ($n=10^4$, 10^5, 10^6, 10^7 cm^{-3} and $G_o=10^4$). Note the high ratios for $n \gtrsim 3x10^5$ cm^{-3}, similar to ratios observed in interstellar shock waves.

In the $n=10^7$ cm^{-3} case warm CO emission is produced in the transition zone and slightly deeper, where the temperature drops from ~1000 K to ~100 K. It should be emphasized that this model is preliminary and not yet a fit to the observations, but it is clear that such high density PDR models show promise in explaining certain observations of warm CO as well as 2μm H_2 emission. If true in M17, it means that the newly-formed massive O stars have ionized and photodissociated large regions of the ambient molecular cloud, exposing a number of sites of ongoing star formation. These dense cores can be probed by their luminous infrared PDR emission, as well as by the radio emission techniques currently in use.

V. Summary

We have reviewed the heating and cooling of molecular clouds and their surfaces. Heating by an external FUV field is shown to be an important mechanism via grain photoelectric emission, H_2 vibrational heating, and gas-grain coupling. The atomic surface regions generally cool by CII(158μm) and OI(63μm) emission. For very dense ($\sim 10^7$ cm^{-3}) cores illuminated by FUV fluxes $\sim 10^4$ times the average local interstellar value, these atomic regions attain temperatures of $\sim$4000 K and gas-grain and OI(6300Å) cooling become important. Deeper into the cloud, the molecular transition zone is often cooled by CO rotational transitions, although H_2 cooling can also be important for the very dense cores cited above. Comparisons are made between theoretical models and the recent observations of warm molecular gas in M17.

References

1. P. Myers and A. Goodman, Ap. J., in press (1987).
2. J. Black, in Interstellar Processes, edited by D. Hollenbach and H. Thronson (Reidel, Dordrecht, 1987), p731.
3. S. Lisano and F. Shu, preprint (1987).
4. C. Leung, Ap. J., 199, 340 (1975).
5. J. Burke and D. Hollenbach, Ap. J., 265, 223 (1983).
6. T. Takahashi, D. Hollenbach and J. Silk, Ap. J., 275, 145 (1983).
7. A. Tielens and D. Hollenbach, Ap. J., 291, 722 (1985).
8. S. Lepp and R. McCray, Ap. J., 269, 560 (1983).
9. T. deJong, Highlights Astr., 5, 301 (1980).
10. B. Draine, Ap. J. Suppl., 36, 595 (1978).
11. B. Draine, Ap. J., 241, 1021 (1980).
12. D. Hollenbach and C. McKee, Ap. J. Suppl., 41, 555 (1979).
13. C. McKee and D. Hollenbach, Ann. Rev. Astron. Astrophys., 18, 219

(1980).
14. C. McKee, D. Chernoff, and D. Hollenbach, in Galactic and Extragalactic IR Spectroscopy, edited by M. Kessler (Reidel, Dordrecht, 1984), p.103.
15. M. Burton, Q. J. R. A. S., in press, (1987).
16. R. Garden, T. Geballe, I. Gatley, and D. Nadeau, M. N. R. A. S., 220, 203 (1986).
17. N. Scoville, D. Hall, S. Kleinmann, and S. Ridgeway, Ap. J., 253, 136 (1982).
18. M. Crawford, R. Genzel, C. Townes, and D. Watson, Ap. J., 291, 755 (1985).
19. J. Bally and C. Lada, Ap. J., 265, 824 (1983).
20. D. Hollenbach and E. Salpeter, J. Chem. Phys., 53, 79 (1970).
21. D. Hunter and W. Watson, Ap. J., 226, 477 (1978).
22. J. Black and A. Dalgarno, Ap. J., 203, 132 (1976).
23. P. Goldsmith and W. Langer, Ap. J., 222, 881 (1978).
24. C. McKee, J. Storey, D. Watson, and S. Green, Ap. J., 259, 647 (1982).
25. D. Neufield and G. Melnick, Ap. J., in press, (1987).
26. C. McKee, D. Hollenbach, and T. Jernigan, in Star Forming Regions, edited by M. Peimbert and J. Jugaku (Reidel, Dordrecht, 1987), p. 202.
27. A. Tielens and D. Hollenbach, Ap. J., 291, 747 (1985).
28. A. Tielens and D. Hollenbach, ICARUS, 61, 40 (1985).
29. T. Phillips and P. Huggins, Ap. J., 251, 533 (1981).
30. A. Harris, J. Stutzki, R. Genzel, J. Lugten, G. Stacey and D. Jaffe, Ap. J. Lett., in press, (1987).
31. D. Hollenbach, T. Takahashi, and A. Tielens, in preparation, (1987).
32. M. Burton, D. Hollenbach, and A. Tielens, in preparation, (1987).
33. M. Felli, E. Churchwell, and M. Massi, Astron. Astrophys., 136, 53 (1984).
34. E. Churchwell, M. Felli, and M. Massi, B. A. A. S., 18, 671 (1986).
35. R. Snell, N. Erickson, P. Goldsmith, B. Ulich, C. Lada, R. Martin and A. Schulz, Ap. J., 304, 780 (1986).
36. L. Mundy, R. Snell, N. Evans, P. Goldsmith, and J. Bally, Ap. J., 306, 670 (1986).
37. N. Evans, L. Mundy, J. Davis, and P. VandenBout, Ap. J., 312, 344 (1987).
38. G. Munch and K. Taylor, Ap. J. Lett., 192, L93 (1974).
39. P. Atherton, T. Hicks, N. Reay, G. Robinson, S. Worswick, and J. Phillips, Ap. J., 232, 786 (1979).

STRUCTURE AND MOTIONS IN MOLECULAR CLOUDS

BRUCE G. ELMEGREEN
IBM Watson Research Center, P.O.Box 218, Yorktown Heights, NY 10598

Abstract: Interactions between dense clumps in molecular cloud complexes are discussed. Two types of clump motions are envisioned: orbital-type motions, where the average relative velocity between clumps approximately equals the observed cloud linewidth, and magnetically-constrained turbulence, where the relative velocities between near-neighbor clumps can be much less than the cloud linewidth. Various observations that could distinguish between these possibilities are proposed. The empirical correlations observed for molecular clouds are attributed to virialized motions in a constant pressure environment. A simple model of supersonic clump interactions is used to identify possible states of turbulent pressure equilibrium.

INTRODUCTION

A clumpy and filamentary structure in giant molecular clouds has recently become evident from large scale CO maps[1,2]. Very little is known yet about the motion or evolution of these clumps, but their discovery alone heralds an era of rapid progress for molecular cloud physics. The clumps apparently contain a large fraction of the mass and kinetic energy of a giant cloud, and the dissipation of their relative motion probably controls the cloud's evolution. Each clump may also be the site of present or future star formation. One is tempted to speculate that clumps and filaments are a natural state for interstellar gas. Their tenuous and filamentary distribution has a curious resemblance to the clumps and filaments in IRAS cirrus clouds[3]. Little is understood about the relative motions of clumps in a cloud. This discussion considers two possible types of motion, and derives the average clump densities and pressures when collisional interactions occur.

MAGNETIC INTERACTIONS

The dense clumps in a molecular cloud could all be moving randomly at the full virial theorem velocity. Then their collisions will have strong shocks, the interclump magnetic field lines should become very tangled, and the cloud will dissipate its kinetic energy rapidly. This could be the case if the unperturbed magnetic field pressure is much less than the turbulent pressure.

Another possibility is that the clumps are condensations on a relatively strong field, and that each clump moves along with its immediate neighbors because of field line entanglements. Then the interconnecting field lines will be only slightly irregular, the individual clump collisions can be weak, and the total dissipation rate in the cloud should be small[4]. This second model corresponds to an oscillation of each clump on its magnetic flux tube, which is constrained to a small region by entanglements with other flux tubes. The cloud complex can still oscillate with a total velocity dispersion comparable to the virial-theorem velocity, but each neighboring pair of clouds will have only a small relative velocity.

The collective behavior of clump motions might be illustrated by a plot of the absolute value of the velocity difference between each pair of clumps in a complex, or in a small region of a complex, versus the projected separation between that pair. The relative velocity is expected to increase with separation if magnetic entanglements are important, but the relative velocity should be independent of separation (barring obvious effects of rotation) if the clump motion is purely orbital. A cross-correlation analysis of clump kinematics by Pérault, Falgarone and Puget[5] suggests that the relative velocity increases with separation, so some type of confinement appears to be present.

Physical cloud-cloud collisions are a more important mechanism for momentum transfer than magnetic interactions when the relative clump motion is at the full virial-theorem velocity. Magnetic interactions become more important when the collision velocity is low. This follows from the ratio of the magnetic collision cross section to the geometric collision cross section[6]:

$$\frac{\sigma_{mag}}{\sigma_{geom}} \sim \left(\frac{v_A/v}{\rho/\rho_c} \right)^{2/3},$$

for cloud density ρ_c, cloud collision velocity v, Alfvèn velocity inside each cloud, v_A, and average complex density ρ. If v is very low, then $\sigma_{mag}/\sigma_{geom}$ can be large, and most of the interactions will be magnetic.

The magnetically interacting volume around each clump is essentially $\sigma_{mag}^{3/2}$. If n_c is the number of clumps per unit volume, then the filling factor of the magnetic influence is

$$n_c \sigma_{mag}^{3/2} = \frac{\sigma_{geom}^{3/2} n_c \rho_c v_A}{\rho v} \simeq \frac{v_A}{v} .$$

Here we have set $\rho = n_c M_c = n_c \rho_c \sigma_{geom}^{3/2}$ for clump mass M_c. Evidently, the filling factor for purely magnetic interactions becomes unity when the relative clump motions are comparable to the Alfvén velocity inside each clump. Also, in this limit, the energy dissipated per magnetic collision becomes very small[4]. Cloud complexes may be able to remain in this state for a long time.

CORRELATIONS

The masses, velocity dispersions, radii and densities of giant molecular clouds are correlated by the following empirical relation, from Falgarone and Puget[7]:

$$\left(\frac{M}{100\ M_\odot} \right)^1 = \left(\frac{c}{0.36 \text{km s}^{-1}} \right)^4 = \left(\frac{R}{\text{parsecs}} \right)^2 = \left(\frac{n}{570 \text{cm}^{-3}} \right)^{-2} .$$

This is a more recent version of the original correlation suggested by Larson[8], Myers[9] and others[10].

This system of three equations can be written in a more intuitively plausible form by considering the virial condition for a cloud that is close to pressure equilibrium with its environment[11]. The following three equations are equivalent:

$$M = K_1 \frac{4\pi\rho R^3}{3} \quad ; \quad Pressure = \rho c^2 = K_2 \quad ; \quad c^2 = K_3 \frac{GM}{2R(1+\beta)} .$$

The constants come from the constants in the empirical relations: $K_1 = 0.7$, $K_2 = 2.1 \times 10^4 k_B$ for Boltzmann constant k_B, and $K_3 = 1.2$ for $\beta = 1$, which is an assumed ratio for the magnetic pressure to the kinematic pressure. The virial relation with $K_3 = 1$ is for an isothermal self-gravitating sphere with a pressure boundary and magnetic fields. The average cloud pressure, K_2, is not too different from the expected value of three times the total external pressure, considering magnetic, thermal and turbulent pressures in the surrounding gas. The factor of three is because an isothermal sphere has an average pressure that is three times the boundary pressure[4]. The closeness of K_1 and K_3 to unity suggests that giant molecular clouds are nearly virialized, roundish, objects, subject to a fairly uniform pressure from the ambient interstellar gas.

This statement about pressure equilibrium has been written in different forms, such as a uniformity in the column density[8] or magnetic field strength[12], with the assumption in this latter case that magnetic fields support the cloud against gravitational forces and that the velocity dispersion comes from strong magnetic waves. These other statements are essentially identical to that of pressure uniformity because pressure in a self-gravitating cloud is proportional to the square of the column density,

$$\text{Pressure} \sim \frac{GM^2}{R^4} ,$$

and because pressure also depends on the magnetic field strength if the cloud is supported by the field [12,13],

$$\text{Pressure} \sim \frac{B^2}{8\pi} \sim \frac{GM^2}{R^4} .$$

A uniform pressure for all clouds is therefore the same as a uniform column density or field strength, if the clouds are strongly self-gravitating and have substantial magnetic support.

The advantage of considering only the pressure as the essential ingredient to the empirical correlations (and not the magnetic field strength for example) is that the pressure condition should be satisfied whether or not there are substantial magnetic fields in the cloud, and whether or not the observed molecular linewidths come from Alfvén waves. The suggested wave nature of the motions[14] is particularly difficult to visualize if the gas is entirely clumped. Alfvén waves travel between the clumps at a high velocity (because the density is low) and they travel inside the clumps at a low velocity (because the density is high), but they may never have a velocity equal to the observed cloud velocity dispersions. The velocity dispersions in this case are more sensibly attributed to clump motions, and not waves in the usual sense. The supersonic linewidths of the individual clumps could be magnetic waves, however.

ON THE MEANING OF PRESSURE

There are many types of pressure in a clumpy magnetic cloud. The macroscopic pressure of the whole ensemble of clumps is identified here with the quantity $\rho_1 c_1^2$ for average cloud density ρ_1 and average cloud velocity dispersion in one dimension, c_1. This pressure may not actually exist anywhere in a clumpy cloud because the gas may nowhere have a density equal to the average value. The total kinematic pressure in a clump is similarly $\rho_2 c_2^2$ for clump density ρ_2 and 1-D velocity dispersion c_2. Part of this clump dispersion is thermal, c_{th}, and part is non-thermal. The non-thermal component could consist of magnetic waves, as discussed below. There is also the pressure in a clump at the time of a collision, $\rho_2 c_1^2$, which, in this case, is written for a collision at the full velocity dispersion in the cloud, and so applicable to the orbital motion model discussed above. This clump collision pressure greatly exceeds any other pressure in the cloud, by approximately the ratio of the clump-to-average density. Evidently the pressures in individual clumps can vary with time by a large factor. There are also magnetic pressures inside the clumps and in the interclump medium; these are written here in terms of the average kinematic pressure in the clumps and in the whole cloud, respectively, as β_2 and β_i. If pressure is essential for determining the basic cloud properties, we should know how these various pressures are related.

Some preliminary research in this direction is underway. The results are illustrated in Figure 1. The left-hand side shows the average values of the density and velocity dispersion ratios as a function of the ratio of the thermal speed to the whole cloud dispersion. Giant cloud complexes are thought to be represented by solutions with

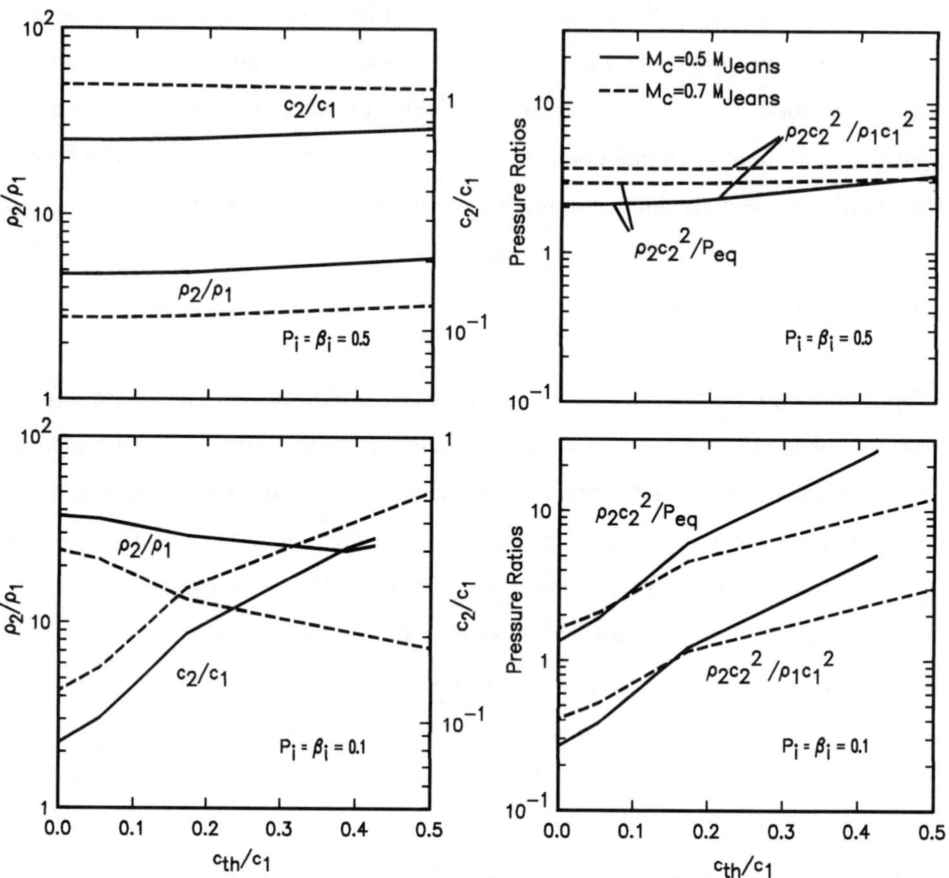

FIGURE 1 - Density, velocity dispersion and pressure contrasts for clumps in a cloud complex. The top two figures are for a higher interclump pressure than the lower figures.

small c_{th}/c_1, while small cloud complexes have large c_{th}/c_1. The right hand side of the figure shows the ratio of the average pressure in a clump to the equilibrium value of the pressure, P_{eq}, calculated as if there were no collisions, and to the macroscopic pressure, $\rho_1 c_1^2$. The average pressure is higher than the equilibrium value because of collisions. The top and bottom diagrams are for different interclump pressures (normalized to $\rho_1 c_1^2$).

The model assumes that clumps are compressed during a collision, and it determines the compressed density from the shock jump conditions for mass, momentum and magnetic flux. The properties of the cloud that enter a shock are assumed to be the average cloud properties, as determined by the solution to the whole system of equations. The energy jump condition is replaced by an adiabatic condition for the amplitude of magnetic waves, as discussed below. After each collision, the clumps are assumed to expand back to low density, with an expansion rate that comes from an equation of motion, including magnetic and pressure forces from the clump and the interclump medium, and self-gravity. If a clump does not soon collide with another clump, then it may expand all the way to pressure equilibrium with the interclump medium, which is described by another equation, also including magnetic and kinematic pressures and self-gravity. Sometimes this expansion will be terminated by another collision. Then the clump density is assumed to return to the shocked value. The collision rate is determined from the average clump radius, and this average clump radius comes from an integral over the clump expansion, weighted by $e^{-t/\tau}$ for collision time τ. The weighting factor is the probability that no collision has occurred in the time t, and so determines the relative number of clumps that reach various radii in their expansion phase. The result of these considerations is a coupled system of non-linear equations, which can be solved numerically by iteration. The solution is presumably the equilibrium state of the clumpy cloud complex.

The velocity dispersion in the clumps is assumed to result from thermal motions and magnetic waves[7] . The wave amplitude is given by the external wave amplitude multiplied by a transmission coefficient. The external wave is assumed to result from clump-clump motions at the orbital speed (using the orbital motion model). The transmission coefficient comes from Appendix B in reference (4). The result is

$$c_2^2 = 40\, I^2(\rho_1/\rho_2)^2 c_1^2 + c_{th}^2$$

for a ratio I (assumed = 0.5) of the total interclump to clump mass. The basic model for this clump velocity dispersion comes from Falgarone and Puget[7], but a transmission coefficient is used to determine the internal wave amplitude, instead of the shock steepening condition used by these authors.

The clump velocity dispersion is assumed to change with compression and rarefaction as the magnetic field strength changes, in proportion to the Alfvén velocity. This gives the condition

$$c^2 = \frac{\rho}{\rho_2}(c_2^2 - c_{th}^2) + c_{th}^2,$$

i.e., the non-thermal part of the clump velocity dispersion scales with the instantaneous density.

The results in Figure 1 illustrate two basic types of solutions for clump properties. When the interclump pressure is low ($P_i = \beta_i = 0.1$), the clumps, on average, find their low pressure state, which is characterized by a high clump density and a near-thermal velocity dispersion. In this state, the external waves do not get into the clump. The average clump pressure can be much larger than the equilibrium value because the denseness of the clumps gives the shocks a high pressure when the clumps collide. The other type of solution is shown for the high pressure case ($P_i = \beta_i = 0.5$). Here the average clump properties correspond to the highest possible clump pressures, which occur when the clumps have a low enough density to let the external waves get inside. The solution in this case has a low clump density and a large internal clump dispersion, which is almost entirely from magnetic waves.

The stability of these two solutions has not been investigated yet in any detail, but the low clump density equilibrium state would seem to be very unstable. This state gets its pressure almost entirely from magnetic waves, which are allowed into the clumps because the clump density is low. If the clump density increases slightly, then the waves will have a slightly more difficult time getting inside, and the internal velocity dispersion will decrease. Because $c_2^2 \propto \rho_2^{-2}$ from the wave transmission coefficient given above, the clump pressure varies approximately as ρ_2^{-1}. Instability usually results when a higher density produces a lower pressure.

CLOUD EVOLUTION

The evolution of a clumpy cloud complex might be imagined to go something like this. The complex is first assembled from clouds in the ambient interstellar medium. This assembly may occur in a pressure front, or as the result of some large scale instability in the ambient gas. The pieces are essentially diffuse interstellar clouds. When their mutual gravity first becomes important, they may move relative to each other at the full orbital velocity, creating strong shocks and rapid dissipation during cloud-cloud collisions. The radiation of energy away from the cloud by Alfvén waves will be rapid in this stage too[4]. This initial motion might then settle into a less dissipative state, in which the collisions are softer, and more the result of magnetic entanglements than physical impacts. This is the turbulent-entrainment state described in §II. Presumably this settling occurs after only one or two internal crossing times, which may be before the incipient cloud is ever observed as a cloud, or recognized to be in the process of forming a cloud. If this early stage of orbital cloud motions and ragged magnetic field line orientations is to be observed, then one might search in clouds without much star formation[1].

Once the cloud settles down from its initial state, clump-clump collisions should begin to establish a type of pressure equilibrium. This equilibrium should be steady in the sense of time-averaged pressures, but large pressure variations can occur for each clump, and among different clumps. The velocity dispersions of the clumps might be highly supersonic at first, consisting of magnetic waves and shearing distortions from clump-clump interactions. The clump densities should be correspondingly low, to admit the waves. After some time, however, the individual clumps should proceed, one-by-one, toward a state of high internal density and low, nearly-thermal velocity dispersions. These clumps can be stable in the midst of collisions provided their masses are less than the Jeans mass for their thermal motions. If their masses happen to be too large for their temperature, then they should collapse to form stars once the wave energy begins to get excluded.

Star formation presumably occurs on a clump-by-clump basis, with interactions contributing little to the gravitational collapse of each clump. Clumps that are too small to collapse at their temperatures might remain stable until they are destroyed by a particularly energetic collision, or by ionization or some other energy associated with star formation. They could also increase their mass and then collapse, by accretion of material from the adjacent flux tubes, or by coalescence with other clumps. Ionization of their periphery might also trigger star formation, because of the increased boundary pressure[15].

REFERENCES

1. Pérault, M., Falgarone, E. and Puget, J.L. 1985, *Astr.Ap.*, **152** , 371.
2. Bally, J., Langer, W.D., Stark, A.A., and Wilson, R.W. 1987, *Ap.J.(Letters)*, **312**, L45.
3. Hauser, M. 1988, this conference.
4. Elmegreen, B.G. 1985, *Ap.J.*, **299**, 196.
5. Pérault, M., Falgarone, E. and Puget, J.L. 1986, *Astr.Ap.*, **157** , 139.
6. Clifford, P. and Elmegreen, B.G. 1983, *M.N.R.A.S.*, **202**, 629.
7. Falgarone, E. and Puget, J.L. 1986, *Astr.Ap.*, **162**, 235.
8. Larson, R.B. 1981, *M.N.R.A.S.*, **194**, 809.
9. Myers, P.C. 1983, *Ap.J.*, **270**, 105.
10. Scalo, J.M. 1987, in *Interstellar Processes*, ed. D.J. Hollenbach and H.A. Thronson (Dordrecht: Reidel), p. 349.
11. Chièze, J.P. 1987, *Astron.Ap.*, **171**, 225.
12. Goodman, A.A. and Myers, P.C. 1987, this conference.
13. Elmegreen, B.G. 1978, *Ap.J.(Letters)*, **225**, L85.
14. Arons, J. and Max, C.E. 1975, *Ap.J.(Letters)*, **196**, L77.
15. Klein, R.I., Whitaker, R.W. and Sandford, M.T. II 1985, *Protostars and Planets II.*, ed. D.C. Black and M.S. Matthews, (Tucson: Univ. of Arizona), p. 340.

THE EVOLUTION OF MOLECULAR CLOUDS

FRANK H. SHU AND SUSANA LIZANO

Astronomy Department, University of California, Berkeley, CA 94720

Abstract. We review the problem of the structure and evolution of molecular clouds, with particular emphasis given to the relationship with star formation. Our basic hypothesis is that magnetic fields are the primary agents for supporting molecular clouds, although damped Alfven waves may play an important role in the direction parallel to the field lines. This picture naturally leads to a conception of "bimodal star formation." We propose that high-mass stars form from the overall gravitational collapse of a supercitical cloud, whereas low-mass stars form from small individual cores that slowly condense by ambipolar diffusion from a more extended envelope until they pass the brink of gravitational instability and begin to collapse dynamically from "inside-out." We also review the evidence that the infall stage of protostellar evolution is terminated by the development of a powerful stellar wind. Recent observations indicate that the wind in low-mass protostars is largely neutral and atomic. We outline a possible origin for such winds as the result of a centrifugally-driven magnetized flow from the equator of a protostar rotating at break-up because it is accreting matter of high specific angular momentum from a surrounding nebular disk.

I. MECHANISMS OF CLOUD SUPPORT

Molecular clouds constitute the principal sites of the births of stars (Zuckerman and Palmer 1974; Burton 1976, Evans 1978; Solomon, Scoville, and Sanders 1979; Blitz and Thaddeus 1980), but they cannot all be collapsing at free fall or star formation would proceed at far too great a rate (Zuckerman and Evans 1974). Although the exact lifetimes of molecular clouds have excited some controversy (see the review of Elmegreen 1985), there exists little disagreement on the issue that they need to be supported against their considerable self-gravity for timescales that exceed at least $\sim 10^7$ yr. Since the Jeans mass M_J of a large cloud, computed at the mean temperature and density, is typically much smaller than the mass of the cloud M_{cl}, large clouds cannot be supported as a whole by thermal pressure. Thermal pressure could be important for the cores (see below), but not for their envelopes.

Without reviewing the history of this subject (see, however, Shu, Adams, and Lizano 1987), we merely note that most proposals suppose either turbulence (e.g., Larson 1981) or magnetic fields (e.g., Mouschovias 1976) to play the dominant role (for a recent review of magnetic field measurements, see Heiles 1987). The large line-widths generally observed in molecular clouds have posed a problem for theoretical understanding ever since Alan Barrett and his collaborators taught us how to decompose the thermal and "turbulent" contributions (Barrett, Meeks, and Weinreb 1964). It has recently been pointed out by Shu (1987) and by Myers (1987) that the observed levels of "turbulence" in molecular clouds may be explicable in terms of nonlinear Alfven waves that permeate molecular clouds supported by magnetic fields.

As Bruce Elmegreen has also noted in his lecture, the energy for the turbulence may originate with many sources: stellar winds, cloud collisions, field entanglements, expanding H II regions, supernovae, etc., but ultimately the disturbances will excite a spectrum of MHD waves. Waves with superalfvenic fluid motions will generate compressive shocks that dissipate them rapidly. Thus, the fluid velocities will generally become subalfvenic (but still supersonic): $v_f \leq v_A$.

In the ^{12}CO line, there is an observational selection toward seeing the highest velocities, $v_f \sim v_A$, because photon trapping tends to shield regions of common (low) velocities (Peter Goldreich, private communication). However, it is easy to show that for magnetically supported clouds, the mean Alfven speed is automatically of the magnitude needed for virial equilibrium, *i.e.*,

$$v_A^2 \equiv \frac{B^2}{4\pi\rho} \sim \frac{GM_{cl}}{R}. \tag{1}$$

Thus, $v_f \sim v_A$ implies that $v_f \sim (GM_{cl}/R)^{1/2}$, i.e., cloud "turbulence" automatically has a tendency to look sufficient for virial equilibrium. Moreover, if B does not vary strongly from region to region (where CO is observed), equation (1) with $\rho \propto M_{cl}R^{-3}$ implies that the regions being supported magnetically should have nearly the same mean column densities, $M_{cl}/\pi R^2 \propto \rho R \propto B \approx$ constant. This yields $v_A \propto \rho^{-1/2} \propto R^{1/2}$. This line of argument then provides a mechanistic basis for understanding the observed correlation (Dame et al. 1985, Solomon and Sanders 1985): $v_f \propto R^\alpha$ with $\alpha \approx 0.5$.

II. BIMODAL STAR FORMATION

The importance of magnetic fields for molecular cloud support can be gauged by a comparison of its actual mass M_{cl} against the critical mass

$$M_{cr} = 0.13G^{-1/2}\Phi, \tag{2}$$

where Φ is the total magnetic flux that threads M_{cl} (see, e.g., Mouschovias and Spitzer 1976). Logically, two regimes of interest exist in the problem of star formation (see, e.g., the discussion of Mestel 1985):

(a) In the *supercritical* regime, $M_{cl} > M_{cr}$, the cloud's self-gravity can overwhelm the magnetic support *even if the fields were to remain frozen in the fluid.* Cloud evolution in this state would be characterized by magnetically diluted collapse (e.g., Scott and Black 1980).

(b) In the *subcritical* regime, $M_{cl} < M_{cr}$, one cannot induce indefinite gravitational collapse (star formation) by *any* amount of increased external load (external pressure) if Φ is conserved (field freezing) because the mass-to-flux ratio M_{cl}/Φ would remain fixed and subcritical. Cloud evolution in this state would likely be driven by ambipolar diffusion (e.g., Nakano 1979, 1981, 1982).

Shu, Lizano, and Adams (1987) have argued that these two cases provide a natural basis for the phenomenon of bimodal star formation, the notion that somehow the formation of low and high mass stars involve distinctly separate mechanisms (Herbig 1962, Mezger and Smith 1977). In our view, OB stars form from portions of molecular clouds that are supercritical and also contain insufficient "turbulence" to provide the necessary support against the self-gravitation of the region. The formation of supercritical regions probably occurs by agglomeration of smaller clouds in the spiral arms of a galaxy (or in any other circumstance under which molecular clouds become closely packed; see the discussion of Shu 1987). In any case, a relatively large supercritical region (a massive molecular cloud core) would tend to collapse as a whole, yielding the well-known tendency of OB stars to form in groups. The molecular gas surrounding the compact H II regions in G10.6−0.4 apparently exhibits such an overall collapse (Ho and Haschick 1986; Keto, Ho, and Haschick 1987). An even more spectacular case is the discovery by Welch et al. (1987) that $\sim 10^5\ M_\odot$ is collapsing dynamically toward the center of the most luminous cluster of OB stars in our Galaxy – the W49A region containing a spectacular "ring" of compact H II regions. A similar phenomenon may also be occurring in W51 (W. J. Welch, private communication).

When a cloud has less mass than M_{cr}, the cloud could theoretically attain a stable equilibrium state even if the nonmagnetic sources of internal support were reduced to zero (see Fig. 1 of Mouschovias and Spitzer 1976). Realistically, however, such a cloud, even if left perfectly alone, would evolve toward a condition of gravitational instability by the process of ambipolar diffusion (Mestel and Spitzer 1956). The dominant constituent of molecular clouds is electrically neutral and is, therefore, not directly affected by magnetic forces. Its tendency to contract gravitationally can be balanced (temporarily) by the frictional force set up when neutrals try to slip past ions and the magnetic fields to which the latter are tied. Nevertheless, with the passage of time, the magnetic flux is gradually expelled from every local dense pocket of gas and dust, and such small molecular cloud cores (Myers and Benson 1983) become the sites for the formation of low-mass stars (Shu 1983, Lizano and Shu 1987a, Myers 1987).

Figure 1 shows an example of the process. Assuming symmetry with respect to rotations about the z axis, we consider an infinite periodic chain of mutually gravitating, identical, regions, spaced a uniform distance apart along z. The calculation for ambipolar diffusion and force balance is carried out for the points interior to the "Roche lobe" of each region, with the magnetic field assumed to have a uniform value B_0 on the boundary. Because no attempt is made to follow the evolution of the material or field outside the Roche lobe (the common envelope), matter is not allowed to cross the Roche surface. The ionization fraction is approximated by an analytic fit to the equilibrium calculations of Elmegreen (1979). We compute the gas kinetic pressure

$$P = a^2\rho, \tag{3a}$$

where ρ is the mass density and $a \equiv (kT/m)^{1/2}$ is the isothermal sound speed, by assuming a constant temperature T. We simulate Alfvenic "turbulence" by the heuristic inclusion of an additional pressure P_{turb} with an associated dispersion velocity squared that satisfies:

$$dP_{turb}/d\rho = K/\rho, \tag{3b}$$

where K is taken to be a constant in accordance with the empirical relationship noted earlier. (See Lizano and Shu 1987b for further details.)

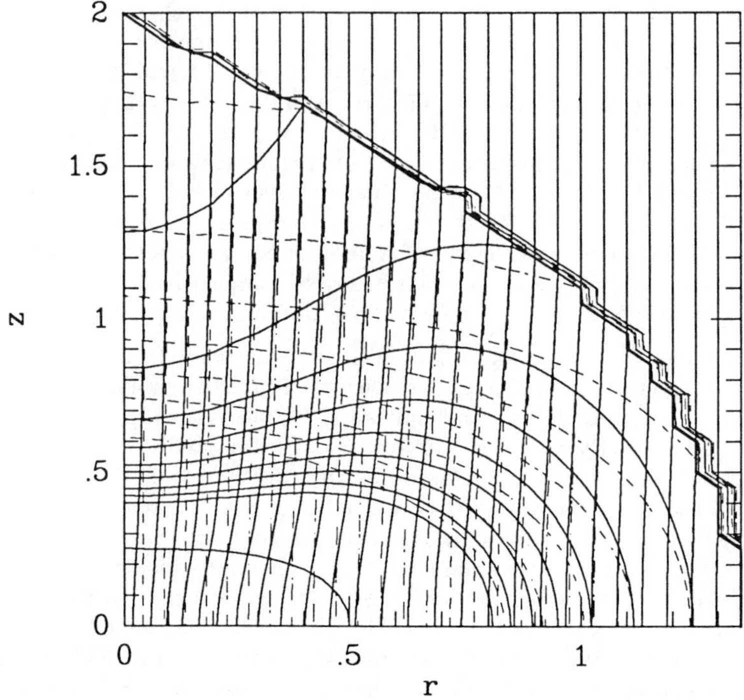

Figure 1. The evolution of a molecular cloud core by ambipolar diffusion. The growing central concentration drags in field lines slightly as indicated by the displacement of the solid vertical curves from the dashed ones. The isodensity contours for the dashed initial state correspond, from top to bottom, to 3, 4, 5, 6, 7, 8, 9, and 10 in units of ρ_0, and the central density is 23. The isodensity contours for the solid curves correspond, from top to bottom to 2, 3, 4, 5, 6, 7, 8, 9, 10, 30 in units of ρ_0, and the central density is 370. The piling up of the contours outside the Roche lobe is an artifact of the plotting routine.

The dimensionless parameters in the specific calculation illustrated in Figure 1 have been chosen so that if $B_0 = 30\ \mu$G, $T = 10$ K ($a = 0.2$ km s^{-1}), and the unit of density ρ_0 corresponds to 10^3 hydrogen molecules per cm^3, then the unit of length approximately equals 0.1 pc (i.e., the spacing 4 between adjacent Roche lobes equals 0.4 pc), and the mass interior to each Roche lobe is approximately 7 $M_\odot$. The constant K has been chosen so that $dP_{\text{turb}}/d\rho$ is 6 times larger than a^2 at a density ρ_0. The dashed curves display a somewhat arbitrarily chosen "initial" state; and the solid curves illustrate the situation a time 1.5×10^6 yr later, when the central density has increased by a factor of 16 over its original value. The magnetic field has diffused outward relative to the neutrals in a Lagrangian sense, but the increasing central concentration of the matter due to its growing self-gravitation has pulled the field lines inward in an Eulerian sense. The central portions of the configuration given by the solid curves have sizes, densities, and velocity widths that resemble the NH_3 cores of Myers and Benson (1983). The evolution of the inner regions at this stage proceeds very quickly, with the central density trying to develop a singular cusp on a time scale $\sim 10^5$ yr. The ensuing dynamical collapse cannot be followed with a quasistatic code.

III. OUTFLOWS AND THE DISRUPTION OF REMNANT MOLECULAR CLOUD CORES

The above description gives an outline of the processes that lead up to the gravitational collapse of molecular cloud cores and the formation of accreting protostars. Although considerable progress has been made in recent years on the problem of protostellar structure and evolution (see the review of Shu, Adams, and Lizano 1987), we shall not discuss it here inasmuch as the topic lies beyond our charge at this symposium. Instead, we wish to jump to the last stage of star formation – the process by which infall is terminated and the newborn star becomes optically revealed. Growing evidence has accumulated that this process is mediated by the onset of powerful stellar winds (see the reviews of Lada 1985, Welch et al. 1985, and Bally 1987). The disruption of the remnant molecular cloud core by the stellar outflow has been captured especially convincingly in the case of L43 (Mathieu et al. 1987).

Until recently, however, the underlying properties of the hypothesized stellar wind have remained largely conjectural. Attempts to find the wind from its signature as an ionized gas have generally set upper limits that lie one to two orders of magnitude below the momentum requirement of the observed bipolar outflow of swept-up molecular gas (Rodriguez and Canto 1983, Levreault 1985, Strom et al. 1986). Nevertheless, recent investigations of extended far-infrared emission associated with the CO lobes of L1551 (Clark and Laureijs 1986; Edwards et al. 1986) leave little doubt that there must exist a powerful source of missing mechanical luminosity coming from the central source IRS 5 (a rotating protostar according to the models of Adams, Lada, and Shu 1987). The stellar photons from L1551 are insufficient to heat the very extended distribution of warm dust; the latter must somehow be heated by a mechanical outward transport of energy, presumably in the form of a *neutral* stellar wind.

Recent observations validate this expectation in the case of the bipolar flow source, HH7-11. Emerging from the central star, SVS 13, is a stellar wind composed of *atomic* hydrogen (Lizano et al. 1987). (A similar situation may also hold for L1551 IRS 5, but the viewing geometry is less favorable – nearly equator-on – so that galactic hydrogen at moderately high velocities supplies more opportunities for confusion.) The H I detected in HH7-11 by the Arecibo telescope reaches speeds as high as ± 170 km s^{-1}, with $\sim 0.015\ M_\odot$ being contained in the fast moving gas represented by the line wings in Figure 2. Since the profile is roughly triangular, the average line-of-sight velocity is 170 km $s^{-1}/3 \approx 60$ km s^{-1}, yielding a travel time across the (projected) radius represented by the Arecibo beam of ~ 0.3 pc/60 km $s^{-1} \sim 5000$ yr. Thus, the rate of mass loss $\dot{M}_w$ being suffered by SVS 13 is ~ 0.015 $M_\odot/5000$ yr $\approx 3 \times 10^{-6}$ $M_\odot$ yr^{-1}. The accumulated atomic hydrogen represented by the line core in Figure 2 amounts (after various corrections) to ~ 0.2 $M_\odot$, implying a total outflow time of 0.2 $M_\odot/3 \times 10^{-6}$ $M_\odot$ yr$^{-1} \sim 7 \times 10^4$ yr. Such a stellar wind would more than suffice as the underlying power that drives the extended CO lobes (see, e.g., Edwards and Snell 1984).

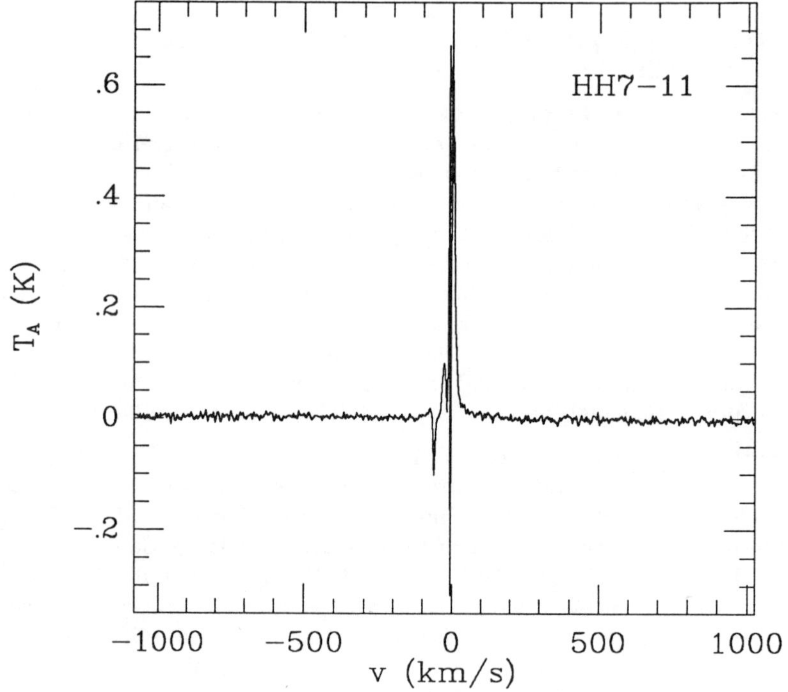

Figure 2. H I profile of HH7-11 taken at Arecibo.

IV. A POSSIBLE MASS-LOSS MECHANISM

An important property of the stellar wind in HH7-11 is the finding (on the Kitt Peak 12-m dish) that the high-velocity gas (up to $\pm$ 150 km s^{-1}) bears CO in a proportion consistent with all of the elemental carbon being contained in the form of that molecule. The combination CO plus *atomic* hydrogen characterizes matter in cool stellar atmospheres (rather than, say, molecular clouds or molecular disks), and places strong contraints on the possible physical mechanisms that can be responsible for accelerating the stellar wind. We have proposed that the mass loss originates from the interface of a magnetized protostar rotating break-up on its equator and an adjoining accretion disk (Shu, Lizano, and Najita 1987). The basic mechanism is that of a centrifugally-driven magnetic wind (Hartmann and MacGregor 1982).

The physical reasoning leading to our conclusion is as follows. Heavy mass loss $\dot{M}_w$ of the magnitude seen in HH7-11 must originate from near the photosphere of a star (to have sufficient density at the sonic transition). However, the sound speed in the photosphere forms a very small ratio compared to the escape speed; therefore, thermal pressure is able to overcome gravity only where the local gravity is relatively weak (cf. an analogous problem in the case of mass-transfer binaries considered by Lubow and Shu 1975). A narrow band where the gravity is locally weak will naturally occur on the equator of a protostar rotating at break-up because the central star is being formed by accretion from a surrounding nebular disk (that is itself supplied by infall from a rotating molecular cloud core). Nevertheless, the centrifugal effects can bring about a reversal in sign of the effective gravity only if the gas can be kept (nearly) corotating with the star. If the gas pushed outward by the thermal pressure preserves its original angular momentum, it will rotate too slowly to remain in orbit at the larger radius, and inertial effects would force it back on the star. Strong magnetic fields (generated, say, by dynamo action after the low-mass protostar has developed an extensive outer convective zone), tied to the deep interior of the star, will try to enforce the requisite corotation of the surface layers. (The coupling between the the neutrals and the small fraction of ions is assumed good enough to treat the gas as a single conducting medium.) The gas pushed out by the thermal pressure in an equatorial band now acquires more than enough angular momentum to maintain itself at the enlarged distance, so it continues to move outward as it is torqued up by the stellar rotation. The process drives the gas eventually to superalfvenic speeds, at which point the motions become essentially ballistic. If the gas speed at the Alfven point exceeds the local escape speed, the flow will develop into a wind that blows to infinity. Detailed model calculations are underway at Berkeley to assess the quantitative merits of the model, as well as to investigate the degree to which the stellar outflow may eventually be channeled magnetically toward the rotational poles.

To complete the overall evolutionary picture, we consider what regulates the numerical value of the mass-loss rate $\dot{M}_w$. In our view, because a protostar is likely to be rotating at break-up during the epoch when it is accompanied by a surrounding nebular disk, incorporation of the material from the disk, which contains a high specific angular momentum, cannot take place on the star *without* a mechanism for magnetically braking the latter in the first place. In a quasi-steady state, the wind must blow at a rate $\dot{M}_w$ that is a significant fraction f of the disk accretion rate $\dot{M}_d$ (which is itself approximately given by the mass infall rate from the surrounding molecular cloud core). A detailed analysis for low-mass protostars on the deuterium-burning birthline shows the fraction f to be approximately given by the formula (Shu et al. 1987):

$$f \approx 1.5/(2.5 + v_{\rm term}^2), \tag{4}$$

where $v_{\rm term}^2$ is the (streamline-averaged value of the) square of the terminal speed of the stellar wind measured in units of the square of the break-up speed $(\Omega_* R_*)^2$ on the equator of the star. For HH7-11, equation (4) yields the prediction $f \approx 0.4$, which is in good agreement with an atomic mass-loss rate $\dot{M}_w = 3 \times 10^{-6}\ M_\odot\ {\rm yr}^{-1}$ and a mass-infall rate (deduced from modeling of the infrared spectral energy distribution, F. C. Adams, private communication) of $\sim 1 \times 10^{-5}\ M_\odot\ \mathrm{yr}^{-1}$.

It has been a privilege to be able to contribute to a symposium volume that honors the career of Alan H. Barrett. His pioneering studies of molecules in space, and the careful follow-up work of his many students, have made possible much of the modern advance in our knowledge concerning dark cloud evolution and its connection with star formation. This work is funded in part by grants from the National Science Foundation and from the NASA astrophysics program which supports a joint Center for Star Formation Studies at UC Berkeley, UC Santa Cruz, and NASA Ames Research Center. S. L. gratefully acknowledges the support of fellowships from the National University of Mexico and the Amelia Earhart Foundation.

REFERENCES

Adams, F. C., Lada, C. J. and Shu, F. H. 1987, *Ap. J.*, **312**, 788.

Barrett, A. H., Meeks, M. L., and Weinreb, S. 1964, *Nature*, **202**, 475.

Blitz, L., and Thaddeus, P. 1980, *Ap. J.*, **241**, 676.

Burton, W. B. 1976, *Ann. Rev. Astr. Ap.*, **14**, 275.

Clark, F. O., and Laureijs, R. J. 1986, *Astr. Ap.*, **154**, L26.

Dame, T.M., Elmegreen, B. G., Cohen, R. S., and Thaddeus, P. 1985, in IAU Symposium No. 106, The Milky Way, ed. H.van Woerden, R. J. Allen and W.B. Burton, (Dordrecht: Reidel), p. 303.

Edwards, S., and Snell, R. L. 1984, *Ap. J.*, **281**, 237

Edwards, S., Strom, S. E., Snell, R. L., Jarrett, T. H., Beichman, C. A., and Strom, K. M. 1986, *Ap. J. (Letters)*, **307**, L65.

Elmegreen, B. G. 1979, *Ap. J.*, **232**, 729.

Elmegreen, B.G. 1985, in *Protostars and Planets II*, ed. D. C. Black and M. S. Matthews (Tucson: University of Arizona Press), 33.

Evans, N. J. 1978, in *Protostars and Planets*, ed. T. Gehrels (Tucson: Univ. of Arizona Press), pp. 152-164.

Hartmann, L., and MacGregor, K. B. 1982, *Ap. J.*, **259**, 180.

Heiles, C. 1987, in *Interstellar Processes*, ed. D. J. Hollenbach and H. A. Thronson (Dordrecht: Reidel), p. 171.

Herbig, G. 1962, *Adv. Astr. Ap.*, **1**, 47.

Ho, P. T. P., and Haschick, A. D. 1986, *Ap. J.*, **304**, 541.

Keto, E. R., Ho. P. T. P., and Haschick, A. D. 1987, *Ap. J.*, **318**, 712.

Lada, C. J. 1985, *Ann. Rev. Astr. Ap.*, **23**, 267.

Larson, R. B. 1981, *M. N. R. A. S.*, **194**, 809.

Levreault, R. M. 1985, Ph.D. Thesis, University of Texas.

Lizano, S., Heiles, C., Rodriguez, L., Koo, B. C., Shu, F. H., Hasegawa, T., Hayashi, S., and Mirabel, I. F. 1987, *Ap. J.*, submitted.

Lizano, S., and Shu, F. H. 1987a, in *NATO/ASI, The Interstellar Medium*, ed. G. Morfill, in press.

Lizano, S., and Shu, F. H. 1987b, in preparation.

Lubow, S. H., and Shu, F. H. 1975, *Ap. J.*, **198**, 383.

Mathieu, R. D., Benson, P. J., Fuller, G. A., Myers, P. C., and Schild, R. E. 1987, preprint.

Mestel, L., and Spitzer, L. 1956, *M. N. R. A. S.*, **116**, 503.

Mestel, L. 1985, in *Protostars and Planets II*, ed. D. C. Black and M. S. Matthews (Tucson: University of Arizona Press), p. 320.

Mezger, P. G. and Smith L. F. 1977, in *IAU Symposium No. 75, Star Formation*, ed. T. de Jong and A. Maeder, (Dordrecht: Reidel), p. 133.

Mouschovias, T. Ch. 1976, *Ap. J.*, **207**, 141.

Mouschovias, T. Ch., and Spitzer, L. 1976, *Ap. J.*, **210**, 326.

Myers, P. C. 1987, in *Interstellar Processes*, ed. D. J. Hollenbach and H. A. Thronson (Dordrecht: Reidel), p. 71.

Myers, P. C., and Benson, P. J. 1983, *Ap. J.*, **266**, 309.

Nakano, T. 1979, *Pub. Astr. Soc. Japan*, **31**, 697.

Nakano, T. 1981, *Prog. Theor. Phys. Suppl. No. 70*, 54.

Nakano, T. 1982, *Pub. Astr. Soc. Japan*, **34**, 337.

Rodriguez, L. F. and Canto, J. 1983, *Rev. Mex. Astr. Astrof.*, **8**, 163.

Scott, E. H., and Black, D. C. 1980, *Ap. J.*, **239**, 166.

Shu, F. H. 1983, *Ap. J.*, **273**, 202.

Shu, F. H. 1987, in *Star Formation in Galaxies*, ed. C. J. Lonsdale Persson (NASA Conference Publ. 2466), p. 743.

Shu, F. H., Adams, F. C., and Lizano, S. 1987 *Ann. Rev. Astr. Ap.*, **25**, 23.

Shu, F. H., Lizano, S., and Adams, F. C. 1987, in *Star Forming Regions*, ed. M. Peimbert and J. Jugaku (Dordrecht: Reidel), 417.

Shu, F. H., Lizano, S., Ruden, S., and Najita, J. 1987, in preparation.

Solomon, P. M. and Sanders, D. B. 1985, in *Protostars and Planets II*, ed. D. C. Black and M. S. Matthews (Tucson: University of Arizona Press), p. 59.

Solomon, P. M., Scoville, N. Z., and Sanders, D. B. 1979, *Ap. J. (Letters)*, **232**, L89.

Strom, K. M., Strom, S. E., Wolf, S. C., Morgan, J., and Wenz, M. 1986, *Ap. J. Suppl.*, **62**, 39.

Welch, W. J., Vogel., S. N., Plambeck, R. L., Wright, M. C. H., and Bieging, J. H. 1985, *Science*, **228**, 1329.

Welch, W. J., Dreher, J. W., Jackson, J. M., Terebey, S., and Vogel, S. N. 1987, *Science*, submitted.

Zuckerman, B., and Evans, N. J. 1974, *Ap. J. (Letters)*, **192**, L149.

Zuckerman, B., and Palmer, P. 1974, *Ann. Rev. Astr. Ap.*, **12**, 279.

III. Grains and Large Molecules

MOLECULAR OBSERVATIONS:

I. MOLECULAR CHAINS AND RINGS
II. THE COLOGNE 3-M RADIO TELESCOPE ON GORNERGRAT

G. WINNEWISSER
I. Physikalisches Institut, Universität zu Köln, Köln.

Abstract Recent advances in the detection of new interstellar molecules have been triggered partly by the commissioning of two new, large millimeterwave telescopes: the IRAM 30 m and the Japanese 45 m. Refined laboratory techniques have resulted in new detections which have helped in assigning strong and ubiquituous but hitherto unidentified interstellar lines.

Since Bill Irvine has given in his presentation on "Dense Clouds", a comprehensive summary of recent interstellar detections, I intend to add a few points which are specific to chain and ring molecules, their abundance and laboratory production. The second part of this contribution will report on recent progress we have made with the installation and commissioning of the Cologne 3-m radiotelescope in the Gornergrat-Süd-Observatorium near Zermatt, Switzerland.

MOLECULAR CHAINS AND RINGS

Since the early detection of OH (1963) by our guest of honor, Alan H. Barrett, and his colleages, the list of interstellar molecules has grown to a total of more than 80 molecular species (Table I). In addition a large number of rare isotopic species have been detected.

The list of interstellar molecules is dominated by organic molecules, i.e. molecules containing one or more carbon atoms. In fact, the heaviest and most complex molecules contain the carbon bond which seems to be the key to interstellar synthesis of gas phase molecules. To date the carbon bond has been detected in the form of carbon chain molecules, such as $HC_{2n}CN$ (n = 0,1,2,3,4,5) and several derivatives, and three ring molecules, SiCC, C_3H_2 and cyclic C_3H. All three ring molecules have been detected in the laboratory only recently. Missing in the detection list of interstellar molecules are branched chains.

Molecular Chains

The known interstellar chain molecules comprise several series with different ligands terminating the chain on one end. The well-known cyanopolyyne series seems to serve as a guide to the chain lengths one can hope to detect in the other series as well. The following series have been found either in interstellar clouds or in the envelope of the late type giant IRC + 10216:

CH, C_2H, C_3H, C_4H, C_5H, C_6H

CN, C_3N

CO, C_3O

CS, C_2S, C_3S

The fact that C_2N and C_2O have not been detected in interstellar clouds is difficult to understand at present. Their laboratory spectra have been measured, and they are asymmetric rotor molecules like CCS.

In addition, the symmetric top species

CH_3CN, CH_3CCCN, $CH_3CCCCCN$ (?)

CH_3CCH, CH_3CCCCH

have been seen.

Essentially all of these species have been found in the carbon-rich source TMC1. A summary of the determined column densities versus the chain length (number of heavy atoms) is reproduced in Fig. 1 from the work of M. Guélin and coworkers (Cernicharo, et al.[1]). It is interesting to note that an even-odd abundance effect with the number of C-atoms seems to exist in the C_nH series.

The members of the C_nH sequence with an even number of C atoms are more abundant than members with an odd number. For the higher members of the series, i.e. C_5H, C_6H, this effect seems to disappear. A detailed review of the carbon chemistry in space has been given by Winnewisser and Herbst[2].

With the exception of C_6H, HC_7CN, and HC_9CN, all these carbon chain molecules have been synthesized in the laboratory. Very recently, the laboratory detections of CCS and CCCS have been announced by Saito and coworkers[3,4]. These molecules have also been synthesized in a flow discharge of CS_2 in our Giessen laboratory by M. Winnewisser and co-workers. One absorption transition is reproduced in Fig. 2. The strength of the laboratory lines leads us to expect that C_4S and probably C_5S can be detected as well. One is also tempted to believe that the carbon series HC_nCP, C_nCP starting with HCP, CP, could possibly be synthesized in the laboratory and detected in space. HCP exists in the laboratory.

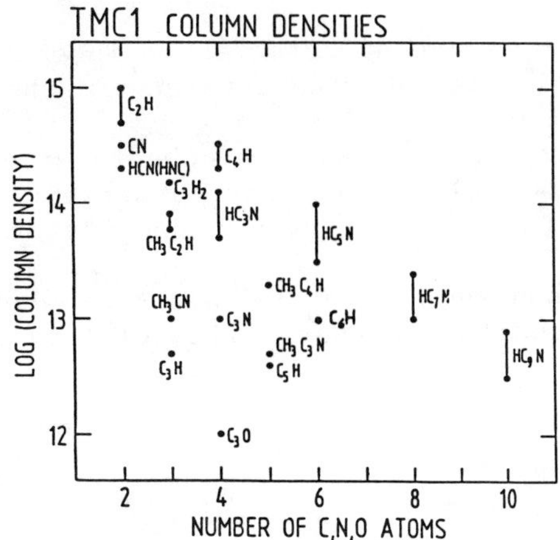

Fig.1. Column densities of carbon chain molecules plotted versus the number of heavy atoms. (from ref. 1).

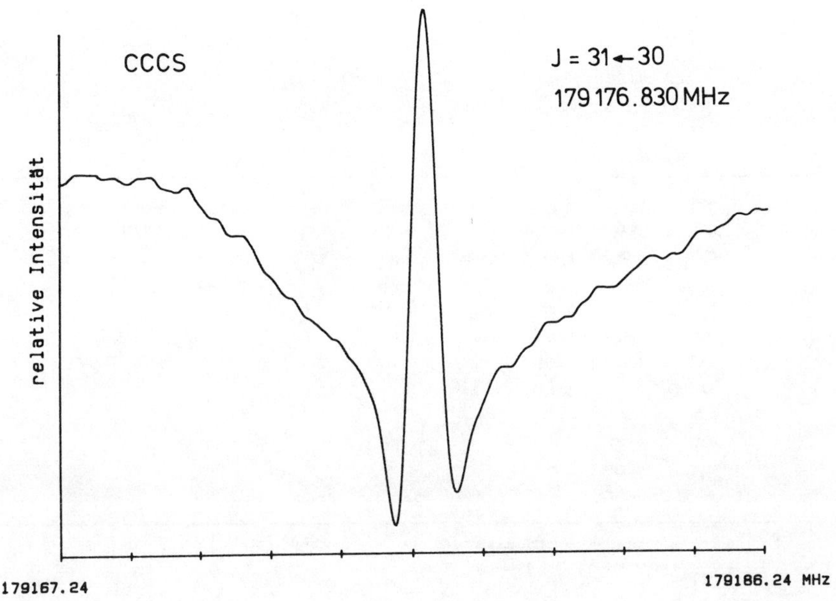

Fig. 2. Laboratory absorption line of CCCS measured in our Giessen spectroscopy laboratory.

Ring Molecules

C_3H_2 was the first true ring detected in interstellar space[6]. The ring molecule SiCC was detected earlier[7] in the circumstellar envelope of IRC + 10216.

Very recently, the gas-phase synthesis of cyclic C_3H in the labortory was announced by Saito and co-workers[8], together with the detection in TMC1 with a column density of 6×10^{12} cm^{-2}. This value is similar to that for linear CCCH. The fact mentioned above, that the abundance of C_nH depends on whether n is even or odd, is probably due to the possibility of forming cyclic C_3H only when n is odd.

Table 1

Interstellar Molecules (july 87)

2	3	4	5	6	7	8	9	10	11	13
H_2	H_2O	NH_3								
OH	H_2S									
SO	N_2H^+									
SO^+	SO_2									
SiO	HNO									
SiS	NaOH(?)									
NO	H_2D^+									
NS										
HCl										
PN										
CH^+	HCN	H_2CO	HC_3N	CH_3OH	HC_5N	$HCOOCH_3$	HC_7N	CH_3C_5N(?)	HC_9N	$HC_{11}N$
CH	HNC	HNCO	C_4H	CH_3CN	CH_3CCH	CH_3C_3N	$(CH_3)_2O$	$(CH_3)_2CO$		
CN	C_2H	H_2CS	H_2CNH	CH_3SH	CH_3NH_2		CH_3CH_2OH			
CO	C_2S	HNCS	H_2C_2O	NH_2CHO	CH_3CHO		CH_3CH_2CN			
CS	SiCC	C_3N	NH_2CN	H_2CCH_2	H_2CCHCN		CH_3C_4H			
CC	HCO	C_3H(lin)	HCOOH	C_5H	C_6H					
	HCO^+	C_3H(ring)	(CH_4)							
	HOC^+(?)	C_3O	(SiH_4)							
	OCS	C_3S	C_3H_2							
	HCS^+	$HOCO_2^+$								
	CCS	(HCCH)								
		$HCNH^+$								

( ) : circumstellar identifications

(?) : uncertain identifications

II. THE COLOGNE 3-m RADIO TELESCOPE ON GORNERGRAT

The new Cologne 3-m radio telescope has recently been commissioned for regular observations. It is now located in the Gornergrat-Süd observatory (alt. 3135 m, latitude 45°59'04"; longitude 07°47'04") near Zermatt, Switzerland. The surface accuracy of the telescope is 30 micron, and current receivers cover the frequency range from 70 to 116 GHz, with several higher frequency receivers scheduled for operation during fall 1987. Back ends include a filter bank, several acousto-optic spectrometers (AOS), and a 500 MHz bandwidth continum system. Table II gives a summary of the present technical status at Gornergrat. The telescope will be used for several scientific aims:

(i) large scale molecular surveys of different molecules,

(ii) detailed surveys using CO as a probe and the data of Columbia survey as guide to areas of special interest,

(iii) molecular observations in the submillimeter wave region.

The Gornergrat has proven to be a very dry site. Values of around 3 mm precipitable water vapor in the summer and well below 1 mm in the winter have been measured. The nearly similar spatial resolutions of the IRAS spectra at 100 and 60 microns and the 3-m telescope in the 80 - 116 GHz range make this combination of instruments ideally suited for investigating large scale structures and physics of interstellar molecular clouds and star-forming regions.

Since you, Alan, liked the site and the telescope so much, and especially the Matterhorn, we include the first 3-mm picture of that famous mountain (Fig. 3). CO spectra from eight clouds covering a broad range of physical conditions indicate the kind of work that can done (Fig. 4).

Fig. 5 presents recent CO mapping results of the high latitude ($l = 111\overset{\circ}{.}6$; $b = 20\overset{\circ}{.}1$) dark cloud L1228. An IRAS outflow-like point source is located in the middle of the area of highest visual opacity. In an accompanying paper by Armstrong et al.[9] in this book, the physical parameters of this source are summarized.

We have recently developed a model aimed at explaining the major observational features of outflow sources[10]. The initial scenario required by the model is that of a central stellar object embedded in a parental cloud of gas and dust forming a halo whose shape deviates slightly from spherical symmetry, an assumption which should be met in practically all realistic cases.

Different gravitational attraction of the nonsymmetrical halo leads to a disk-like structure and initiates a mass inflow towards the central object. This radial disk-like inflow draws the main bulk of its mass from a basically non-rotating halo surrounding a rotating disk structure. On its way towards the central stellar object, the inflowing matter traverses with freefall velocity a mass gap between the inner rim of the rotating disk and the outer fringes of the central object. Upon reaching the neighborhood of the equator of the stellar object, the inflow divides itself into

Fig. 3. Antenna temperature at 3-mm wavelength as a function of azimuth and elevation in the direction of the Matterhorn obtained at the Gornergrat-Süd-Observatory.

Table 2: Technical Data of the 3-m Telescope

Total weight	3.5 metric tons
Main reflector	
Shape	Paraboloid of rotation
Diameter	3 m
f/D ratio	0.4
Construction	4 panels (GfK technique)
Surface accuracy	30 μm (r.m.s.)
Weight	225 kg
Subreflector	
Shape	Hyperboloid of rotation
Diameter	38 cm
Construction	Aluminum
Pointing adjustment	Servo drive
Telescope movement	
Rotation axes	elevation over azimuth
Max. slew speed	1°/s
Max. slew speed during observations	0.2°/s
Max. wind during observations	18 m/s
Pointing accuracy	better than 0.001°
Receiver	
Tuning range	70–116 GHz
Bandwidth	max. 0.5 GHz
Noise temperature	230–270 K
Filter spectrometer	
Bandwidth	256 MHz
Resolution	1 MHz
Acousto-optical spectrometer MR	
Bandwidth	250 MHz
Resolution	200 kHz
Acousto-optical spectrometer HR	
Bandwidth	64 MHz
Resolution	31.25 kHz
Continuum backend	
Bandwidth	500 MHz

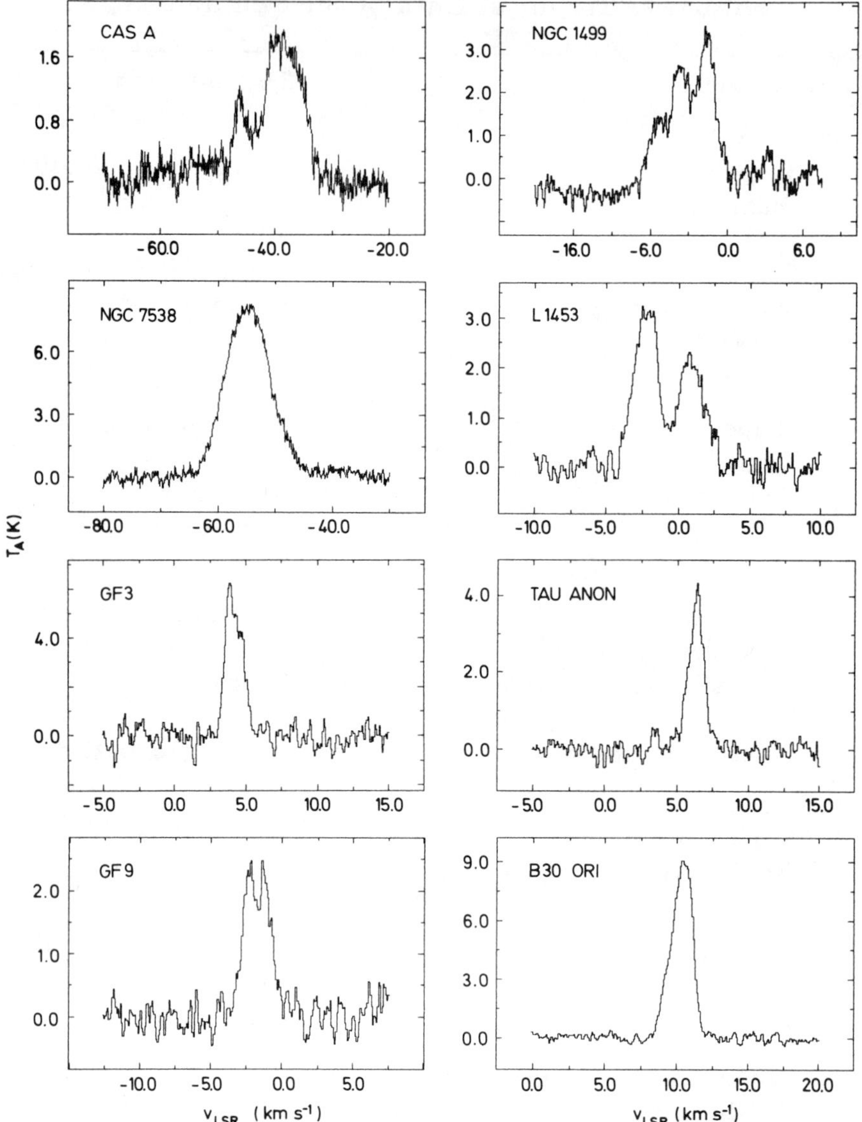

Fig. 4. CO spectra taken with the high resolution AOS (30 kHz/ channel, 2048 channels). Only a small number of channels are displayed. CAS A is a supernova remnant, NGC7538 a molecular cloud associated with an HII region. GF3 and GF9 are globular filaments, NGC 1499 is the California nebula, L1453 a dark cloud complex, TAU-Anon is a diffuse dark cloud and B30 a dark cloud in the ring of lambda-Ori.

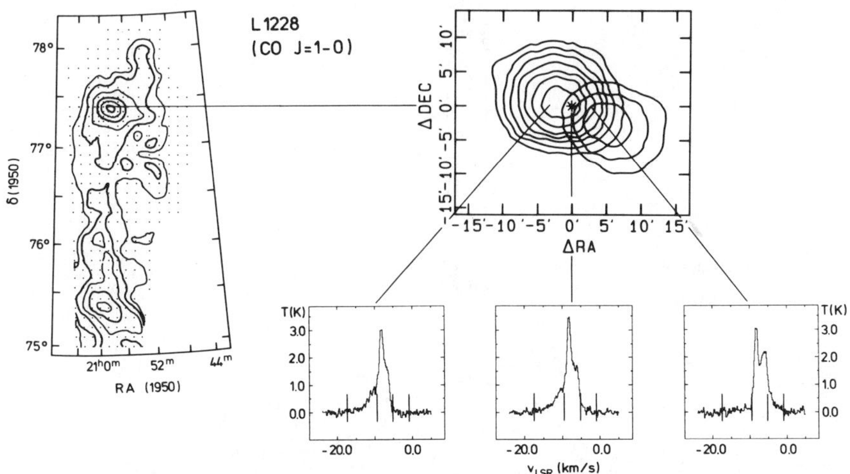

Fig. 5. Contour map of the total integrated CO intensity (left). The lowest contour level is 4 K km/s , the interval is 2.2 K km/s. The map on the right represents the redshifted (-5.2 to -0.44 km/s; 3 contour levels) and the blueshifted (-17.9 to -9.15 km/s) emission. The spectra are shown for three positions. The IRAS source is indicated by a star (L. Haikala).

two parts: the inner part with mass flux $\dot{M}_1$ penetrates a small ring-like strong shock region and mixes finally with the surface layer of the stellar object, while the outer part of the mass flux $\dot{M}_2$, which is small compared to $\dot{M}_1$, is deflected, by passing the shock region and moving towards the poles. The model is shown schematically in Fig. 6. The kinetic energy carried by $\dot{M}_1$ is released in the form of radiation at a zone around the equator. A small part of the radiation energy is absorbed above the "radiation zone" by the circumventing material $\dot{M}_2$, which becomes the outflowing material at the poles. Details of the calculations are given by Yue et al.[10]

The radial inflow is a transient phenomenon lasting only about (2-3) x 10^4 years if we assume a typical radial scale of 10^4 AU for the non-rotating halo which supplies the radial disk flow. When the non-rotating halo is exhausted, the radial inflow and the well-collimated bipolar outflow stop, leaving behind a rotating faint disk around the stellar object.

Thus, the four major observational features associated with bipolar outflows,

(i) its bipolar geometry,

(ii) the high velocity of the outflowing material of order of 100 km/s at very large distance (10^3 - 10^4 AU),

(iii) the high degree of collimation and

(iv) the time scale of about 10^4 years,

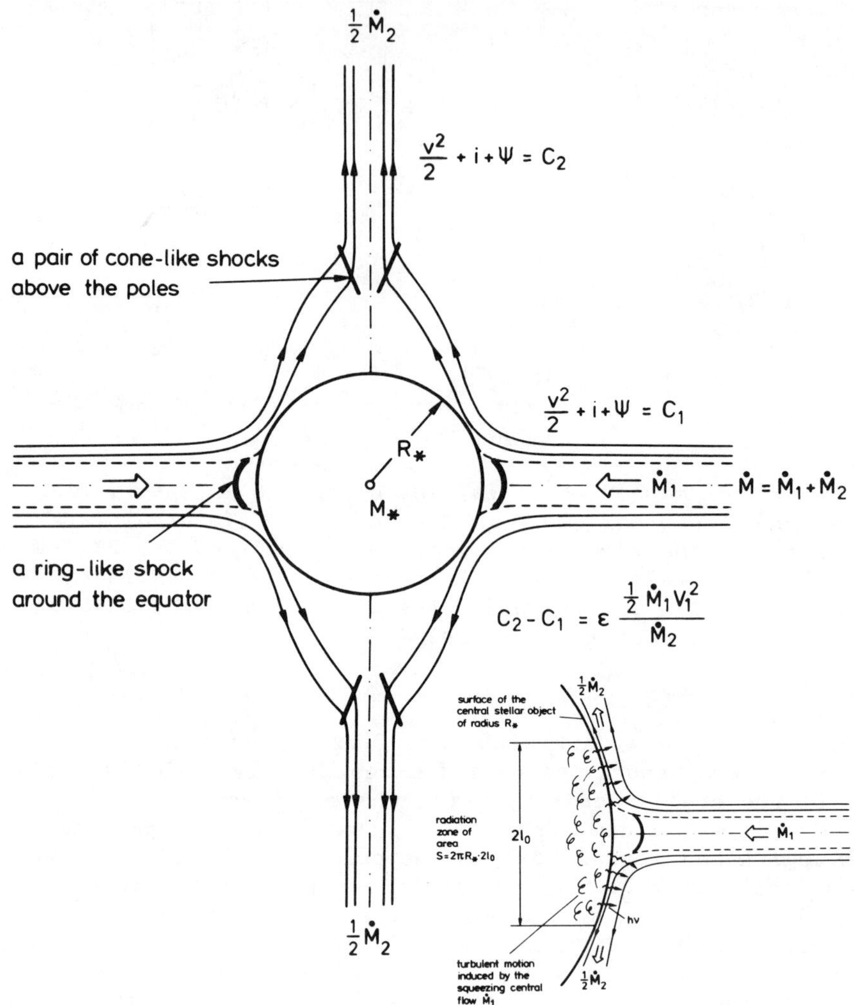

Fig. 6. Bipolar jets are induced by the radial disk-like inflow surrounding the equatorial region of the central stellar object. The radial inflow of a flux $\dot{M}$ is divided into two parts: (i) the central major part of mass flux $\dot{M}_1$ flows along the plane of the surrounding disk through a strong ring-like shock region and eventually mixes with the surface material of the stellar object. (ii) the outer mass flow $\dot{M}_2$ is hindered from falling onto the central object by an increased vertical pressure gradient which redirects this flow towards the polar regions. There it forms a collimated bipolar jet in each direction after passing through a pair of cone-like shocks above each of the poles. A small fraction of the released energy by $\dot{M}_1$ is absorbed by $\dot{M}_2$ in the form of radiation. The picture is not drawn to scale.

can all be explained by the proposed dynamical model. The interaction with the local environment into which the bipolar flow is finally ejected determines its observed geometrical shape.

REFERENCES

1. J. Cernicharo, M. Guélin, K.M. Menten, C.M. Walmsley, Astron. Astrophys. (in press 1987).
2. G. Winnewisser and E. Herbst, Topics in Current Chemistry, 139, 121 (1987).
3. S. Saito, K. Kawaguchi, S. Yamamoto, M. Ohishi, H. Suzuki, and N. Kaifu, Ap. J. (in press 1987).
4. S. Yamamoto, S. Saito, K. Kawaguchi, N. Kaifu, H. Suzuki, M. Ohishi, Ap. J. (in press 1987).
5. N. Kaifu, H. Suzuki, M. Ohishi, T. Miyaji, S. Ishikawa, T. Kasuga, M. Morimoto, and S. Saito, Ap. J. (in press 1987).
6. P. Thaddeus, J.M. Vrtilek, andC.A. Gottlieb, Ap. J. (Letters), 299, L63, (1985).
7. P. Thaddeus, S.E. Cummins, R.A. Linke, Ap. J. (Letters), 283, L45 (1984).
8. S. Yamamoto, S. Saito, M. Ohishi, H. Suzuki, S. Ishikawa, N. Kaifu, A. Murakami, Ap. J. (in press 1987).
9. T. Armstrong, L. Haikala, and G. Winnewisser, this volume.
10. Z.Y. Yue, B. Zhang, and G. Winnewisser, Astron. Astrophys. (submitted).

THE SIZE AND COMPOSITION OF INTERSTELLAR GRAINS

JOHN S. MATHIS
Washburn Observatory
The University of Wisconsin-Madison

Abstract

The observed interstellar extinction allows a measurement of the volume of interstellar grains (Purcell 1969). The result depends on the fraction of grains which are electrical conductors. For nonconducting grains, the required volume is about five times that of the silicates containing all of the available silicon. The conclusion is that the bulk of interstellar grains must be made of the common elements H, N, C, or O. Some form of solid carbon, which does not have strong bands, seems necessary to provide the necessary volume.

The following topics are discussed: (a) The forms of solid carbon: amorphous, hydrogenated amorphous, "glassy", or graphitic. (b)The nature of the 2175 Å bump, along with two stars which show well-determined short-wavelength bump positions (Cardelli and Savage 1987). (c) The difficulties of assigning graphite as the only carrier of the bump. (d) The properties of the "far-ultraviolet rise" of the extinction for $\lambda^{-1} > 6\ \mu m^{-1}$. (e) The properties of infrared emission bands, and reasons for identifying the bands with aromatic hydrocarbons of some sort, possibly the molecules with 20 - 60 carbon atoms called "polycyclic aromatic hydrocarbons". (f) The properties of red and infrared emission, and possible connections with hydrogenated amorphous carbon.

I. Introduction

Within the past three years there has been a great deal of activity in research regarding the nature of interstellar grains. The primary reasons have been (a) the identification of certain materials through their spectroscopic signatures, especially in the near-infrared region of the spectrum; (b) precise studies of differences among interstellar extinctions in various directions; and (c) laboratory studies of possibly important constituents of interstellar dust. This activity has led to both increased understanding and confusion regarding the precise nature of the grains. The confusion arises because the complexity of the true situation is appreciated better than it was previously.

There have been several recent reviews of the nature of grains. Among them are those of Tielens and Allamandola (1986a,b), which are especially complete as regards the chemistry of the grains. Mathis (1986,1987) reviewed theories of grains and the ultraviolet (UV) aspects of the observations of interstellar dust, respectively. Greenberg (1986) discussed the nature of dust in dense clouds.

The well-determined interstellar extinction law, including its variations from direction to direction, has not yet provided a unique determination of the properties of the materials in the grains. There are a variety of rather different theories which can claim to fit the observations of the average extinction as well as the polarization. Even spectral bands (mainly in the near-infrared, 3 - 20 μm) are somewhat ambiguous as to the identity of the material giving rise to each feature. Most bands are caused by the bending or stretching of a particular bond (such as the C-H), rather than by a specific molecule. However, the frequencies of the functional groups do depend somewhat on the environment of the bond which produces the absorption. One can sometimes make definite assignments of molecules on the basis of the details of the band shapes which are observed.

There is one general property of the grains in the diffuse interstellar medium (ISM) which Purcell (1969) showed can be determined directly from the known interstellar extinction law: the volume of grains per H atom, which I will call V. The relation arises from considering the entire ISM, gas plus grains, as a medium with an electric susceptibility. The fundamental Kramers-Krönig relations relate V to an integral of the observed extinction, over all frequencies. About half of the integral arises from the frequencies less than 6E13 Hz ($\lambda > 5\ \mu$m), so the integral is uncertain (factor of 2?) because of the difficulties in the determination of the infrared extinction. The unobservable ultraviolet (UV) region, wavelengths < 912 Å, contributes only 5% of the integral.

One finds that $V \approx 4.8E\text{-}27/(0.3 + 0.7F)$ cm^3 (H atom)$^{-1}$, where F is the volume fraction of grains which are electrically conducting rather than dielectric at low frequencies (a few Ghz, which is the rate of grain spinning relative to the galactic magnetic field). Greenberg (1974) obtained a similar result. In practice, F is the fraction of grains which are graphite (an electrical conductor) rather than silicate or some other nonconducting material. Note that an appreciably smaller volume of material is required if the grains are electrically conducting.

If all of the cosmic abundance of silicon is in silicates, their volume is about one-fifth of that required above if there are no electrically conducting grains. Carbon grains, if present, can contribute an addition volume of 1.3 times that of the silicates. If the carbon is graphite (which, as discussed below, is not entirely true), one needs only 1.9 times the volume of the silicates because graphite is a conductor. In this case, the graphite and silicates alone are more than enough to produce the required volume of grains. If the carbon is not an electrical conductor (i.e., is amorphous), the combined volumes of silicates and carbon are not enough to provide the volume needed, by about a factor of two, even if all of the carbon is in grains. In view of the present uncertainty of the amount of long-wavelength extinction, this factor of two is only marginally significant. Note that Purcell (1969), who considered (necessarily) only the optical part of the spectrum, obtained a lower limit on δ of only one-fifth as large as the value quoted above.

It is clear that the estimate of the volume of grains places an important constraint on their composition: the very common elements (H, O, and C, if we omit the noble gases He and Ne) must make up the bulk of the grains. Silicates or other materials containing trace elements in every molecule cannot be solely responsible for the observed interstellar extinction.

We must be careful to keep in mind the large difference between the dust in the diffuse ISM, which I will call "diffuse dust", and the dust found inside cold, dark molecular clouds. Observationally, diffuse dust can be defined by its optical extinction properties: the ratio $A(\lambda)/E(B - V) \approx 3.1$, while it is much larger in clouds. The UV extinction law deep within clouds is almost unknown because the UV extinction is so large that no observations can be made. There are many infrared absorption bands which are produced by molecular ices of CO, water, ammonia, formaldehyde, CN radicals, alcohols, and others (see Tielens and Allamandola 1986 for a review).

The only strong absorption features in diffuse interstellar dust are the well-known 9.7 μm and 19 μm silicate features and the λ2175 "bump". There is also a quite weak C-H stretch band at about 3.4 μm, discussed in §IIE below. Some material which has no sharp absorption features seems necessary.

The depletion of carbon in the ISM cannot yet be directly observed (see Jenkins 1987 for a recent review of depletions) because the element is mainly singly ionized. The available lines of C II are either too strong, in which case they are strongly saturated, or too weak for detection with the observational equipment available to date (i.e., IUE or Copernicus). Note that any present "measurements" of interstellar carbon column densities must be taken with extreme caution.

II. Continuous Interstellar Extinction and Its Interpretation

The extinction law is conventionally normalized by dividing the observed extinction of each individual star by E(B - V), its color excess between the B and V filter bands, so that the law is by definition zero at V and unity at B. This choice of normalization can have serious effects on the appearance of the law.

For wavelengths longer than the red, the extinction law appears to be almost the same in all directions, even including dust in dense regions and possibly molecular clouds (Rieke and Lebofsky 1985), provided that the law is normalized between two wavelengths > 0.6 μm. With the conventional normalization between B and V, the extinction , divided by E(B-V), is very different in dense clouds and in the diffuse ISM. This variation makes the use of E(B-V) an unfortunate choice for normalizing the extinction law.

The ultraviolet portion of the extinction can be thought of as containing three components: (a) the huge λ2175 "bump", (b) an increase in extinction with λ^{-1} beginning about 1700 Å, called the "FUV rise", and (c) a linear rise extending across the IUE spectral range ($3\ \mu m^{-1} < \lambda^{-1} < 9\ \mu m^{-1}$). The decomposition of the extinction into these three components is artificial, since a linear combination of each is also possible. However, there are surely separate carriers of the bump and of the FUV rise.

A. The Ultraviolet Bump: The properties of the 2175 Å bump have been recently explored by Massa and Fitzpatrick (1986) and Fitzpatrick and Massa (1986; FM). Their conclusion was that the central position of the bump is surprisingly fixed in wavelength: (2175 ± 9) Å. In FM they found extreme variations, among some 45 stars, of only ±17 Å. Their sample included some lines of sight which show very peculiar extinction in general, such as towards Herschel 36 in the H II region M 8, or the Trapezium in the Orion Nebula. However, recent work (see below) has shown that there are exceptions to this result. Another property of the 2175 Å bump is that the width does vary among various lines of sight. Within FM's sample, the FWHM varied by almost a factor of two. The strength of the bump also varied accordingly.

Very recently, Cardelli and Savage (1987) have discussed two stars whose 2175 Å bumps lie outside of the range quoted above: HD 62542, a seemingly normal B5 V star associated with a dark cloud near the Gum Nebula, and HD 29647, a mercury-manganese star in Taurus whose peculiar extinction had been noted by Snow and Seab (1980). Cardelli and Savage (1987) determined the extinction very carefully. By fitting the profile of the bump with an analytic function in a least-square sense, they found that the central positions for the bump for the two stars are 2110 Å for HD 62542 and 2128 Å for HD 29647. The bump in both stars is also broader than any in the FM sample, but the bumps are very weak. Their areas are 0.6 times the average area in the FM sample, which is not the smallest in the FM sample because of the large widths.

B. The Far-UV Rise: The UV extinction law also has a rapid rise with increasing wave number from 6 μm^{-1} to the largest wave numbers (about 9 μm^{-1}) available to the IUE satellite. The two stars with the peculiar position of the bump also have the most extreme amount of the FUV rise. The dust in the general Large Magellanic Cloud as well as in the 30 Doradus region (Fitzpatrick 1986) has a weak bump (at the normal wavelength), and a strong FUV rise as well. However, there are stars which have a weak bump and a weak FUV rise.

Greenberg and Chlewicki (1983) and FM found that the FUV rise has a specific spectral shape. The strength of the rise is not correlated to that of the bump, suggesting that the grains responsible for the FUV rise are not the same as those producing the bump. The FUV rise could be the short-wavelength side of a large feature similar to the 2175 Å bump but with a central wavelength about 800 Å. Small graphite grains should show such a feature near 800 Å, but too narrow to account for the FUV rise.

C. Interpretations of the Continuous Extinction

Forms of Solid Carbon: "Graphite", "Amorphous" and "Hydrogenated Amorphous": There has been a lot of confusion in the literature regarding the meanings of the terms "graphite" and "amorphous carbon." The question is well addressed in Tielens and Allamandola (1986b). In truly amorphous carbon, the C atoms are arranged at random, with an average spacing of about 2 Å. If one is given the locations of three atoms, no prediction can be made regarding the positions of any others. However, carbon has strong bonds with itself, and this situation is highly unstable. Instead, regions within the solid will form clusters of benzene rings in which the arrangement is in the form of a plane hexagon, with spacings of 1.415 Å around the hexagon. Further ordering will produce larger sheets of hexagonal structure, up to several tens of Ångstroms or more. In a sufficiently large grain, layers of these sheets will form, stacked in a specific way with 3.35 Å between layers. Except at the edges of the sheets, the layers are weakly bonded to one another. At the edges of the domains of ordered layers, there are randomly oriented bonds and also tetragonal (diamond) bonding which hold the sheets together. In the highest ordered situation, there are large ("infinite") sheets of hexagonal rings, held weakly together by van der Waals forces between the sheets. This material is graphite. From the positions of three of the atoms in one sheet of graphite, one can predict the locations of all other atoms in the solid.

In practice, then, there is a continuum of possible structures for solid carbon (even besides diamonds), ranging from completely amorphous to perfectly structured. One can define a correlation length, the distance over which one can predict the position of the C atoms from knowing the positions of those near the origin. "Amorphous" carbon, as the term is usually used astronomically, means a correlation length of some tens of Å (many benzene rings), and is not truly amorphous. Very little annealing is required to produce it from truly amorphous carbon. "Soot", produced from burning hydrocarbons in insufficient air, has such a partially ordered structure. It shows a weak x-ray diffraction pattern of the graphite structure. "Glassy" carbon mean a material which has somewhat longer correlation length than amorphous, but does not have enough order to be termed "graphite".

The diamond structure is also of astronomical interest. In it, the carbon atoms are located at vertices of tetrahedra. Diamonds of 50 Å size have actually been identified in meteorites (Lewis et al., 1987). Thus, diamond is one of the probable materials of which the grains are composed. Duley (1984) found that the condensate from laser-irradiated graphite had a considerable amount of tetrahedral symmetry. His "amorphous carbon" is, therefore, probably intermediate between diamond and the amorphous carbon as defined above.

Hydrogenated Amorphous Carbon (HAC): Amorphous carbon, as described above, can bond H atoms on its outer boundaries and within the structure, and is likely to be highly reactive with H atoms which hit its surface (Duley and Williams 1981). The resulting hydrogenated amorphous carbon (HAC) shows the C-H bands at 3.4 μm and 11.3 μm, as well as the C-C bands at 6.29 μm (Borghesi et al. 1987). HACs can possibly be identified with the "Quenched Carbonaceous Condensate" which has been studied in the laboratory (Sakata et al 1983, 1984). This material produced by electric discharges through hydrocarbon gases such as methane. Presumably it has a very disordered structure, but shows many of the infrared emission bands, as well as an absorption at near 2200Å. This ultraviolet feature seems to be broader than the interstellar bump.

The Case Against Graphite Producing the Bump and the Continuous Extinction: For several years, it seemed very reasonable to assume that interstellar extinction was produced by a mixture of uncoated graphite and amorphous silicate grains. Graphite has an absorption which should peak at the energy corresponding to the maximum difference in density of states. A distribution of grain sizes can reproduce the extinction over the entire observed range of 0.1 - 20 μm very well (Mathis et al. 1977; Draine and Lee 1984; Rowan-Robinson 1986). These models cannot now be considered to be correct. The problem is that the central wavelength of the graphite absorption depends on particle size (Gilra 1972; Savage 1975). In large grains, the bump is broadened and shifted to longer wavelengths. The width should be well correlated with the central wavelength. This effect is not observed (FM). One possibility is to have very small graphite particles (size $\ll \lambda/2\pi = 350$ Å) produce the bump, in which case the wavelength dependence of extinction is independent of the size. This was suggested by Hecht (1986), who attributed the bulk of the solid carbon to HACs. The lack of hydrogen for the small particles causing the bump might be because they are heated sufficient by a single photon to anneal themselves into graphite and expel the hydrogen. The variation of the width of the bump could be caused by either size or temperature effects or (more likely) by impurities.

The discovery of at least two stars with bumps differing from the standard position (Cardelli and Savage 1987; see above) is most interesting. Coatings might produce a shift of the bump to shorter wavelengths only if the coating material has optical properties resembling a metal, with a small, or even negative, real part of the dielectric constant. It is also interesting that the observed position of the bump in these two stars is in fact where it is predicted to be from the optical constants of Taft and Philipp (1965) for small particles-- the position of the bump as observed was explained as being shifted to longer wavelengths because of the mix of finite-sized grains. Perhaps the "normal" bump, at 2175 Å, might be at a shifted wavelength. Such a shift can be caused by a variety of effects, such as holes or impurities within the material, as well as by coatings on the grains.

E. Infrared Absorption Bands: The 3.4 μm band is important because it is caused by the C-H stretch which several possible grain materials should have: hydrogenated amorphous carbon (HAC), the polycyclic aromatic hydrocarbons (PAH) molecules, and organic refractory mantles left from the chemical reactions of free radicals of the icy mantles which grains have inside of molecular clouds. The band is so weak that it is seen only towards IRS 7 (A(V) = 30 mag.), in the direction of the galactic center (Butchart et al. 1986). The central optical depth is about 0.3. As far as I can tell, there is a complete lack of the 3.4 μm C-H stretch in the spectrum of VI Cygni No. 12 (Gillett et al. 1975), roughly < 0.03, although A(V) = 10 mag. This is about one-third as much as would be expected if the 3.4 μm band/A(V) ratio were the same for it and IRS 7. Is local dust deficient in the strength of the C-H stretch? I would hesitate to conclude that on the basis of comparing only two objects.

The 9.7 μm band is seen in both emission and absorption in many objects (see Savage and Mathis 1979 for references), providing that they are oxygen-rich (O/C > 1). In particular, it is observed in the circumstellar envelopes of M stars. Its great breadth is fitted very well by laboratory constants of amorphous, but not crystalline, silicates (Krätschmer and Huffman 1979). The identification with silicates, rather than with SiO or other oxides, is strengthened by the presence of a band peaked about 19 μm (Forrest, McCarthy, and Houck 1979; McCarthy et al. 1980). The shapes of these bands were recently used by Draine and Lee (1984) to determine the optical constants of "astronomical silicate". Their identification has recently been strengthened by the measurement (Aitken et al. 1987) of the polarization as well as the extinction in the object AFGL 2591, in which the observed profiles of both bands are very well reproduced by the "astronomical silicate" results.

The interstellar polarization of the 19 μm band (Aitken et al. 1987) apparently rules out the suggestion (Duley and Najdowsky 1983) that the 9.7 μm band is caused by tiny MgO particles as well as possibly from SiO and FeO, since MgO is too symmetrical (it has a cubic structure) to produce polarization, SiO has no band at 20 μm, and FeO has too little absorption on the short side of 19 μm (Brehat et al. 1966) to fit the observations.

III. Infrared Emission Bands and Other Grain Diagnostics

The Unidentified Infrared Emission Bands (UIBs): There are intense emission bands at 3.3, 6.2, 7.7, 8.6, and 11.3 μm in the spectrum of a wide variety of objects (Cohen et al. 1986 and refs. therein): among others, the very active galaxy M82, many C-rich planetary nebulae, H II regions, and in a wide variety of "reflection" nebulae. The wide variety of conditions under which they can be produced strongly suggests that a single process is responsible for the emission. The bands are emitted with generally similar strengths in this wide variety of objects and radiation fields, suggesting that absorption of a single photon by a grain (or large molecule) is likely to cause the entire process. However, there is some variation of the relative strengths of the UIBs from place. This variation can be caused by either differing compositions of the specific particles which are producing the emission, or else by differences in the excitation energy per mode (Barker et al. 1987).

There is an emission continuum often seen along with the UIBs, extending from about 5400 Å redward through at least 7000 Å, but which has a strength which can vary significantly from place to place even within the same reflection nebula.

This continuum is very strong in HD 44179 (the "Red Rectangle"). A continuum also underlies the UIB emission and extends to 13 μm (Cohen et al. 1985).

The UIBs and/or the associated continuum contribute significantly to the 12 and 25 μm IRAS filter signals in both reflection nebulae (Castelaz et al. 1986; Cohen et al. 1987) and (presumably) to the diffuse "cirrus" clouds, which have much stronger short-wavelength IRAS emissions that would be expected from the cold dust contributing to the 60 and 100 μm signals. There is a wide variation from cloud to cloud in the 12 μm strength relative to the extinction or to the 100 μm emission (de Vries 1985). The excess 12 and 25 μm emission from "cirrus" is correlated with the presence of CO clouds (Weiland et al. 1986).

The UIBs can be produced by either HACs or a smaller, more ordered form of aromatic hydrocarbon, the "polycyclic aromatic hydrocarbons", or PAHs. Either of these might be able to explain the UIBs and the red continuum (Duley 1985; d'Hendecourt et al. 1986).

Polycyclic Aromatic Hydrocarbons ("PAHs"): Duley and Williams (1981) suggested that two of the UIBs are characteristic of aromatic molecules. Léger and Puget (1984) pointed out that these bands can be accounted for by PAHs, which are molecules of graphitic-like structure (rows of adjacent benzene rings), with hydrogens on some of the vacant sites at the edges. Independently, Allamandola et al. (1985) proposed a similar explanation; they showed that the UIB emission in the Orion Bar could be fitted by the spectrum of automobile exhaust, which is a mixture of PAHs and soot, or HAC. The evidence for PAHs has become stronger since that time (see Léger and d'Hendecourt 1986; Tielens and Allamandola 1986). The UIBs are fitted well (but not perfectly, as we shall discuss) in both wavelength and intensity by a mixture of PAHs of various sizes, up to about 50-60 carbons each. PAHs can explain the surprisingly bright 12 μm emission from the "cirrus" discovered by IRAS (Puget et al. 1985). About 6% of the cosmic carbon in PAHs would explain the intensities of the UIBs in reflection nebulae (Léger and Puget 1984) . Such molecules are stable against dissociation by UV radiation because the photon's energy is shared by the lattice, but Duley and Williams (1986) have emphasized that chemical reactions can destroy PAHs rather quickly in the diffuse cloud environment.

There is an extensive discussion by Omont (1986) regarding the nature of PAHs, as well as an entire volume (Léger et al. 1986) devoted to their properties as well as those of solid carbon.

The mechanism of producing the emission is the radiative de-excitation of the PAH molecule following the absorption of a single UV photon, say of ten eV. This photon produces an excitation of the neutral PAH to an excited electronic level, whence it cascades down with the emission of the bands. The spacing of the rotational levels is quite even, so many infrared photons are emitted at almost the same wavelength. The evidence for the process occurring is (a) the coincidence of the several bands with laboratory measurements of a variety of PAHs, averaged over those species, or with automobile exhaust which contains PAHs. (b) The similarity of the UIBs in various objects and at various radiation fields. (c) The energetics are reasonable from single photons with plausible efficiencies.

The PAHs are probably not mainly neutral in space; they can be positively and negatively charged. The comparison of laboratory spectra and the observed UIBs is often done by considering neutral PAHs. This comparison may be necessary because of the paucity of studies of the spectra of ionized molecules, but it is not really appropriate. Laboratory studies of both positively and negatively ionized PAHs would be useful.

While the agreement of the infrared spectrum of the PAHs and celestial emission bands is very suggestive, I have some reservations about assuming that they are a major component of interstellar extinction. For one thing, they have very strong absorptions in the UV spectral region (4 - 10 eV; 0.3 μm > λ > 0.12 μm). For free molecules like PAHs, these absorptions are quite narrow, yet there are no spectral features at all in the UV spectrum other than the 2175 Å bump. Can there always be such a mixture of PAHs of various types, such that their individually sharp spectra merge together into a rather featureless blend? Materials in solid particles, or perhaps coated onto grains, are more likely to achieve this condition than free molecule ions. It would be surprising if PAHs are responsible for the 2175 Å bump, with its width varying from object to object, but with a central wavelength that is almost invariant. Similarly, the IR emissions of individual PAHs are quite narrow, and do not fall on precisely the wavelengths of the observed emission bands. Perhaps one can avoid these objections if there is always a stochastically established distribution of PAHs, with a sufficiently smooth spectrum in both the infrared and UV, but I doubt it. Donn et al. (1987) have emphasized that when the IR emissions of many (neutral) PAHs are averaged, the resulting bands are not precisely coincident with the unidentified infrared emission bands. Furthermore, the emission presumably occurs because a single PAH molecule is excited by an UV photon of several volts' energy, and this energy is then internally redistributed among the excited states of the molecule. The temperature equivalent to this level of excitation is hundreds of degrees Kelvin. In this case, one should see emission from the excited states of the molecule; there should be emission bands in the optical part of the spectrum as well as the IR.

I emphasize that there seems to be little doubt that some sort of hydrogenated aromatic hydrocarbon in responsible for the infrared emission bands. The problem is, that there are many possibilities of varying degrees of disorder, including HACs and the materials produced by discharges through simple hydrocarbons.

IV. Summary and Prospects

I think that the present activity in research on grains has resulted in a major improvement in our understanding of their nature, although the answer seems to be that they are more complicated than I at one time believed. A mixture of amorphous silicates and several forms of carbon, some with a small amount of associated hydrogen, can account for most of the properties of the diffuse interstellar dust: the amount of extinction, the infrared spectral bands (at least semiquantitatively), and possibly the 2175 Å bump.

It seems to me that there is a good likelihood that there is a regular progression in sizes of carbon particles, starting from polycyclic aromatic hydrocarbons (PAHs), including ions. I would say that a very few percent (say, < 2%) of the carbon is in this form. Somewhat larger and more complicated, but not large (say, 10 - 50 Å), hydrogenated amorphous carbon (HAC) grains could provide some of the UIBs and some of the red and IR continuum which is seen in varying amounts from place to place (while the PAHs produce mostly the rather narrow bands). The excitation of the HAC particles of this size might be a problem. The HACs, or amorphous carbon form the large ones because the H atoms on the edges do not affect their spectral properties. These particles will not show the 2175 Å bump because they will not have the continuous density of states possessed by graphite. The case for very extensive mantles on the grains is weak because of the lack of the C-H bond in VI Cyg No. 12. However, there are many unanswered questions. What is the precise origin of the 2175 Å bump: if it is small graphite flakes, with impurities to broaden it by various amounts, how can two (or more!) stars have bumps at shorter wavelengths? What causes the FUV rise, which is not silicaceous but is not correlated with the bump? Can PAHs survive chemically in the ISM? There are certainly plenty of mysteries for future research!

The work leading to this paper has been partially supported by NASA.

References

Aitken, D. K., Roche, P. F., Smith, C. H., James, S. D., and Hough, J. 1987, *M.N.R.A.S.*, submitted.

Allamandola, L. J., Tielens, A. G. G. M., and Barker, J. R. 1985, *Ap. J. (Lett.)*, **290**, L25.

Barker, J. R., Allamandola, L. J., and Tielens, A. G. G. M. 1987, *Ap. J. (Lett.)*, **315**, L61.

Borghesi, A., Bussoletti, E., and Colangeli, L. 1987, *Ap. J.*, **314**, 422.

Brehat, F., Evrard, O., Hadni, A., and Lambert, J.-P. 1966, *C. R. Acad. Sci. Paris*, **263**, 1112.

Butchart, I., McFadzean, A. D., Whittet, D. C.B., Geballe, T. R., and Greenberg, J. M. 1986, *Astr. Ap.*, **154**, L5.

Cardelli, J. A., and Savage, B. D.. 1987, *Ap.J.*, submitted.

Castelaz, M. W., Sellgren, K., and Werner, M. W. 1986, *Ap. J.*, **313**, 853.

Chlewicki, G., et al. 1987, *Astr. Ap.*, **173**, 131.

Cohen, M., et al. 1986, *Ap. J.*, **302**, 737.

Cohen, M., Tielens, A. G. G. M., and Allamandola, L. J. 1985, *Ap. J. (Lett.)*, **299**, L93.

d'Hendecourt, L., Léger, A., Olofsson, G, and Schmidt, W. 1986, *Astr. Ap.*, **170**,91.

de Vries, C. P. 1985, *Mitt. Astr.Ges.*, **63**, 158.

Désert, F. X., Boulanger, F., Léger, A., Puget, J. L., and Sellgren, K. 1986, *Astr. Ap.*, **159**, 328.

Donn, B., Khanna, R., and Salisbury, D. 1987, *B. A. A. S.*, **18**, 1030.

Draine, B. T., and Lee, H.-M. 1983, *Ap. J.*, **285**, 89.

Duley, W. W. 1984, *Ap. J.*, **287**, 694.

- - - -. 1985, *M. N. R. A. S.*, **215**, 259.

Duley, W. W., and Najdowsky, I. 1983, *Ap. Sp. Sci.*, **95**, 187.

Duley, W. W., and Williams, D. A. 1981, *M. N. R. A. S.*, **196**, 269.

- - - -. 1986, *M. N. R. A. S.*, **219**, 859.

Fitzpatrick, E. L. 1986, *A. J.*, **92**, 1068.

Fitzpatrick, E. L., and Massa, D. 1986, *Ap. J.*, **307**, 286 (FM).

Forrest, W. J., McCarthy, J. F., and Houck, J. R. 1979, *Ap. J.*, **233**, 611.

Gillett, F. C.,Jones,T.W., Merrill,K.M., and Stein,W.A. 1975, *Astr. Ap.*, **45**, 77.

Gilra, D. P. 1972, in *Scientific Results from the OAO-2*, ed. A. D. Code (NASA, Sp-310), p. 295.

Greenberg, J. M. 1974, *Ap. J. (Lett.)*, **189**, L81.

- - - -, 1986, in _Light on Dark Matter_, F. P. Israel, ed. (Reidel: Dordrecht), p. 177.

Greenberg, J. M., and Chlewicki, G. 1983, *Ap. J.*, **272**, 563.
Hecht, J.H. 1986, *Ap. J.*, **305**, 817.
Jenkins, E. B. 1986, in *Intersetllar Process*, ed. D. A. Hollenbach and H. A. Thronson (Reidel: Dordrecht), p. 533.
Joyes, P. 1986, in *Polycyclic Aromatic Hydrocarbons and Astrophysics* , eds. A. Léger, L. d'Hendecourt, and N. Boccara (Reidel: Dordrecht), p. 15.
Krätschmer,W., and Huffman, D.R. 1979, *Ap. Sp. Sci.*, **61**,195.
Léger, A., and d'Hendecourt, L. 1986, in *Polycyclic Aromatic Hydrocarbons and Astrophysics* , eds. A. Léger, L. d'Hendecourt, and N. Boccara (Reidel: Dordrecht), p. 223.
Léger, A., and Puget, J. L. 1984, *Astr. Ap.*, **137**, L5.
Léger, W., d'Hendecourt, L., and Boccara, N. 1986, *Polycyclic Aromatic Hydrocarbons and Astrophysics*, Reidel (Dordrecht).
Lewis, R. S., Wacker, J. F., Anders, E., and Steel, E. 1987, *Nature*, **326**,160.
Marchand, A. 1986, in *Polycyclic Aromatic Hydrocarbons and Astrophysics*, eds. A. Léger, L. d'Hendecourt, and N. Boccara (Reidel: Dordrecht), p. 31.
Massa, D., and Fitzpatrick, E. L. 1986, *Ap. J. Suppl.*, **60**, 305.
Mathis, J.S. 1986a, in Light on Dark Matter, F. P. Israel, ed. (Reidel: Dordrecht), 171.
----, 1986b, in *Interrelationships Among Circumstellar, Interstellar, and Interplanetary Dust*, Proceedings of a Conference held at Wye, Maryland, NASA Conf. Publ. 2403, p. 29.
----. 1987, Chapter 22 in *Scientific Accomplishments of the IUE*, eds. Y. Kondo et al. (Reidel: Dordrecht) p. 517.
Mathis, J. S., Rumpl, W., and Nordsieck, K. H. 1977, *Ap. J.*, **217**, 425.
McCarthy, J. F., Forrest, W. J., Briotta, D. A.., and Houck, J. T. 1980, *Ap. J.*, **242**, 965.
Omont, A. 1986, **164**, 159.
Puget, J. L., Léger, A., and Boulanger, F. 1985, *Astr. Ap.*, **142**, L19.
Purcell, E. M. 1969, *Ap. J.*, **158**, 433.
Rieke, G. H., and Lebofsky, M. J. 1985, *Ap. J.*, **288**, 618.
Rowan-Robinson, M. 1986, *M. N. R. A. S.*, **219**, 737.
Sakata, A., Wada, S.,Okutsu, Y., Shintani, H., and Nakada,Y. 1983, *Nature*, **301**, 493.
Sakata, A., Wada, S., Tanabe, T. and Onaka, T. 1984, *Ap. J.*, **287**, L51.
Savage, B. D. 1975, *Ap. J.*, **199**, 92.
Savage, B. D., and Mathis, J. S. 1979, *Ann. Rev. Astr. Ap.*, **17**, 73.
Schild, R. E. 1977, *A. J.*, **82**, 337.
Snow, T. P., and Seab, C. G. 1980, *Ap. J. (Lett.)*, **242**,L83.
Taft, E. A., and Philipp, H. R. 1965, *Phys. Rev.*, **138**, A197.
Tielens, A. G. G. M., and Allamandola, L. J. 1986a, *The Composition, Structure, and Chemistry of Interstellar Dust*, NASA Technical Memorandum 88350, presented at the summer school on Interstellar Processes, Jackson, Wyoming, July 1986.
Tielens, A. G. G. M., and Allamandola, L. J. 1986b, *Evolution of Interstellar Dust, in Physical Processes in Interstellar Clouds*, Proceedings of a meeting held in Irsee, Germany, August, 1986, Eds. G. Morfill and M. Scholer.
Weiland, J. L., Blitz, L., Dwek, E., Hauser, M. G., Magnani, L., and Rickard, L.J. 1986, *Ap. J. (Lett.)*, **306**, L101.
Witt, A. N., Schild, R. E., and Kraiman, J. B. 1984, *Ap. J.*, **281**, 708.

"INFRARED CIRRUS": NEW LIGHT ON THE INTERSTELLAR MEDIUM

MICHAEL G. HAUSER
Laboratory for Astronomy and Solar Physics, Goddard Space Flight Center, Greenbelt, Maryland 20771, U.S.A.

Abstract The ability of the Infrared Astronomical Satellite to measure emission from interstellar dust on angular scales from a few arcminutes to the whole sky has provided a powerful new tool for interstellar medium studies. This emission, referred to as 'infrared cirrus', shows in detail the global distribution of the material long known to obscure shorter wavelength observations of our Galaxy and beyond. It also provides new insights on topics ranging from the nature of interstellar grains to the interpretation of the infrared energy distributions of galaxies. The high galactic latitude cirrus is correlated both with atomic gas and with diffuse molecular clouds in the solar vicinity. The latter clouds are particularly interesting since they resemble neither normal diffuse molecular clouds nor dark clouds in their composition. The status of both global and specific cirrus cloud studies is reviewed, and implications for dust and cloud properties are discussed.

INTRODUCTION

The terminology 'infrared cirrus' was introduced just over three years ago by Low *et al.*[1] to describe complex extended sources of infrared emission apparent in the data from the Infrared Astronomical Satellite (IRAS).[2] The term initially referred to emission identified both with interplanetary and interstellar dust sources, but the only identified interplanetary 'cirrus' source to date is commonly referred to as 'zodiacal dust bands'. The subsequent literature has commonly used 'cirrus' to denote emission from tenuous dust clouds scattered throughout the interstellar medium[3] (ISM). I will adopt the latter usage in this review, with emphasis on the component seen at high galactic latitudes.

The IRAS measurements of infrared cirrus provide a powerful new tool for study of the ISM. This arises in part because IRAS was sensitive to very small amounts of interstellar material. For example, assuming 20 K dust of typical abundance and radiative properties, IRAS data smoothed to 1/2-deg sampling readily reveal clouds with atomic hydrogen column density $N_{HI} \sim 10^{19}$ cm^{-2}, velocity-integrated CO brightness $W_{CO} \sim 10^{-2}$K km/sec, or extinction $A_V \sim 0.01$ mag.[4] The IRAS data are also of particular value because they cover essentially the entire sky,[2] and because they provide new information on cloud mass and energetics, grain properties, and local radiation fields.

Other motivations for study of the infrared cirrus, though of less direct relevance to this Symposium, might be mentioned. An understanding of the contribution of cirrus clouds to the total infrared emission of our Galaxy facilitates interpretation of the energy distributions of external galaxies[5]. Direct mapping of the total dust column density responsible for optical extinction bears directly on determination of the polar extinction,[6] and hence on the extragalactic distance scale. Measurement of the angular distribution of the dust provides a check on determinations of the correlation function of external galaxies.[7] Finally, characterization of the brightness and angular distribution of the cirrus emission itself is essential to understanding natural limitations to the sensitivity of future space infrared observations.

As a result of the many motivations suggested above, there is a burgeoning infrared cirrus literature. My aims are to review the observational data and to summarize some of the implications for the interstellar medium. Prospects for future cirrus measurements are also briefly mentioned. Additional information and references may be found in the recent review by Beichman.[8]

CIRRUS OBSERVATIONS

Morphology: individual clouds

The presence of cold dust in the ISM was, of course, well-known prior to IRAS, and it came as no surprise that IRAS data revealed clumpy, extended 100 μm emission. The strength of the associated 60 μm emission, implying dust equilibrium color temperatures ~ 34 K in the clouds initially studied by Low *et al.*,[1] was somewhat surprising, but heating of graphite grains by typical interstellar radiation fields offered a plausible explanation. The infrared cirrus was early recognized to be often associated with HI clouds.[1,9-12] Numerous investigators have shown that within a cloud there is a rather close correlation between 100 μm brightness, I(100), and HI column density.[9,11-13] Variations in the $I(100)/N_H$ ratio from cloud to cloud are measures of differences in local radiation field or dust-to-gas mass ratios. Cirrus emission has also been noted to be correlated with other indicators of dust, such as extinction,[14] faint high latitude optical reflection of general galactic starlight,[15] and classical reflection nebulae.[16]

One very surprising feature of the cirrus emission has been its brightness at 12 and 25 μm wavelengths, where equilibrium thermal emission from ~ 30 K grains would be totally negligible. The IRAS data often show similar cloud morphology at these wavelengths as they do at the longer wavelengths.[9,16-18] As discussed below, this 'warm cirrus' component has very interesting, though not fully understood, implications for the character of the dust.

A second surprise, particularly in the case of high latitude cirrus, has been the discovery that some of the emission arises in dominantly molecular, rather than atomic, gas regions. This has been shown both by direct correlation of recently discovered, local (high latitude) molecular clouds[19,20] with IRAS data,[18] and by CO searches guided by IRAS cirrus data.[13] These tenuous molecular clouds are typically seen as cores of larger regions of cirrus and HI emission, and have molecular abundances more similar to those of classical dark clouds than of diffuse clouds.[21] They may hold clues to gas phase transitions between the atomic and molecular states.

A summary of the correlation between infrared, optical, and radio measures of cloud content for some representative cirrus regions is given in Table 1. Clouds which are predominantly atomic are listed separately from those with substantial molecular gas. The ratio $A_V/I(100)$ is shown only for clouds where extinction has been measured from star counts (so that it is independent of gas phase measurements). Clearly more such measurements are needed, but the mean value for the 3 clouds shown is $(.08\pm.04)$ mag/(MJy/sr). For the regions in Table 1, which generally lack strong local heating sources, the mean ratio of I(100) to total hydrogen column density, N_H, is (1.4 ± 0.7) and (0.4 ± 0.2) for the atomic and molecular cirrus clouds respectively, both in units of MJy/sr per 10^{20} H cm^{-2}. The relatively modest dispersion, at least within a particular class of cloud, suggests that the IRAS 100 μm brightnesses are a useful quantitative tracer of the mass. The apparent difference between the means for the atomic and molecular clouds may indicate systematic calibration error in the determination of H_2 column density from CO measurements for these relatively low optical depth molecular clouds. de Vries _et al._[13] have, in fact, argued on the basis of HI, CO, and 100 μm measurements of a cloud complex in Ursa Major that the infrared measurements are a more reliable gas mass tracer for tenuous molecular clouds than masses derived from CO measurements using the $N(H_2)/W_{CO}$ ratio appropriate for giant molecular clouds.

Morphology: global distribution

The large scale infrared cirrus distribution has been mapped in

TABLE 1 Correlation of cirrus with extinction and gas column density

Source	$A_V/I(100)$ [mag/(MJy/sr)]	$I(100)/N_H$ $\frac{\text{MJy/sr}}{10^{20}\ \text{H cm}^{-2}}$	Ref.
Atomic gas regions			
ESO plates 025 & 487	.11		14
A		1.6	1
B		.9	"
C		.3	"
D		1.0	"
field at b~60°		1.4	7
field at l~125°,|b|<10°		.6	11
field at l~225°,|b|<10°		.4	"
near Magellanic Stream		.7	10
average for b>50°		.8	12
average for b<-50°		.7	"
Auriga		.5	"
Lupus		2.1	"
Orion		1.1	"
Ursa Maj		1	
Molecular gas regions			
MBM 16*	.04	.6	18
MBM 20*	.1	.6	"
MBM 30*		.3	"
MBM 26*		.2	22
MBM 54-55*		.2	"

*Cloud from the survey of Magnani, Blitz, and Mundy, ref. 20.

several ways. A 100 μm cirrus map obtained by filtering the IRAS scan data to remove low angular-frequency structure shows that the global distribution of cirrus clouds outside of the galactic plane (|b|>15°) exhibits many of the same features as the HI distribution.[3] Good has recently prepared all-sky cirrus maps in all four bands by fitting a spline curve to the lower envelope of the data from each scan and creating a map of the residuals from the spline curves.[23] While this technique leaves prominent structure within about ten degrees of the ecliptic plane due to the zodiacal dust bands, correlated interstellar features are evident at all four IRAS survey wavelengths. These features persist, though with decreasing density, to the galactic poles.

An alternative approach to isolation of infrared emission from the ISM has been taken by Boulanger and Perault,[12] who have carefully modeled and then subtracted the interplanetary dust emission from the IRAS data, producing maps of the residual infrared emission over the sky. They have used these maps to study individual cloud regions, and also have averaged their maps over galactic longitude to study the variation of cirrus emission out of the galactic plane with galactic latitude. The resulting brightness distributions are very similar to the velocity-integrated HI distribution averaged in the same fashion, varying roughly linearly with cosec(|b|) at all wavelengths for $|b| \sim 10^{o}$-90°.

The global distribution of cirrus clouds associated with molecular gas is less well known at present, since we lack a complete survey of high-latitude molecular clouds. However, work in progress by Blitz and Gir[24] shows that the known molecular cirrus clouds tend to lie on prominent filaments and loops of material, suggesting that shocks may play an important role in the evolution of these clouds.

Energy Distribution

As indicated above, one of the most surprising properties of the infrared cirrus is the strong emission at the short IRAS wavelengths. This is illustrated in Fig. 1, which shows the luminosity per unit mass for a number of representative regions. The short wavelength emission, typically ~ 25% of the total emission in these cases, clearly cannot be thermal equilibrium radiation from the same grains producing the long wavelength emission.

The regions represented in Fig. 1 are characterized by having low-to-modest UV optical depth and no strong local heating

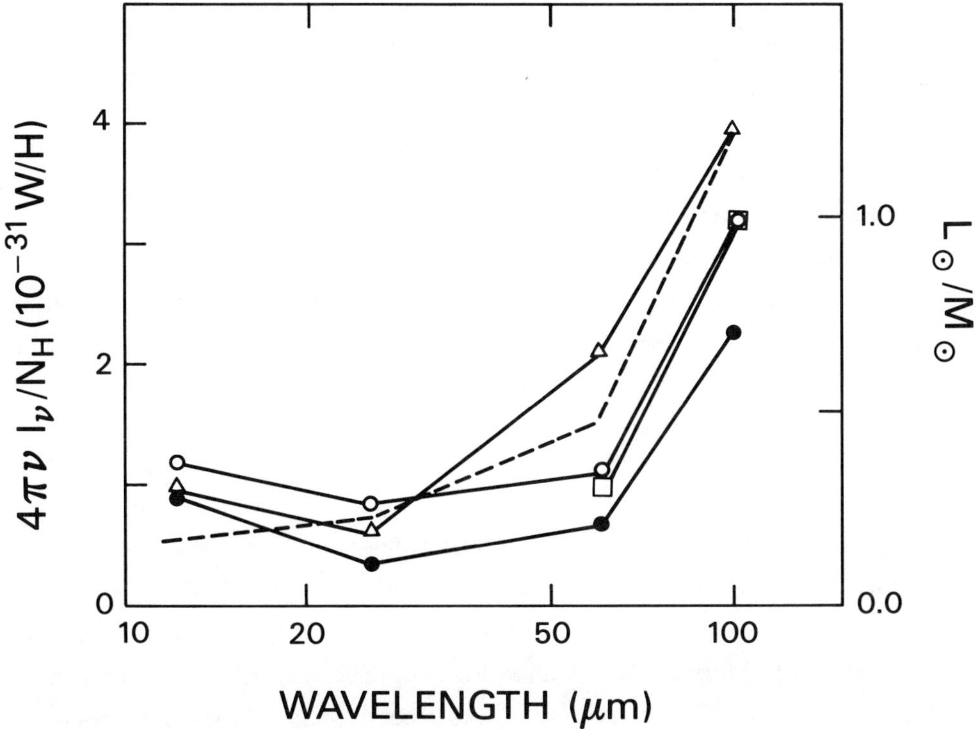

FIGURE 1 Energy distribution of selected infrared cirrus regions. Open symbols represent IRAS data for atomic gas-cirrus regions: squares, cloud B (ref. 1); triangles, cloud at b ~ 60° (ref. 9); circles, average for |b| > 50° (ref. 12). Closed circles show the distribution for a typical molecular gas-cirrus cloud (cloud 20 of ref. 20, as given in ref. 18). The dashed line is the calculated distribution for a model including transient heating (ref. 27; see text).

sources, i.e., all are predominantly heated by their local interstellar field. The ratio of 60 μm to 100 μm intensity implies a color temperature ~ 30 K, whereas the 12 μm to 25 μm intensity ratio implies T ~ 300 K. Several recent studies have explored the variation of infrared color with radiation density. Castelaz et al.[16] showed that in the Pleiades, with a radiation field of order 10 times the interstellar field in the solar vicinity, I(12)/I(25) still implies a color temperature close to 300 K. However, I(60)/I(100) implies an appreciably elevated temperature, T ~ 40-50 K. They further showed that, whereas I(60)/I(100) decreases with increasing distance from the exciting star 23 Tau, I(12)/I(25) remains quite constant. Such studies have been extended to even more intense fields in the vicinity of very luminous stars in the ρ Oph (Ryter *et al.*[25]) and California nebula (Boulanger *et al.*[26]) regions, showing a definite anti-correlation of I(60)/I(100) with I(12)/I(25) as radiation intensity increases to 100 times local interstellar values.

DISCUSSION

The energy distribution of the infrared cirrus provides new insight into the size distribution and character of interstellar dust. Low *et al.*[1] initially interpreted the long-wavelength IRAS emission as predominantly thermal equilibrium emission from small graphite grains, and this has been supported by detailed models.[27,28] Though Harwit, Houck, and Stacey[29] noted that part of this radiation might arise from atomic fine-structure lines, present evidence, such as the strong correlation of I(100) with N_H, does not favor this interpretation.[11] Draine and Anderson[27] suggested that the short wavelength cirrus emission could arise from very small grains transiently heated by single UV photons, much as the near-infrared continuum of reflection nebulae was explained by Sellgren.[30] Sufficiently small grains, e.g., graphite grains smaller than 20 Å, can reach temperatures of several hundred degrees K by absorbing a single photon. Assuming that the grains cool from each heating event by thermal emission, such temperatures are adequate to produce the observed 12 and 25 μm emission as well as part of the 60 μm emission. The dashed line in Fig. 1 illustrates the expected energy distribution from one of the Draine and Anderson models of grain-size distribution (power law spectral index of −3.5 down to 100 Å radius, power law index of −4 from 3 Å to 100 Å) in a radiation field equal to the local interstellar field.[31] This model generally accounts for the observed energy distribution, though the parameters can be adjusted to improve the match to the data in any particular case (e.g., see Weiland et al.[18]).

A variation on this explanation, modeled initially by Puget, Leger, and Boulanger[32] and subsequently by various authors,[33,34] accounts for the short wavelength emission, especially in the 12 μm IRAS band, as line emission from large polycyclic aromatic hydrocarbon (PAH) molecules excited transiently by single UV photons.[35,36] The so-called 'unidentified infrared lines' at wavelengths of 3.3, 6.2, 7.7, 8.6, and 11.3 μm, which appear in the spectra of regions such as reflection nebulae[37] and which partially fall in the IRAS 12 μm band, have been identified with the bending and stretching modes of such molecules. Sensitive spectroscopic observations are needed to confirm or deny the presence of these lines in the low surface brightness IRAS cirrus clouds, and hence to clarify the character of their emitting material.

Regardless of the detailed physics of the very small grains, the transient stimulation hypothesis receives further support from the variation of infrared colors with radiation density. In the Pleiades,[16] the decrease of I(60)/I(100) with increasing distance from 23 Tau supports the interpretation of this radiation as largely thermal equilibrium emission. The constancy of I(12)/I(25) to distances ~ 1 pc from the star is the expected behavior for transient heating, where the emission spectrum depends more upon the character of the 'grains' and the energy of individual photons than upon the photon flux. At the still higher energy densities studied by Ryter et al.[25] and Boulanger et al.,[26] I(12)/I(25) decreases substantially with increasing energy density, suggesting progressive destruction of the particles responsible for the 12 μm emission. This behavior is, in any event, quite unlike thermal equilibrium emission.

It is thus quite clear that the 'cold' (60 and 100 μm) and 'warm' (12, 25, and some 60 μm) cirrus emission typically arise from different components of the ISM via different radiation mechanisms. The short wavelength infrared cirrus measurements have made us keenly aware of a very small grain component of the ISM, with many important implications for grain properties, evolution, and cloud chemistry just beginning to be explored.[38] Derivation of basic cloud physical properties must also bear in mind the multiple infrared emission mechanisms. Optical depths and dust column densities calculated from the observed 60 and 100 μm intensities with the assumption of an isothermal cloud in equilibrium (i.e., ignoring the small grain contribution at 60 μm) can be substantially too low.[18]

The generally good correlation of infrared cirrus and gas-phase measurements (Table 1) show the IRAS maps to be a valuable resource for further exploration of the ISM. As an example of work in progress, Desert[39] has constructed a catalog of some 500 'infrared excess' cirrus clouds, i.e., clouds with substantially more 100 μm emission than expected on the basis of their HI emission. Such clouds can be expected either to have substantial molecular content, strong local heating sources, or anomalous dust-to-gas ratios. About half of the high latitude molecular clouds found by Magnani, Blitz, and Mundy[20] (and only a few of the several hundred non-detections in their survey) appear in Desert's catalog, evidence supporting the utility of this approach.

SUMMARY AND FUTURE MEASUREMENT PROSPECTS

The infrared cirrus component of the IRAS sky maps is providing a pervasive and revealing look at the ISM. The surprisingly broad infrared energy distribution of the cirrus has called attention to a very small grain and/or large molecule population not adequately recognized previously. Further studies promise to enhance our understanding of both the character of the grains and their evolution. The predominant cirrus emission at high galactic latitudes, both in discrete complexes and the large scale average, appears to be attributable to dust in atomic gas clouds heated by the local interstellar field. Interesting exceptions, however, are clouds with cores of substantially molecular material. These clouds promise to be an important new astrophysical laboratory, perhaps shedding light on processes responsible for gas-phase transitions between the atomic and molecular states, and on the formation and/or destruction of the larger molecular clouds. Infrared data are now an important tool for ISM studies, not only for the giant molecular clouds and HII regions as was recognized prior to IRAS, but also for the pervasive, tenuous atomic and molecular clouds.

The IRAS cirrus measurements have advanced interstellar medium studies from previous epochs, when astronomers discovered dusty clouds by blindly 'bumping into' them as they groped through the ISM at short wavelengths, to an era of full technicolor vision. It is scarcely surprising that important new insights have emerged. Future space astronomy programs in the infrared will continue this progress. The Diffuse Infrared Background Experiment on NASA's Cosmic Background Explorer (COBE) mission, now scheduled for launch in 1989, will provide all-sky maps in 10 infrared bands from 1 to 300 μm with a 0.7° beam.[40] This will further clarify the overall energetics of the cirrus clouds, and, since one band nearly coincides with the 3.3 μm 'PAH' feature, may also help to clarify the nature of the very small grains. Highly sensitive infrared spectrometers and photometers on future observatories such as ESA's Infrared Space Observatory (ISO) and NASA's Space Infrared Telescope Facility (SIRTF) will permit definitive studies during the next decade.

ACKNOWLEDGMENTS

I have been greatly assisted in the preparation of this review by the insights, and in many cases the sharing of pre-publication results, of many colleagues, especially L. Blitz, F. Boulanger, F. Desert, E. Dwek, T. N. Gautier, J. Good, L. Magnani, F. Verter, and J. Weiland. The cheerful and efficient assistance of the IPAC librarian, R. Hernandez, was an essential aid to compilation of the literature. This work has been supported in part by NASA's IRAS Extended Mission Program.

REFERENCES

1. F. J. Low et al., Ap. J. (Letters), 278, L19 (1984).
2. G. Neugebauer et al., Ap. J. (Letters), 278, L1 (1984).
3. T. N. Gautier, III, in Light on Dark Matter, edited by F. Israel (D. Reidel Publ. Co., Dordrecht, 1986), pp. 49-54.
4. T. N. Gautier, III and M. G. Hauser, in The Milky Way Galaxy, edited by H. van Woerden et al. (D. Reidel Publ. Co., Dordrecht, 1985), pp. 219-222.
5. G. Helou, Ap. J. (Letters), 311, L33 (1986).
6. M. G. Hauser et al., Ap. J. (Letters), 278, L15 (1984).
7. E. J. Groth, B.A.A.S., 18, 1034 (1986).
8. C. A. Beichman, Ann. Rev. Astron. Astrop., 25, 521 (1987).
9. F. Boulanger, B. Baud, and G. D. van Albada, Astr. Ap., 144, L9 (1985).
10. R. X. McGee, R. F. Haynes, R. J.-M. Grognard, and D. Malin, M.N.R.A.S., 221, 543 (1986).
11. S. Terebey and M. Fich, Ap. J. (Letters), 309, L73 (1986).
12. F. Boulanger and M. Perault, preprint submitted to Ap. J., 1987.
13. H. W. de Vries, A. Heithausen, and P. Thaddeus, Ap. J., 319, 723 (1987).
14. C. P. de Vries and R. S. Le Poole, Astr. Ap., 145, L7 (1985).
15. A. Sandage, Astron. J., 81, 954 (1976).
16. M. W. Castelaz, K. Sellgren, and M. Werner, Ap. J., 313, 853 (1987).
17. C. P. de Vries, Astr. Ap., 150, L15 (1985).
18. J. L. Weiland, L. Blitz, E. Dwek, M. G. Hauser, L. Magnani, and L. J Rickard, Ap. J. (Letters), 306, L101 (1986).
19. L. Blitz, L. Magnani, and L. Mundy, Ap. J. (Letters), 282, L9 (1984).
20. L. Magnani, L. Blitz, and L. Mundy, Ap. J., 295, 402 (1985).

21. L. Magnani, L. Blitz, and J. G. A. Wouterloot, preprint submitted to Ap. J., 1987.
22. F. Verter, private comm.
23. J. Good, private comm.
24. L. Blitz and B. Gir, private comm.
25. C. Ryter, J. L. Puget, and M. Perault, to be published in Astr. Ap., 1987.
26. F. Boulanger, C. Beichman, F. X. Desert, G. Helou, and M. Perault, preprint submitted to Ap. J. (Letters), 1987.
27. B. T. Draine and N. Anderson, Ap. J., 292. 494 (1985).
28. M. Rowan-Robinson, M.N.R.A.S., 219, 737 (1986).
29. M. Harwit, J. R. Houck, and G. J. Stacey, Nature, 319, 646 (1986).
30. K. Sellgren, Ap. J., 277, 623 (1984).
31. J. S. Mathis, P. G. Mezger, and N. Panagia, Astr. Ap., 128, 212 (1983).
32. J. L. Puget, A. Leger, and F. Boulanger, Astr. Ap. (Letters). 142, L19 (1985).
33. F. X. Desert, F. Boulanger, A. Leger, J. L. Puget, and K. Sellgren, Astr. Ap., 159, 328 (1986).
34. F. X. Desert, in Light on Dark Matter, edited by F. Israel (D. Reidel Publ. Co., Dordrecht, 1986), pp. 213-216.
35. A. Leger and J. L. Puget, Astr. Ap. (Letters), 137, L5 (1984).
36. L. J. Allamandola, A. G. G. M. Tielens, and J. R. Barker, Ap. J. (Letters), 290, L25 (1985).
37. K. Sellgren, L. J. Allamandola, J. D. Bregman, M. W. Werner, and D. H. Wooden, Ap. J., 299, 416 (1985).
38. J. S. Mathis, in Light on Dark Matter, edited by F. Israel (D. Reidel Publ. Co., Dordrecht, 1986), pp. 171-176.
39. F. X. Desert, private comm.
40. J. C. Mather, Opt. Eng., 21, 769 (1982).

MAGNETIC ALIGNMENT OF GRAINS

ROGER H. HILDEBRAND
University of Chicago, Chicago, Illinois, U.S.A.

Abstract This paper presents a review of mechanisms that can account for alignment of dust grains not only in diffuse clouds but also in dense clouds where the conditions are less favorable. It is now becoming feasible to investigate the magnetic fields and the alignment of grains in dense clouds using OH Zeeman splitting and near- and far-infrared polarimetry. The results thus far favor alignment of superparamagnetic grains by the Davis-Greenstein mechanism.

INTRODUCTION

Since the discovery by Hiltner[1] and Hall[2] that starlight is polarized we have had 38 years to find a satisfactory explanation. From the beginning, the correlation with reddening left little doubt that the effect is due to partial absorption of the starlight by dust grains that are somehow aligned with respect to something. We can now say that the dominant mechanism is magnetic alignment of spinning grains, but that is far from a complete solution to the problem.

I will review briefly the mechanisms that have proved to be inadequate and the choices that remain, and will then summarize the results of observations showing that grains are aligned, in some cases at least, in dense cool clouds where the alignment must succeed under conditions vastly different from those along the paths to the visible stars. Those observations bear on the solution and, with or without a solution, enable us to map magnetic fields where no stars are visible.

ALIGNMENT BY NON-MAGNETIC PROCESSES

Hiltner's maps[3] showed polarization for hundreds of stars over wide areas of the sky. One did not need to know how the grains became aligned to see that they were aligned. The polarization vectors showed a pattern which Chandrasekhar and Fermi[4] interpreted as tracing a magnetic field directed along the spiral arms of the Galaxy. They assumed that the undulations were hydromagnetic waves and used the angular dispersion of the polarization vectors about the plane of the Galaxy and estimates of the density of the gas and the velocity of the turbulent motion to derive a value for the magnitude of the field. Their conclusion that a field of a few microgauss is directed along the spiral arms has been confirmed by observations of synchrotron emission, Zeemann splitting, and, especially, by Faraday rotation of the emission from nearby pulsars[5].

It is appropriate to ask whether the observed polarization in the arms of the Galaxy could be due to something other than magnetic alignment. In an early examination of that question Gold[6] considered the consequences of a bulk motion of gas with respect to the grains. Each impact of a gas atom on a grain produces a change in angular momentum which is perpendicular to the relative velocity and, preferentially, perpendicular to the long axis of the grain. The result of the random walk produced by a succession of impacts is that the grain tends to spin with its greatest average projection along the direction of the streaming gas. Since the starlight transmitted by a cloud of aligned grains is polarized with the E-vector in the plane containing the shortest average projection, this process cannot account for the gross features of the observed polarization unless the stream is perpendicular to the plane of the Galaxy and passes undeflected through the disk. As we will see presently, the relative velocity of the gas and dust must be greater than the thermal velocity, $\sim$1 km/sec, to overcome the disorienting effect of photon emission and considerably greater to overcome the effect of suprathermal rotation. Since we have not found such a systematic high velocity flow, and since we do see a magnetic field which would deflect even a slightly ionized cloud falling into the disk, we must look for other explanations.

The stream of photons provided by the Galactic radiation field does not provide a solution: The momentum flux is inadequate (less than that of 10 hydrogen atoms/cm^3 moving at 1 km/sec.), and, since the photons tend to move in the plane of the Galaxy, the polarization would be in the wrong direction.

As Harwit[7] has pointed out, one must consider not only the linear momentum of the photons but also their intrinsic angular momentum. The linear momentum of a photon changes the angular momentum of a grain typically by an amount $\delta L = x(h\nu)a/c$, where a is half the long dimension of the grain and x is a factor less than one, but of order one, depending on the shape of the grain (e.g. $x = 1/\sqrt{3}$ for a thin rod). The intrinsic angular momentum of an absorbed photon changes the angular momentum of the grain by an amount $\delta L = h/(2\pi)$. From this comparison it is evident that the effect of the linear momentum will predominate when $\lambda < 2\pi a$ as is generally true in diffuse clouds. Nevertheless the intrinsic angular momentum should not be ignored. A grain emits several hundred far-infrared photons for every UV photon and every gas atom it encounters. Each photon emitted changes the angular momentum of the grain by $h/(2\pi)$ in some random direction. The effect is sufficient to destroy alignment due to streaming of gas at intercloud densities unless the streaming is above thermal velocities. (For more on non-magnetic processes see Purcell and Spitzer[8].)

ALIGNMENT OF FERROMAGNETIC GRAINS

When we turn to magnetic effects, it is natural to begin by considering ferromagnetic grains aligned like compass needles. As Spitzer and Tukey[9] have shown, a rotational kinetic energy $\sim kT$

is sufficient to prevent alignment by that process even if the grains are made of pure iron and are as large as 1 μm (i.e. kT >> ΔU = (magnetization) x (field strength) x (grain volume)). For rotational kinetic energies above kT the difficulty of aligning permanent magnets would become even greater. If the "compass needles" were not so easily knocked about, they would tend to line up along the field and the transmitted starlight would be polarized with the E-vector normal to the plane of the Galaxy.

THE DAVIS-GREENSTEIN MECHANISM

It has been necessary to set aside four processes because they are too feeble or work in the wrong direction. What remains, or at least what remains that anyone has thought about, is the Davis-Greenstein mechanism[10], in which paramagnetic relaxation of spinning grains removes components of rotation perpendicular to the magnetic field.

Random impacts of gas molecules are sufficient to make the grains spin at about 10^5 rad/sec. In the absence of aligning torques the impacts are also sufficient to destroy any pre-existing alignment in about 10^5 years at a gas density of $\sim 10/cm^3$. In a much shorter time, only about one year (much less in the case of suprathermal rotation), the spin axis of a grain will come to coincide with the principal axis of maximum rotational inertia as a result of internal processes: as Purcell[11] has pointed out, internal damping (due to mechanical stresses or to the Barnett effect) will reduce the rotational kinetic energy, $J^2/(2I)$, by increasing the moment of inertia, I, to its maximum value. The alignment of the spin axis with the magnetic field then proceeds by paramagnetic relaxation at a rate proportional to the magnetic susceptibility and proportional to the square of the field strength.

The problem with the Davis-Greenstein mechanism as originally conceived is that the time required for alignment in the field of the Galaxy, > 10^6 years, is longer than the time for the impacts of the gas to destroy the alignment. That difficulty can, in principle, be removed either by suprathermal rotation as proposed by Purcell[11], or by superparamagnetic damping as proposed by Jones and Spitzer[12]. We shall re-examine the time scales when we come to alignment in dense clouds.

Suprathermal Rotation

Suprathermal rotation is rotation at kinetic energies much greater than would result from purely stochastic excitation. If the grains are exposed only to random impulses from gas molecules, then the rotational kinetic energies will be (3/2)kT where T is a temperature between that of the gas and the grains, typically 10 - 100 K. If the grains are exposed to UV photons which produce photoelectrons, or if they absorb hydrogen atoms which are ejected as molecules, then the kinetic energies can be of order 1 eV (T $\approx$ 10,000 K). But if the excitation is not stochastic, if the impulses turn a grain repeatedly in the same direction, then the kinetic energy can be very much greater then k times any temperature in the system.

One may reasonably expect such systematic effects due to irregularities in the grain surface. For example, if particular sites distributed asymmetrically over the surface of the grain favor inelastic collisions, then a small difference in temperature between the grain and the gas is sufficient to spin up the grain to thousands of times kT. Particular sites that favor photoemission or catalyze molecule formation can also drive the grains to suprathermal energies.

However likely suprathermal rotation may be, one cannot conclude that it will ensure alignment of the grains. Since the magnetic damping torque and the angular momentum both increase in proportion to the angular velocity, the time required for alignment by the Davis-Greenstein mechanism is not reduced by suprathermal rotation. What is changed is that the time available for alignment is limited not by gas damping but rather by the time the grain continues to be driven in the same direction. The effectiveness of suprathermal rotation therefore depends on the survival of the surface features responsible for the systematic impulses for a time at least comparable to the alignment time. The effect of "crossovers" from one spin direction to the other has been examined in detail by Spitzer and McGlynn[13].

Superparamagnetic Damping

Superparamagnetic damping is the paramagnetic dissipation of rotational kinetic energy, as in the Davis-Greenstein mechanism, where the imaginary part of the susceptibility, χ'', greatly exceeds that found in most substances. Spitzer and Tukey[9] have considered the possibility that collisions between grains, sometime in the course of their evolution, might produce small clumps of materials such as iron, iron oxide, or spinel. That possibility has been pursued by Jones and Spitzer[12] and recently by Mathis[14].

If a ferromagnetic particle within a grain is below a certain critical size, typically a few hundredths of a micron, then the particle will consist of a single domain and the grain will have a paramagnetic susceptibility with a value of χ'' up to 10^6 times the usual value. Jones and Spitzer[12] conclude that orientation of such grains at the temperatures, gas densities, and magnetic fields to be found in the arms of the Galaxy is physically plausible. The difficulties of alignment would be further removed if the grains experienced both suprathermal rotation and superparamagnetic damping.

ALIGNMENT IN DENSE CLOUDS

In the absence of observational data one could not predict whether grains should be aligned in dense clouds: the characteristic times for the various processes can be only partially anticipated. The times of particular interest are:

t_d = gas damping time $\approx$ (1/2) time for a grain to collide with its own mass of gas $\approx (2/3)\rho a / [nmv] \approx 5 \times 10^{14}/(n\sqrt{T})$ sec. where a and ρ ($\sim$ 0.1 μm, and $\sim$ 2 gm/cm^3) are the radius and density of the grain and n, m, and $v = [8\ kT/(\pi m)]^{1/2}$ are the number density, mass, and average speed of the colliding molecules (See Purcell and Spitzer[8] for dependence on grain shape);

t_a = time to align the angular momentum, J, with principal axis of greatest rotational inertia, $\approx$ (1 year) x $(10^5/\Omega)^2$ where $\Omega \approx 10^5$ rad/sec for thermal and $\sim 10^9$ rad/sec for suprathermal rotation[11] ($t_a \propto a^2$ for Barnett effect);

t_c = crossover time[13] = time between reversals of the spin direction;

t_m = time to accrete a monolayer taking into account the abundances of heavy elements as discussed by Spitzer[15] $\approx 3 \times 10^{11}/(n\xi)$ sec where n is the number density of hydrogen and ξ is the sticking probability; and

t_r = magnetic relaxation time = time to align J with B = $I/(B^2VK) \approx 2\rho a^2/(5B^2K)$ where I and V are the moment of inertia and volume of the grain, and $K = \chi''/\Omega$ is the ratio of the imaginary part of the magnetic susceptibility to the angular velocity $[K \propto (1/T)]$. For ordinary paramagnetic materials, $K \approx 2 \times 10^{-13}$ sec/rad (at 10 Kelvin) and $t_r \approx 3 \times 10^{14}/[B(\mu G)]^2$ sec. For superparamagnetic materials[12] $K \approx 10^{-8}$ to 10^{-12}

sec/rad (again at 10 Kelvin) and t_r would be correspondingly shorter.

Among these times, t_d and t_a are the least troublesome: one should be able to find t_d to within an order of magnitude and t_a is short enough to be negligible even for the lowest values of Ω allowed by Brownian motion. Spitzer and McGlynn[13] estimate that the disorientation of a grain will be nearly complete in two of three crossovers, hence t_r must not be much larger than t_c. It is perhaps reasonable to assume that t_c is comparable to t_m, but t_m depends on the sticking probability, ξ, which depends on local conditions[16]. If ξ is as large as 0.1 for molecules other than H, H_2, and noble gasses, then t_m will be comparable to t_d and no significant improvement in alignment will result from suprathermal rotation. If the time available for alignment is either $\sim t_d$ or $\sim t_m$, and if the grains (either suprathermal or superparamagnetic) are aligned by the Davis-Greenstein mechanism, then the ratio t(required)/t(available) should be proportional to n/B^2 (ignoring the relatively weak dependence on T).

Measurements and Conclusions

In Table 1 and Figure 1 we show the results of far-infrared measurements[17-21] of the thermal emission from aligned grains in several molecular clouds. The data on Sgr A are preliminary; the final results will be presented in a forthcoming paper by Werner *et al*[20]. These measurements have the characteristics that they are independent of optical depth (for $\tau << 1$); they do not require the presence of background or embedded sources, as is necessary for measurements of polarization by selective absorption; they are free of contamination by scattering or modification by the combined effects of emission and absorption; and they sample the bulk of the dust throughout the depth of a cloud.

In two cases we are able to compare our results with those from near-infrared measurements. In Orion, at the Kleinmann-Low

TABLE I Polarization measurements[a]

Object	λ (μm)	P (%)	ψ (E-vector)	Refs.
Orion				
KL (ctr at BN)	270	1.7 ± 0.2	25° ± 6°	17,18,19
KL (ctr at BN)	100	1.7 ± 0.2	22° ± 4°	21
400 μm pk 1!5 s of BN	270	1.7 ± 0.3	29° ± 7°	17,18,19
OMC-2 (400 μm pk)	270	<0.7 (2σ limit)		19
Sgr A (Preliminary results)				
N pk of 100 μm emiss.	100	2.0 ± 0.5	88° ± 6°	20
C (radio pt. source)	100	1.0 ± 0.4	88° ± 11°	20
S pk of 100 μm emiss.	100	0.8 ± 0.4	87° ± 13°	20
Mon R2[b]	100	1.2 ± 0.2	8° ± 5°	21
S140[b]	100	<0.9 (2σ limit)		21
NGC 2024[b]	100	<0.9 (2σ limit)		21
S255/257[b]	100	<1.1 (2σ limit)		21
W49[b]	100	<0.9 (2σ limit)		21

[a]Beam diam. = 55" for Sgr A. 60" for other sources.

[b]Far-infrared peak.

Nebula, we find a position angle approximately perpendicular to that measured by Knacke and Capps[22] at 11 μm and 20 μm, as is to be expected since in this case the near-infrared polarization is due to selective absorption and the far-infrared to emission. In Sgr A we find a position angle approximately equal to that inferred by Aitken et al.[23] for the radiation emitted at 12-13 μm from the sources IRS1, IRS5, IRS8, and IRS10 (all along the northern arm of the radio continuum and all within our central beam).

It is evident that grains are aligned in at least some molecular clouds. The results are not consistent with alignment by streaming gas. At Orion-KL the observed polarization is perpen-

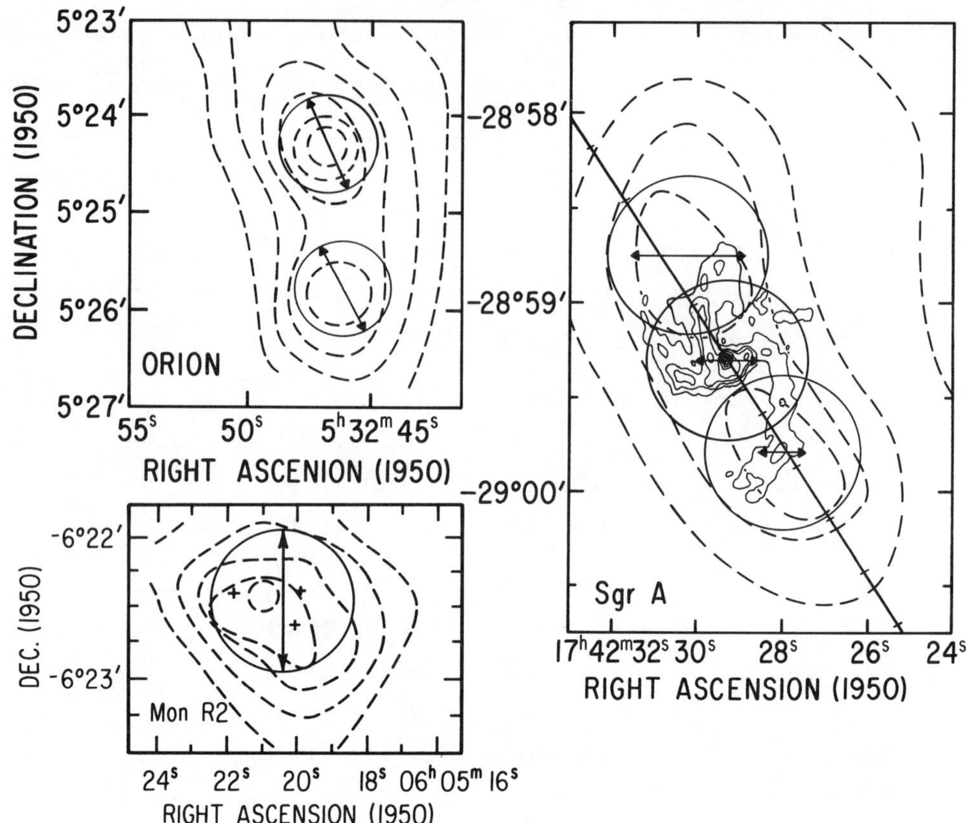

FIGURE 1 Far-infrared polarization in Orion (270 μm), Mon R2 (100 μm), and Sgr A (100 μm). The dashed contours show thermal emission at 400 μm for Orion[30] and at 100 μm for Mon R2[31] and Sgr A[32]. The beam diameters are 60" (Orion and Mon R2) and 55" (Sgr A). In the Sgr A map the solid contours show 2 cm radio emission[33]. The plane of the Galaxy is shown by the diagonal line (tick marks at 1 pc intervals). The arrows show the direction of the E-vector of the observed radiation. The lengths of the arrows in the Sgr A map are proportional to the degrees of polarization. In each case we infer that the direction of the magnetic field as projected on the sky is perpendicular to the E-vector.

dicular to the high velocity flow measured by Erickson et al.[24]. Moreover the polarization is just as great at the 400 μm peak south of KL, where there is no known high velocity flow. In Mon R2 the polarization is at about 45° to the large scale outflow[25].

The OH Zeeman splitting measured in a sample of twelve clouds by Troland and Heiles[26] and in Orion (125 μG) by Troland, Crutcher, and Kazes[27], and the estimates of magnetic support for over one hundred clouds by Myers and Goodman[28] all indicate that the magnetic field, B, increases with gas density, n, less rapidly than $\sqrt{n}$ (starting from intercloud densities; one may find $B \propto \sqrt{n}$ above a few hundred/cm^3). Hence any magnetic alignment mechanism that is inadequate or marginal in the intercloud medium will be inadequate in a cloud. On this basis one can conclude that if grains are aligned by paramagnetic relaxation, then that process must not be marginal in the intercloud medium. This conclusion is subject to the qualification that the Zeeman measurements give only the line-of-sight component of the field and that the beams used in the Zeeman measurements have been considerably larger (∿ 3.5' x 19') than those used in far-infrared polarimetry (∿ 1' diam.). It is obviously desirable to make polarimetric maps over regions as large as those beams and to make comparisons of Zeeman splitting with far-infrared polarization in more regions of the sky.

The argument that a process necessary for magnetic alignment in clouds must be more than marginally effective in the intercloud medium applies with special severity to suprathermal rotation. We find readily measurable polarization in the 400 μm peak of Orion where the conditions for suprathermal rotation are presumably less favorable than, say, at KL or Sgr A and much less favorable than in the surrounding medium - fewer UV photons, less atomic hydrogen, and a smaller difference between the temperatures of the gas and dust. We conclude that either suprathermal rotation is not necessary or it is very effective outside the cloud.

It is tempting to conclude that the grains are superparamagnetic. If they are, then one need not be concerned about the difficulties that confront other explanations for the alignment. Such a conclusion, however, is only as good as the list of possibilities that have been considered.

Finally, we note that the direction of the field in Sgr A as inferred from our results lies approximately in the plane of the dust ring. If that field is toroidal, then the angular momentum of the grain must rotate as rapidly as the ring in order for the grain to hold its alignment with the field. On that basis the time for the alignment should not be long compared to the time for the ring to move through a radian ($\sim 6 \times 10^{11}$ sec. at the position of the northern and southern beams[29] shown on the Sgr A map of Fig. 1).

ACKNOWLEDGEMENTS

I am grateful to M. Werner, J. A. Davidson, M. Morris, G. Novak, and S. R. Platt for permission to present the preliminary results on Sgr A and to G. Novak and S. R. Platt for many discussions of far-infrared polarimetry. This work has been supported by NASA grant No. NSG 2057.

1. W. A. Hiltner, Science, 109, 165 (1949)
2. J. S. Hall, Science, 109, 166 (1949).
3. W. A. Hiltner, Astrophys. J., 114, 241 (1951)
4. S. Chandrasekhar and E. Fermi, Astrophys. J., 118, 113 (1953).
4. R. N. Manchester, Astrophys. J., 188, 637 (1974).
6. T. Gold, Mon. Not. R. Astron. Soc., 112, 215 (1952).
7. M. Harwit, Nature, 226, 61 (1970).
8. E.M. Purcell and L. Spitzer, Jr., Astrophys. J., 167, 31 (1971).
9. L. Spitzer, Jr. and J. W. Tukey, Astrophys. J., 114, 187 (1951).
10. L. Davis, Jr. and J. L. Greenstein, Astrophys. J., 114, 206 (1951).

11. E. M. Purcell, Astrophys. J., 231, 404 (1979).
12. R. V. Jones and L. Spitzer, Jr., Astrophys. J., 147, 943 (1967).
13. L. Spitzer, Jr. and T. A. McGlynn, Astrophys. J., 231, 417 (1979).
14. J. S. Mathis, Astrophys. J., 308, 281 (1986).
15. L. Spitzer, Jr., Physical Processes in the Interstellar Medium (John Wiley & Sons, New York, 1978) Chap. 9, pp. 205-208.
16. J. R. Burke and D. J. Hollenbach, Astrophys. J., 265, 223 (1983).
17. R. H. Hildebrand, M. Dragovan, and G. Novak, Astrophys. J. Lett. Ed., 284, L51 (1984).
18. R. H. Hildebrand, M. Dragovan, and G. Novak, Airborne Astronomy Symposium, edited by H. A. Thronson, Jr. and E. F. Erickson (NASA Conference Publication 2353, Moffett Field, CA 1984) pp. 134-139.
19. M. Dragovan, Astrophys. J., 308, 270 (1986).
20. M. Werner, J. A. Davidson, M. R. Morris, G. Novak, S. R. Platt, and R. H. Hildebrand, (In preparation).
21. G. Novak, Ph.D. Thesis (In preparation).
22. R. F. Knacke and R. W. Capps, Astrophys. J., 216, 271 (1977).
23. D. K. Aitken, P. F. Roche, J. A. Bailey, G. P. Briggs, J. H. Hough, and J. A. Thomas, Mon. Not. R. Astron. Soc., 218, 363 (1986).
24. N. R. Erickson, P. F. Goldsmith, R. L. Snell, R. L. Berson, G. R. Huguenin, B. L. Ulich, and C. J. Lada, Astrophys J. Lett. Ed., 261, L103 (1982).
25. J. Bally and C. J. Lada, Astrophys. J., 265, 824 (1983).
26. T. H. Troland, C. Heiles, Astrophys. J., 301, 339 (1986).
27. T. H. Troland, R. M. Crutcher, and I. Kazes, Astrophys J. Lett. Ed., 304, L57 (1986).
28. P. C. Myers and A. A. Goodman, Astrophys. J., submitted (1987).
29. R. Genzel, D. M. Watson, M. K. Crawford, and C. H. Townes, Astrophys. J., 297, 766 (1985).
30. J. Keene, R. H. Hildebrand, and S. E. Whitcomb, Astrophys. J. Lett. Ed., 252, L11 (1982).
31. H. A. Thronson, Jr., I. Gatley, P. M. Harvey, K. Sellgren, M. W. Werner, Astrophys. J., 237, 66 (1980).
32. E. E. Becklin, I. Gatley, and M. W. Werner, Astrophys. J., 258, 135 (1982).
33. R. D. Ekers, J. H. van Gorkom, U. J. Schwarz, and W. M. Goss, Astron. Astrophys., 122, 143 (1983).

IV. New Techniques for Studying Interstellar Matter

MILLIMETER AND SUBMILLIMETER INTERFEROMETRY

D. DOWNES
Institut de Radio Astronomie Millimétrique, Grenoble, France

Abstract Millimeter interferometers have become highly useful astronomical instruments during the current decade. For the next 5–10 years, the main arrays will be those of the Berkeley-Illinois-Maryland collaboration, Cal Tech, Nobeyama and IRAM. The principal technical challenges are i) the sensitivity, which limits the maximum useful antenna spacing, and hence the angular resolution, ii) the restricted field of view, which might be expanded by mosaicing, under-illumination or multi-beaming, and iii) atmospheric effects, which introduce noise in the phase measurements. There are proposals to add more antennas to the existing millimeter arrays, to build new millimeter and sub-millimeter interferometers, and there are strong incentives to extend the coverage of these arrays to higher frequencies.

INTRODUCTION

In 1981, Alan Barrett wrote to a number of U.S. radio astronomers, warning that the funding of the 25-m millimeter telescope project of the NRAO was endangered, and that something had to be done. Partly as a response to his letter, the U.S. National Science Foundation set up a committee, chaired by Professor Barrett, to advise on what to do next. One of the committee's recommendations was the construction of a large millimeter array, with an instantaneous collecting area of at least 1000 m^2. The scientific justification for this recommendation was fairly obvious: such an array would give high-resolution information on regions of cold (10–1000 K) matter in the universe, which could not be easily obtained by other techniques. The objects of interest included regions of high extinction ($A_V \sim$ 10–200 magnitudes), planetary atmospheres, outflows from forming stars, mass loss from circumstellar envelopes and giant molecular clouds in other galaxies. The advantages which millimeter arrays have over centimeter arrays are the much greater number of molecular spectral lines–more than 600 mm lines have been detected in space, and the larger linewidths, in units of frequency, which allow a higher sensitivity in brightness temperature, for the same angular resolution and system performance. In the continuum, the millimeter arrays can observe solar system objects, stellar winds, compact H II regions, galaxies and quasars in thermal and synchrotron emission, and in dense molecular clouds, they can detect the thermal emission from dust grains.

To show the progress toward the goal set by the Barrett committee, I briefly review the history, current status and future plans for mm and sub-mm interferometers.

MILLIMETER ARRAYS: HISTORY, STATUS AND FUTURE

Massachusetts Institute of Technology / Haystack Observatory

Some of the initial efforts in short-wavelength interferometry were made during the late 1960's and early 1970's, at wavelengths of 17.5 and 13.5 mm, with two 2.4-m antennas on the roof of the Research Laboratory of Electronics, at M.I.T., and later with three 5.5-m dishes at Haystack.[1] This early work was partially interrupted by the discovery of water masers, when the K_u-band receivers for the interferometer were quickly transfered to larger antennas for VLBI, giving the Haystack group the lead in H_2O maser research.

Observatoire de Bordeaux

The first astronomical interferometer operating at a wavelength shorter than 1 cm was the Bordeaux solar interferometer,[2] which obtained its first fringes at 8.6 mm in January 1973, from an active center on the sun. The instrument consisted of two 2.5-m antennas on an east-west baseline of 64.4 meters, which gave a fringe separation of 28″for sources at zero declination at transit. The experience in developing this interferometer was of help in the early proposals for the millimeter interferometer for Plateau de Bure.

Cambridge University

The Cambridge 5-km telescope was used to observe Cygnus-A, W3(OH) and a few other sources at a wavelength of 9 mm during the winters of 1978–80.[3, 4] The eight 13-m antennas were equipped with receivers at 31.4 GHz, in order to achieve the highest possible resolution (0.3″) on the 5-km baseline. Only the strongest sources were mapped, and further millimeter work at Cambridge was concentrated on preparations for the James Clerk Maxwell telescope.

University of California, Berkeley

The first interferometer for the shorter millimeter wavelengths was the UC Berkeley interferometer at Hat Creek,[5] which originally began operating at 1.3 cm in the early 1970's and was then converted to a wavelength of 3.4 mm, giving its first results on the SiO maser line in 1979. At that time, the interferometer had two 6-m dishes (Fig. 1), movable along tracks 300m EW and 200m NS. It has been expanded since then to 3 x 6-m antennas, which can be operated remotely from the Berkeley campus.

FIGURE 1 Two of the three 6-m antennas of the UC Berkeley millimeter interferometer at Hat Creek. (Photo M. Birkinshaw).

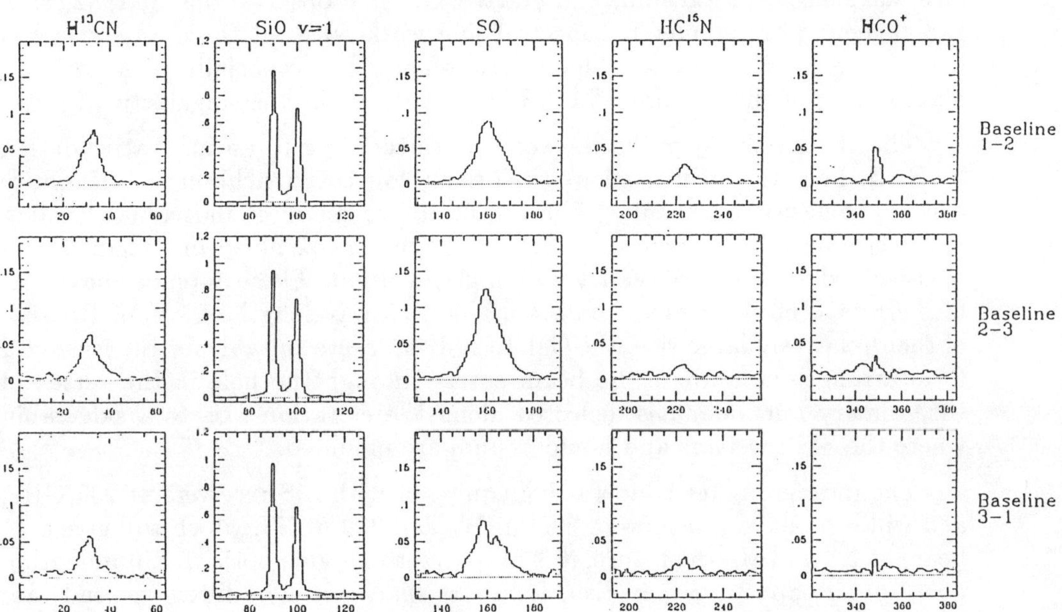

FIGURE 2 Five spectral lines at 3 mm, from Orion, observed on three baselines with the Berkeley interferometer.[6]

The Hat Creek interferometer currently has a 3-level spectral line correlator which allows 256 complex spectral channels per baseline, for each sideband. These channels may be split into four independent bands within the 500 MHz IF bandwidth. For example, to obtain the spectra shown in Fig. 2, the correlator had 64 complex channels (amplitude and phase) in each of four bands, centered on different frequencies, for both upper and lower sidebands of the first local oscillator. This configuration enabled the array to observe five different spectral lines on three interferometer baselines simultaneously.[6]

The Hat Creek array will double in size in the near future, as a result of the Berkeley–Illinois–Maryland collaboration. The expanded array will have 6 x 6-m dishes, yielding 15 simultaneous baselines, enough to make initial maps in a single day. The correlator will be expanded in the next few years to accomodate five times the number of baselines which it currently handles, and the antennas will also be equipped with receivers for the CO(2–1) line at 230 GHz. The three-university collaboration will provides investment money of about $4 million for the expanded interferometer, and operating money of about $1 million/year.

California Institute of Technology

The Cal Tech Millimeter Interferometer at Owens Valley, consisting of 3 x 10.4-m antennas, (Fig. 3) obtained its first fringes in 1980, in the continuum at a wavelength of 2.6 mm, and started CO line observations in 1982, as a two-element interferometer. Spectral line work with all three antennas has been going on since 1984.[7] This interferometer has especially pioneered the observations of the CO line at 115 GHz in galactic and extragalactic objects.

The Owens Valley antennas were built according to an innovative design by R. B. Leighton, which allowed relatively low-cost machining of the entire primary reflector as a unit. The reflectors are made of individual hexagonal panels on a back structure which could be disassembled in Pasadena and reassembled in Owens Valley with high precision. These antennas have surface errors $< 60\ \mu$m r.m.s., and are usable to a wavelength of 1 mm. Because of their relatively large size, the Cal Tech dishes were suitable for the Nasmyth focal arrangement, where the beam enters through the hole in the vertex of the primary mirror and is deflected along the elevation axis to a side cabin where the SIS receivers and liquid helium are mounted.

The interferometer is now being equipped with SIS receivers at 230 GHz, and will soon have baselines of 200 m EW and 220 m NS, which will give a 1" beam at 1.3 mm. In the future, it is proposed to provide optics for illuminating two 5-m sub-apertures per dish, for short spacings. The filter backends are to be replaced with a digital cross-correlating spectrometer allowing a total bandwidth up to 500 MHz, and either four separate bands on three baselines, with the 3 x 10.4-m full aperture array, or else two separate bands on six

FIGURE 3 Two of the 10.4-m antennas of the Cal Tech mm interferometer at Owens Valley. Photo by the author.

simultaneous baselines for the 6 x 5-m sub-aperture array. It is also intended to increase the overall system bandwidth to $\sim$ 1 GHz. In the past, a proposal was also made to increase the number of antennas in the Cal Tech array.

Institut de Radio Astronomie Millimétrique

The French-German IRAM interferometer, being built on Plateau de Bure, France, will have 3 x 15-m dishes, located at 2550 m altitude, and as such, will have the largest antennas and the highest site of the mm interferometers built so far.[9, 10] The antenna technology is different from that used in radio astronomy dishes up to now, in that the panels and most of the reflector support structure are made of carbon fibre, rather than metal, in order to ensure thermal stability.[11] The antennas are designed to have a surface accuracy better than 50 μm, for operation in the range 80 to 350 GHz, and are movable to 26 stations along rails 288 m EW and 160 m NS. An additional, fixed antenna in this same series has been built at the site of the European Southern Observatory at La Silla, Chile. The accuracy of individual panels (on a scale of approximately 1 m) ranges from 7 μm for the inner panels to 18 μm for the outermost ones. At present, the first antenna has been completed (Fig. 4), and initial measurements with tape measure and theodolite indicate an initial surface accuracy of 60–80 μm r.m.s. over the entire 15-m dish. Further panel adjustements will be made when the second antenna is ready so that the surface can be checked with interferometric holography. Testing of the first

FIGURE 4 The first of three 15-m antennas for the IRAM interferometer on Plateau de Bure, France. (Photo IRAM/MAN).

dish in single-dish mode began in 1987 with a cooled Schottky receiver at a wavelength of 3.4 mm, and two filter spectrometers of 512 x 1 MHz and 256 x 100 kHz channels. The second and third dishes of the interferometer will be finished in fall 1987 and winter 1988, respectively, and will be equipped with SIS receivers operating in the band 80–116 GHz. For the future, IRAM has submitted proposals to expand the array to four or six antennas.

Nobeyama Radio Observatory

The Nobeyama Millimeter Interferometer, with 5 x 10-m dishes (Fig. 5) started operating at 22 GHz in summer 1984, and is to conduct initial tests at 43 and 115 GHz in winter 1987–88. This interferometer has the longest baselines among the mm arrays so far constructed, 560 m EW and 520 m at 33 ° from NS, and the largest number of baselines (ten) which can be formed simultaneously.[8] In principle, it could also be used together with the Nobeyama 45-m telescope to obtain short spacing information, and for interferometry of compact objects at very high sensitivity. The Nobeyama capabilities are enhanced by an innovative spectral correlator and by impressive computing power. For the future, it is planned to equip the Nobeyama antennas with receivers at higher frequencies, and there have been some discussions about increasing the number of antennas in the array to ten or fifteen.

FIGURE 5 Antennas of the 5 x 10-m array of the Nobeyama Radio Observatory. (Photo NRO).

Australia Telescope

CSIRO Radiophysics intends to equip the Australia Telescope, in a later stage of its operation, with receivers at a wavelength of 3 mm, which will illuminate at least the inner 15 meters of the five movable 22-m dishes of the compact array at Culgoora. The AT will then be able to do the first millimeter interferometry in the southern hemisphere.

National Radio Astronomy Observatory

The Barrett committee in its main recommendation, asked the National Radio Astronomy Observatory to start a design study for a large millimeter array (MMA). Ideas about this array are given in the series of MMA Memoranda and Newsletters, circulated by the NRAO in recent years. The early proposals were to have two separate instruments, a compact array of about 21 "small" (3–5 m) dishes, possibly mounted on a single structure of about 25-m diameter, to provide a large field of view for mapping extended, low-brightness objects, and another array, similar to the VLA, consisting of about 15 to 27 "large" (6–13 m) dishes, to provide high sensitivity and high resolution. The MMA is regarded as a project for the 1990's, after much of the work on the VLBA is completed.

SUBMILLIMETER INTERFEROMETERS

Smithsonian Astrophysical Observatory

The Smithsonian is the only group so far which has proposed a submillimeter array, to consist of six 6-m dishes, located on a mountain site, and to cover the submillimeter windows centered at 0.87, 0.65, 0.43, and 0.35 millimeters.[17] The estimated cost of the SAO project is of the order of $40 million (1987).

The justification for building a submillimeter array is that this is one of the last parts of the radio spectrum where interferometry has not yet been done, it will provide high angular resolution because of the short wavelengths, and it contains a great deal of the energy emitted by astronomical objects. The Einstein A-coefficient for molecular transitions varies with frequency, ν, as ν^3, so for an optically thin spectral line, the brightness temperature of the line varies as ν^2, and hence the flux density in the line varies as ν^4, and the integral of the line flux over frequency varies as ν^5. For optically thick lines, the flux density at the line peak will vary as ν^2, and the frequency-integrated flux as ν^3. For continuum emission from dust, the flux density will vary as ν^3 or ν^4, depending on the emissivity of the dust grains.

In view of these strong increases with frequency, a submillimeter array may detect many sources, even if it has higher receiver and system temperatures than millimeter arrays. Furthermore, for millimeter and submillimeter interferometers, it may be useful to choose the baselines as a function of observing frequency in order to make maps of a source at the same angular resolution in several spectral lines. If the synthesized beamwidth is kept constant in this manner, and if the velocity resolution is also kept constant, then the sensitivity in brightness temperature varies as $\nu^{2.5}$. Here again, there is an important motivation for extending the coverage of millimeter interferometry to higher frequencies and into the submillimeter range, even if receivers are noisier.

There are, however, some important considerations for the design and operation of submillimeter arrays. Due to the greater extinction of the atmosphere, it will be difficult to obtain coverage of the (u,v) plane with a one-dimensional array, even for sources at high declination. The observing limit will be at most two air masses. Hence a two-dimensional array is essential. The short wavelengths imply narrow fields of view and poor coverage of low spatial frequencies, unless the dish sizes are kept small (6 meters maximum). The short duration of excellent observing conditions for submillimeter wavelengths creates a need for fast imaging. Hence there should be at least six antennas in the array, in order to get 15 simultaneous baselines, which is probably the minimum number required for a reasonable map in one night. All of these factors have been taken into account in the Smithsonian proposal.

ASPECTS OF ARRAY DESIGN

There are two ratios which are lower for millimeter and submillimeter arrays than they are for centimeter arrays, namely,

$$\frac{\text{Maximum Baseline}}{\text{Minimum Baseline}} \quad \text{and} \quad \frac{\text{Field of View}}{\text{Size of typical sources}}$$

Resolution and Minimum Spacing Problems

In the first of these ratios, the minimum baseline is usually slightly larger than the size of the individual dishes, while the maximum usable baseline for typical sources studied may be limited by the sensitivity. For the same integration time, wavelength, bandwidth and noise limit, the maximum practical baseline, B_{max}, scales as $D(N/T_{sys})^{0.5}$, where D is the dish diameter, T_{sys} is the system temperature, and $N = n(n-1)/2$ is the number of interferometer pairs formed with n antennas in the array.

Unlike most synthesis telescopes in the centimeter range, the millimeter arrays until now have had smaller and fewer antennas, higher system temperatures, more problems with the atmosphere, and weaker sources for both calibration and observing. For the most part, the thermal line sources which are observed in the millimeter range have brightness temperatures of only several degrees, and are weak in comparison with the non-thermal sources and optically thick H II regions mapped with interferometers in the centimeter range.

In practice, for the millimeter arrays until now, the minimum detectable brightness temperature in the synthesised beam reaches a few K at a baseline of about 100 m, and for fixed integration time, scales as the square of the baseline. However, the thermal spectral-line sources being studied also have brightness temperatures of this order, so it has been difficult so far to obtain useful signal-to-noise ratios on these sources at baselines much greater than 100 m. Indeed, most of the publications from the Berkeley and Cal Tech interferometers so far report maximum (projected) baselines which are typically about 40 – 70 m, although these will be probably be extended to 100 m in the near future.

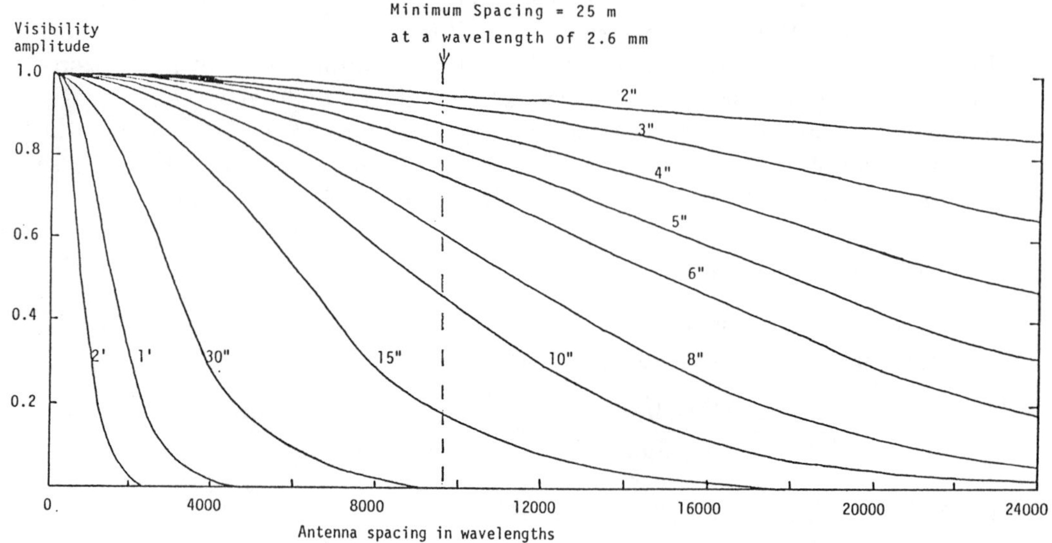

FIGURE 6 Visibility amplitude vs. antenna spacing, for Gaussian sources with half-power widths from 2″to 120″. *Dashed vertical line:* At a spacing of 25 m, an interferometer at 115 GHz will detect only half the flux from 9″sources, and nothing from components > 20″.

The synthesised beamwidths corresponding to the typical baselines used so far at 3 mm are typically 4″– 11″, but these values are uncomfortably close to the sizes of structures for which information is missing from the maps, because of the minimum baseline. The graph in Fig. 6 shows the visibility amplitude for circular gaussian sources of full width to half power from 2″to 120″. At a wavelength of 2.6 mm, a minimum spacing of 25-m, chosen to avoid mutual shadowing of antennas of, say, 15-m diameter, means that 9″structures will have their visibilities reduced by a factor of two, and structures > 20″will be completely missing from the interferometer maps.

In principle, the missing zero-spacing information can be recovered from single-dish maps. For the current mm interferometers, useful combinations have been and will be made with maps from the Onsala 20-m, the IRAM 30-m and the Nobeyama 45-m telescopes. Other possibilities to get back the short-spacing information are a) scanning the antennas during the synthesis[12], which is being tried at Hat Creek, and b) sub-illumination of the antennas, which is being tried at Cal Tech.

Field of View Problems

The second ratio which makes life more difficult for millimeter arrays than for centimeter arrays, is the limited field of view, relative to the size of typical molecular-line sources which one would like to study. Many molecular clouds and nearby galaxies extend over several arc minutes, but the primary beamwidths of current millimeter arrays are only about an arc minute at a wavelength of 3 mm, and correspondingly smaller at the CO(2–1) line at 1.3 mm.

Ways to overcome this problem are multi-beaming to observe several fields simultaneously, or else, if there is only one receiver per antenna, to observe different fields sequentially. The former method requires construction of more receivers and expanded correlators for the arrays; the latter method costs integration time. In both cases, the mosaic of adjacent fields is combined in the data processing. Another possibility might be to combine data taken with different arrays. The Hat Creek array, for example, might do a complete synthesis of a large field at shorter baselines, while the IRAM array might concentrate on the hot spots in that field at longer baselines.

Anomalous Refraction

Another effect which may have stronger consequences for millimeter and submillimeter arrays than for centimeter arrays, because of the smaller primary beams, is anomalous refraction. This phenomenon has been detected at 3 and 1.3 mm with the 30-m telescope on Pico Veleta and at 22 GHz with the 100-m telescope at Effelsberg.[13] It is characterized by displacements of the sources being observed by up to 20″for up to 30 sec of time. The effect is correlated with outside air temperature and time of day, tending to be strongest in the afternoon, and generally disappearing in the winter on cold sites. The anomalous refraction is probably caused by variations in the "wet" component of the refractive index, and corresponds to typical changes of 0.5 mm in the electrical pathlength over baselines of 30 to 100 m. The corresponding variations in the precipitable water vapour content are < 0.1 mm, so the effects on the atmospheric optical depth or the sky brightness are hardly noticeable.

The main effects of the anomalous refraction on mm and sub-mm interferometry will be increased noise in the phase measurements, leading to a broadening of the synthesized beam, and perhaps more serious, a reduction in fringe amplitude, due to the sources moving out of the narrow primary beams at these wavelengths. The electrical length changes are consistent with the phase noise measurements made with interferometers, namely:

$$\Delta l = 0.02\ B^{0.7} \qquad (5\text{ GHz} - -\text{VLA}^{14}),$$
$$\Delta l = 0.01\ B^{0.8} \qquad (22\text{ GHz} - -\text{Nobeyama}^{15}),$$
$$\Delta l = 0.003\ B^{0.8} \qquad (89\text{ GHz} - -\text{Hat Creek}^{16}),$$

where Δl is the variation in electrical length, in mm, and B is the baseline, in meters. However, the single-dish measurements give the impression that one is seeing individual packets of moist air, rather than integrating through a spectrum of different-sized cells at different distances from the telescope.

As the number of antennas in the millimeter and submillimeter arrays increases, it may become useful to apply techniques from optical interferometry to overcome phase errors caused by the the atmosphere. Particularly promising is the bi-spectrum method,[18, 19] in which triple products of the complex visibility are formed on short time scales, then ensemble averaged to improve sensitivity, and then used to derive phases. Calculations indicate that this method may allow a factor of 3–10 improvement over current phase closure or self-calibration methods.[20]

SUMMARY

In the progress toward the large millimeter array recommended by the Barrett committee, one may use some of the experience from the existing arrays with the problems discussed above to consider the effect on one particular aspect of the array design, namely the antenna size.

The minimum spacing problem favors having many small dishes (< 6 m), whereas cost-sensitivity arguments[21] tend to favor fewer, larger dishes. The cost considerations are modified by the mode of operation. Mosaicing, with one receiver per antenna, implies sharing the integration time among fields, and hence an overall sensitivity varying as dish diameter D, rather than D^2, thereby pushing the minima of the cost curves toward smaller dishes. Multi-beaming, however, brings the cost minima back toward the larger dishes. The desire for maximum baseline lengths and highest angular resolution favors bigger antennas, while the wish for a large field of view favors smaller dishes. Experience indicates that pointing problems are under control, and not a factor for deciding on dish size. Anomalous refraction would favor the smaller dishes, because of their larger primary beams.

Table 1 summarises the properties of current and proposed millimeter and submillimeter arrays, which represent different approaches in solving these problems. For the future, probably both types of arrays will be useful, those with wide field of view, and those with high sensitivity on longer baselines. Given the active interest in expanding the existing millimeter arrays, in constructing new and more powerful ones, and the rewards to be obtained by moving to higher frequencies, the young field of millimeter and submillimeter interferometry obviously has a very promising future.

REFERENCES

1. G. D. Papadopoulos and B. F. Burke, Radio Science, 6, 667 (1972).
2. J. Delannoy, J. Lacroix, and E. J. Blum, Proc. IEEE, 61, 1282 (1973).
3. P. Alexander, M. T. Brown, and P. F. Scott, Monthly Notices Roy. Astr. Soc., 209, 851 (1984).
4. P. F. Scott, Monthly Notices Roy. Astr. Soc., 194, 25P (1981).
5. W. J. Welch, J. H. Dreher, W. Hoffman, D. D. Thornton, and M. C. H. Wright, Astron. Astrophys., 59, 379 (1979).
6. W. J. Welch, and D. D. Thornton, in Proceedings of the International Symposium on Millimeter and Submillimeter Radio Astronomy, U.R.S.I., Granada, Spain, (1985), p. 53.
7. C. R. Masson, G. L. Berge, M. J. Claussen, G. M. Heiligman, R. B. Leighton, K. Y. Lo, A. T. Moffet, T. G. Phillips, A. I. Sargent, S. L. Scott, D. P. Woody, and A. Young, in Proceedings of the International Symposium on Millimeter and Submillimeter Radio Astronomy, U.R.S.I., Granada Spain, (1985), p. 65.
8. M. Ishiguro, K. I. Morita, T. Kasuga, T. Kansawa, H. Iwashita, Y. Chikada J. Inatani, H. Suzuki, K. Handa, T. Takahashi, T., in Proceedings of the International Symposium on Millimeter and Submillimeter Radio Astronomy, U.R.S.I., Granada, Spain, (1985), p. 75.
9. L. N. Weliachew, in Proceedings of the International Symposium on Millimeter and Submillimeter Radio Astronomy, URSI, Granada, Spain, (1985), p. 85.
10. L. N. Weliachew, and D. Downes, in ESO-IRAM-Onsala Workshop on (Sub) Millimeter Astronomy, eds. P. Shaver and K. Kjär, ESO, Garching, (1985), p. 51.
11. J. Delannoy in ESO-IRAM-Onsala Workshop on (Sub) Millimeter Astronomy, eds. P. Shaver and K. Kjär, ESO, Garching, (1985), p. 25.
12. R. D. Ekers and A. H. Rots, in Image Formation from Coherence Functions in Radio Astronomy, ed. C. van Schooneveld, IAU Coll. No. 49, D. Reidel, Dordrecht (1979), p. 61.

13. W. J. Altenhoff, J. W. M. Baars, D. Downes, and J. Wink, Astron. Astrophys., in press (1987).
14. J. W. Armstrong and R. A. Sramek, Radio Science, 17, 1579 (1982).
15. T. Kasuga, M. Ishiguro and R. Kawabe, IEEE. Trans. Ant. Prop., AP-34, 797 (1986).
16. J. H. Bieging, J. Morgan, W. J. Welch, S. N. Vogel and M. C. H. Wright, Radio Science, 19, 1505 (1984).
17. J. M. Moran, in Proceedings of the International Symposium on Millimeter and Submillimeter Radio Astronomy, URSI, Granada, Spain, (1985), p. 89.
18. A. Lohmann, G. Weigelt, B. Wirnitzer, Applied Optics, 22, 4028 (1983).
19. F. Roddier, Optics Communications, in press, (1987).
20. T. J. Cornwell, Astron. Astrophys., 180, 269 (1987).
21. D. Downes, MMA Memorandum, 22, NRAO, Socorro (1984).

TABLE I Existing and proposed millimeter and submillimeter arrays.

Array	Status	Number of Antennas	Instantaneous Collecting Area = n $D^2/4$	Shortest Usable Wavelength	Relative Speed for full synthesis
Hat Creek	in operation	3 x 6 m	85 m^2	1.3 mm	1.0
Hat Creek	funded	6 x 6 m	170 m^2	1.3 mm	5.0
Owens Valley	in operation	3 x 10.4 m	255 m^2	1.3 mm	1.0
Owens Valley	proposed	6 x 10.4 m	510 m^2	1.3 mm	5.0
Nobeyama	in operation	5 x 10 m	393 m^2	1.3 mm	3.3
Nobeyama	for future	45 m + 5 x 10 m	1983 m^2	1.3 mm	5.0
Nobeyama	discussed	(10 to 15) x 10 m	785 to 1178 m^2	1.3 mm	15.0 to 35.0
Plateau de Bure	in construction	3 x 15 m	530 m^2	0.8 mm	1.0
Plateau de Bure	proposed	4 x 15 m	707 m^2	0.8 mm	2.0
Plateau de Bure	proposed	6 x 15 m	1060 m^2	0.8 mm	5.0
Australia telescope	later phase	5 x 15 m	884 m^2	2.6 mm	3.3
Smithsonian	proposed	6 x 6 m	170 m^2	0.35 mm	5.0
NRAO	early MMA idea:	20 x 10 m + 24 x 3 m	1740 m^2	$\sim$ 1 mm	63.3

SUBMILLIMETER AND FAR-INFRARED DETECTORS

T. G. PHILLIPS
California Institute of Technology, Pasadena, CA, U.S.A.

Abstract A discussion is given of the relative merits of direct and heterodyne detection techniques in the submillimeter and far-infrared bands, for various astronomical situations. The nature of superconducting tunnel junction detectors (SIS) is described and their performance in heterodyne spectroscopic experiments is examined.

INTRODUCTION

It is well known that at short radio wavelengths, such as in the millimeter band, in high resolution spectroscopy experiments the best sensitivity is obtained using radio style heterodyne receivers, whereas at infrared or optical wavelengths direct detection with bolometers or photodetectors is the superior method. Somewhere in the submillimeter or far-infrared we might expect that there would be a crossover between the two techniques. In fact the precise crossover point depends on the wavelength, resolution, bandwidth and background emission of the experiment and the nature of the observation must be understood before the better technique can be identified.

The astronomical conditions can be quite variable. The spectroscopy of the gas and dust of the interstellar medium is a fascinating topic. It comprises emission and absorption due to the rotation of both simple, light and complex, heavy molecules or molecule ions, plus atomic and atomic ion fine structure features, thermal spectra from dust grains and infrared absorption bands from grain surfaces and physisorbed molecules on grains. A synthetic spectrum of a dense interstellar medium cloud is shown in figure 1.

The cloud is assumed to be homogeneous with a gas and dust temperature of 30 K. The 30 K black body curve bounds the emission. Underlying this is the dust continuum

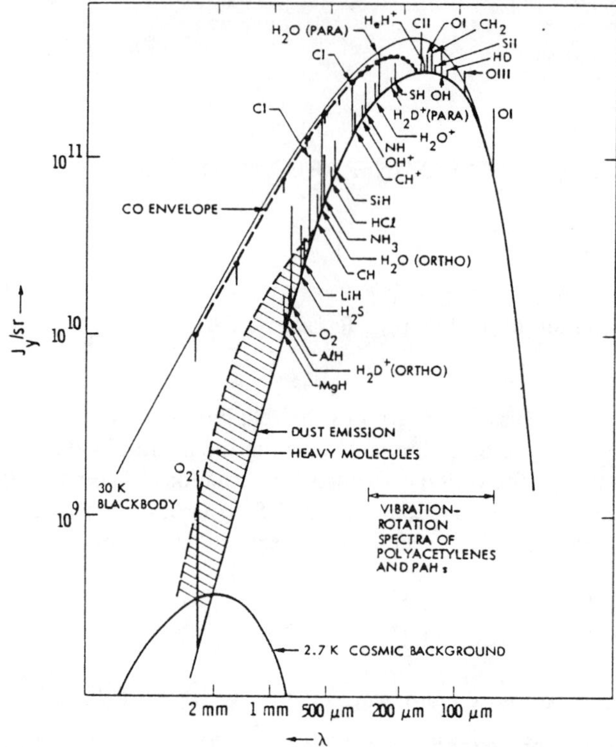

FIGURE 1 A spectrum showing the extent of information contained in the anticipated profile of an interstellar cloud.

which is optically thick and radiates like the blackbody at short wavelengths, but falls well below this at long wavelengths. Emission from abundant molecules like CO reaches the blackbody curve, but most lines are deexcited or optically thin and are less strong. The molecular linewidths may be as narrow as the thermal width and demand spectroscopic resolution ($\Delta\nu/\nu$) of 10^{-7}.

For ground based astronomy, the thermal emission background may be provided by the warm telescope structure, or more probably the atmospheric emission.

Figure 2 shows the transmission of the atmosphere from a high mountain site such as Mauna Kea in Hawaii. At wavelengths longward of 1mm the atmosphere may contribute as little as 10 K to the background temperature, but in the 450 or 350 μm windows may be equivalent to 150 K or more.

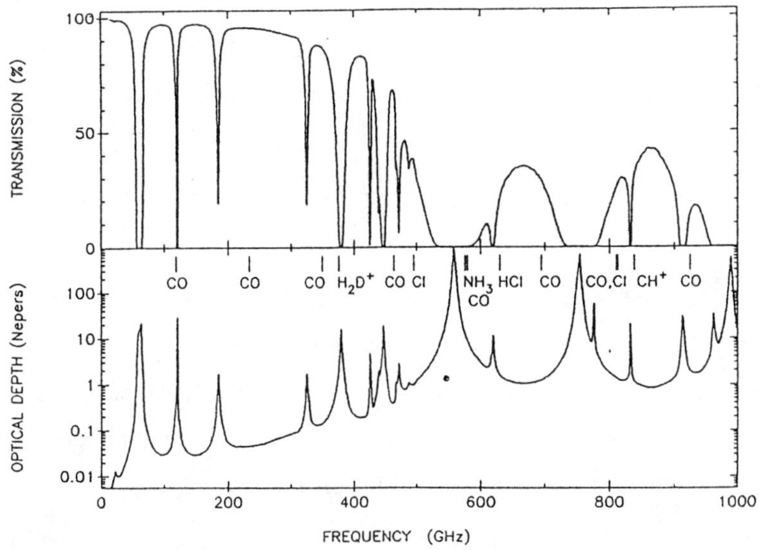

FIGURE 2 Atmospheric transparency in the millimeter and submillimeter bands from Mauna Kea, Hawaii.

HETERODYNE VS DIRECT DETECTION

From the above we see that we must expect to accomodate resolutions of $\sim 10^{-1}$ (continuum) to 10^{-7} (cold cloud molecular lines), and background from 3 K (space) to 150 K. To compare the effectiveness of heterodyne and direct techniques we can ask, what is the signal to noise ratio obtained looking through the same telescope at a source of temperature T_S, using the two techniques?

For direct detection (say using a bolometer), for a source whose spectrum is flat over the filter bandwidth, and for integration time Δt,

$$S/N = \frac{P_S}{NEP}\sqrt{\Delta t} \qquad (1)$$

where P_S is the incident power arriving through the front end filter, bandwidth $\Delta\nu$. In the

Rayleigh-Jeans limit, $P_S = 2kT_S \Delta\nu$. The NEP is the bolometer noise equivalent power in $W/Hz^{1/2}$ units.

For heterodyne detection, the Dicke radiometer equation gives:

$$S/N = \frac{T_S}{T_N}\sqrt{\Delta\nu\Delta t} \tag{2}$$

where $\Delta\nu$ is now the backend spectrometer channel width and T_N is the receiver noise temperature. In both methods we have not worried about reductions in S/N due to switching techniques. If we assume for the moment that the NEP and T_N values are fixed by the detector performance, and that the $\Delta\nu$ represents the resolution required and is the same in both cases, then equations (1) and (2) give:

$$\frac{(S/N)\text{heterodyne}}{(S/N)\text{direct}} = \frac{NEP}{2kT_N}\frac{1}{\sqrt{\Delta\nu}} \tag{3}$$

Typical values for 1mm wavelength operation are NEP = 10^{-15} W $Hz^{-1/2}$ and T_N = 1,000 K, so the signal to noise ratios are equal when $\Delta\nu \sim 1$ GHz. In other words, in the millimeter or submillimeter bands we would expect to use bolometers for bandwidths greater than 1 GHz (i.e. continuum experiments) and heterodyne receivers for high resolution spectroscopy. In fact the NEP and T_N values are likely to be dependent on the frequency, bandwidth and background[1,2] and the analysis can be more complicated.

The NEP and T_N are determined by the combined noise from the photon background and the detector:

$$NEP = \sqrt{\frac{4\epsilon h\nu kT_B\Delta\nu}{\eta(1-\epsilon)^2} + NEP_{det}^2} \tag{4}$$

$$T_N = \epsilon T_B + \frac{h\nu}{\eta(1-\epsilon)k} \tag{5}$$

where T_B is the background temperature, ϵ the emissivity of the background emitter, ν

the frequency of the measurement, η the detector quantum efficiency. The second term in T_N is the quantum noise from simultaneous measurement of amplitude and phase in a heterodyne experiment, which should be the dominant noise for a high frequeny radio style detector.

These two expressions can be used in equation (3) to estimate the relative performances of direct and heterodyne detectors for a given case. For instance figure 3 shows the result for a large space-borne far-infrared telescope (LDR) as a function of frequency and resolution.

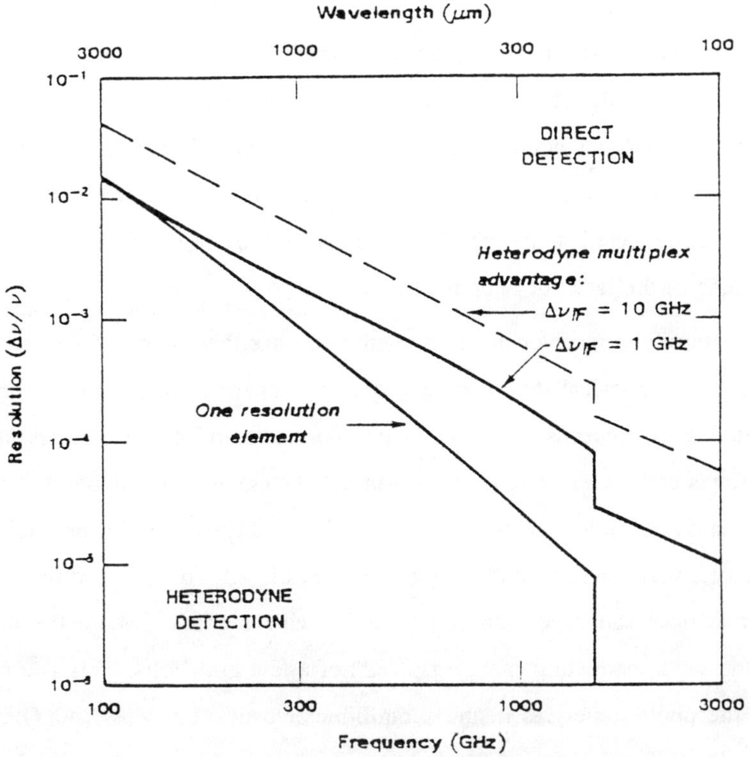

FIGURE 3 Comparison of direct and heterodyne sensitivities for the case of the NASA Large Deployable Reflector.

The overall result is that direct detection is preferable for resolutions of 10^{-2} or worse, but for wavelengths longward of about 200 μm heterodyne systems are better for the spectroscopic resolutions of 10^{-4} or better. Naturally the results depend on the number of backend channels available to the heterodyne system (heterodyne multiplex advantage). Multiple elements in the focal plane are equally advantageous to both types of detection.

SIS DETECTORS

A novel type of heterodyne detector has been developed in the past decade, which exhibits many desireable features and makes it the device of choice, given today's technology. It is the superconducting tunnel junction in the form of a lithographically formed superconductor-insulator-superconductor (SIS) sandwich. This was developed for radioastronomy[3,4] in the late 1970's and shown to be a quantum device with interesting and calculable properties not available to a classical device[5]. Reviews of the theoretical and practical situation up to about 1983 can be found in the literature[6,7]. Here we describe some of the basic technical and physical reasons for the use of the particular device and mention some of the latest developments.

For heterodyne receiver mixer elements one has the choice of fast bolometers or photoconductors, classical diodes or quantum non-linear devices. In the last category are SIS diodes and Josephson mixers. The bolometers and photodetectors have limited response times and therefore limited I F bandwidths (small heterodyne multiplex advantage). The classical diodes cannot have gain and typically have I-V characteristics whose non-linearity involves a semiconductor gap voltage of $\sim$ 1 Volt. By contrast the quantum non-linear devices can have gain and have I-V characteristics where the non-linearity involves the superconducting energy gap (2Δ) of a few milliVolts. This is a much better match to the photon energies in the submillimeter band (1 mV $\sim$ 240 GHz). A major reason for the popularity of SIS receiver devices is that the structure is formed by planar lithography and the microcircuit can easily be contacted and can be modified or embellished with extra local tuning elements.

Figure 4 shows the energy level structure of an SIS junction. The bias voltage can be increased to a value, $V = 2\Delta/e$, to permit tunneling of single quasiparticles from filled states on the left hand side to empty states on the right hand side.

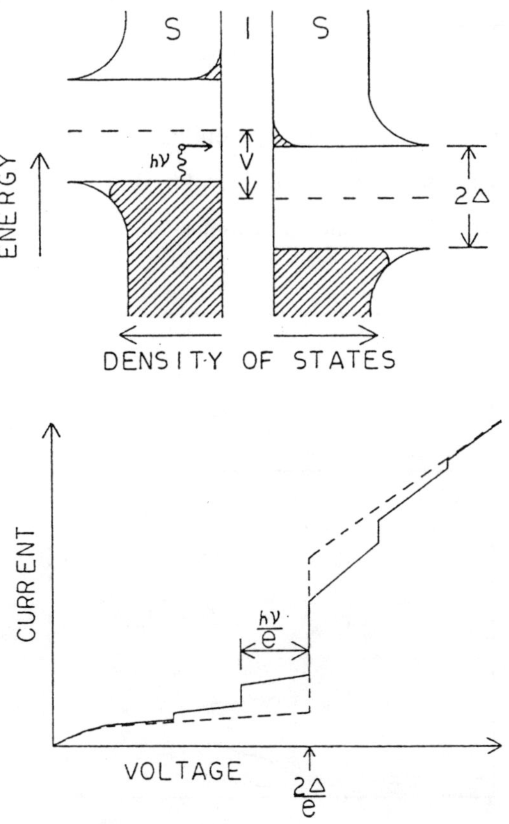

FIGURE 4 Upper - The photon assisted tunneling process in an SIS junction. Lower - The anticipated I-V characteristics without (dashed) and with (full lines) photons present.

At zero voltage bias pairs of electrons can tunnel across the barrier in the D C Josephson effect. However, at large voltage bias values, near that of 2Δ/e, the quasiparticle tunnel current can be activated by the presence of photons. The process of photon assisted tunneling was first observed in these devices by Dayem and Martin[8], but was not used for radioastronomy detection until modern lithography made possible the construction of small area devices[9,3].

The expected I-V characteristic for a device is shown in the lower part of figure 4.

The dashed curve represents current as a function of bias voltage for the device in the dark, the hard line shows the steps of increased current at a lower voltage ($\Delta V = h\nu/e$) due to tunneling assistance by a single photon, or $\Delta V = n\, h\nu/e$ for assistance by n photons simultaneously.

Actual I-V characteristics can be very similar to those computed from theory[7]. An example of the characteristics of a small area Pb alloy junction fabricated by the two angle evaporation process[9,10] are shown in figure 5.

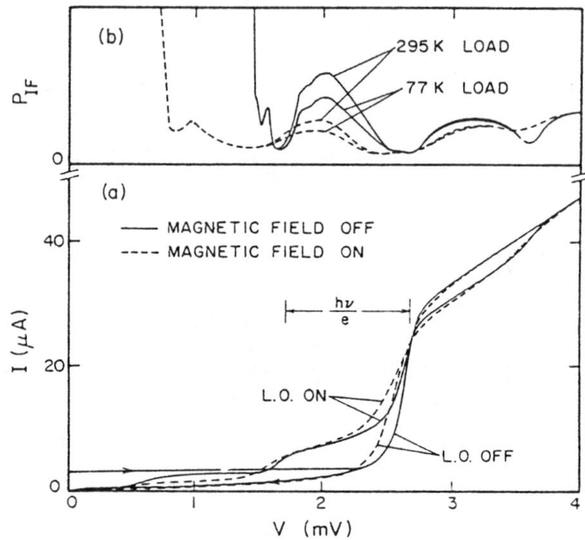

FIGURE 5 Lower - Actual I-V characteristics for an SIS junction with 230 GHz photons present. Upper - I F power for two values of the signal load.

In this example the applied photon field is at about 230 GHz or 1mV, so only one photon assisted step is observed. The figure also shows the I F power as a function of bias voltage for conditions in which the receiver sees an ambient and a liquid nitrogen temperature load. The very large noise power amplitude at low bias voltages is due to sampling of the unstable regions of bias associated with the Josephson effect. Also shown is the effect of a magnetic field on the device which suppresses Josephson currents and increases the useful available range of bias voltages. This suppression is generally required at high frequencies when the photon steps become wide and the local oscillator power becomes larger.

There have been several recent advances in SIS detector technology. These have been in three areas; new materials, new radiation coupling schemes and new lithographic tuning structures.

The Pb alloy based junctions are convenient to manufacture and can have very small ($< 1\ \mu m^2$) area. Figure 6 is a scanning electron micrograph of such a junction fabricated by R. E. Miller of AT&T, Bell Labs[10].

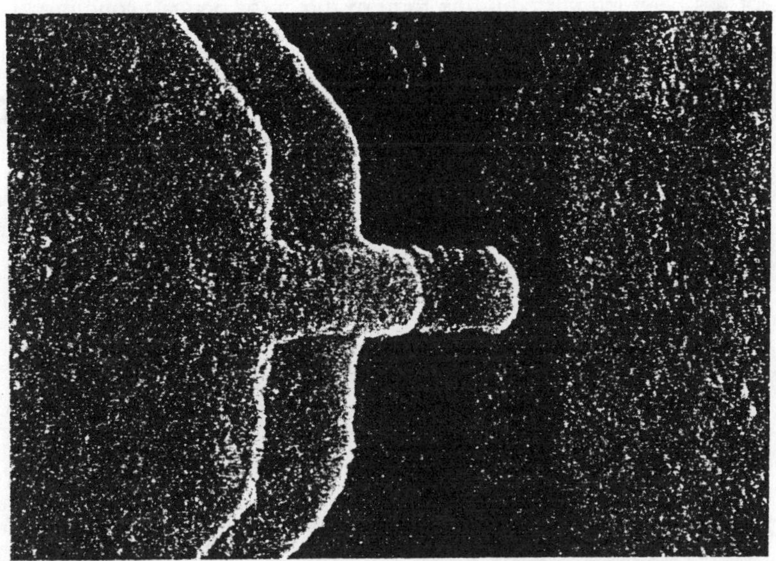

FIGURE 6 Pb alloy SIS junction fabricated by dual angle evaporation by R. E. Miller of AT&T Bell Labs.

They can be somewhat unstable thermally and chemically and gradually suffer gradual degredation. New developments are in the area of Nb based junctions which are physically hard and chemically stable. They are, however, more difficult to manufacture to small area specifications. Excellent results have been achieved by Kerr and collaborators using Nb - Al_2O_3 - Nb junctions[11,12], usually in the form of series arrays of elements to reduce the effects of the somewhat larger capacities of these devices. The major advantage of Nb junctions is the better durability. Junctions can also be constructed from materials of greater gap energies (2Δ) which would make them suitable for high frequency operation if the areas can be kept below 1 μm^2. Encouraging results have been obtained with NbN (Niobium nitride) as an electrode material, with MgO for the barrier[13,14]. Gap voltages as high as 5.2 mV have been obtained, compared with about 2.6 mV for Pb alloys.

Traditionally, SIS junctions have been mounted across full or reduced height rectangular waveguide, to couple to the TE_{01} mode. Variations on this have included circular waveguide[6] and waveguide antenna to stripline conversion mounts for series arrays[15]. A new concept has been introduced in the form of direct (or optical) coupling of radiation to a micro-antenna containing the SIS junction at its center[16,17]. Figure 7 shows the mounting configuration in which such an antenna, in the shape of a 'Bow-Tie' in this case[17], is mounted at the focus of a quartz hyperhemisphere.

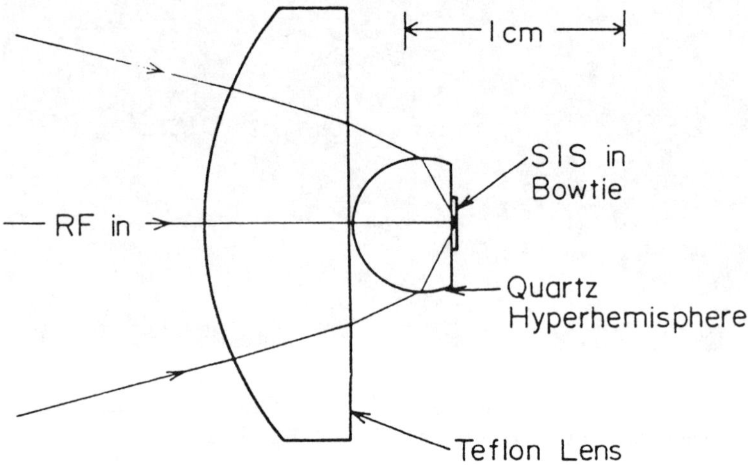

FIGURE 7 The SIS optical mode coupling scheme for Bow-Tie and other antenna structures.

The use of the quartz lens helps to couple to the radiation in a forward direction. Other forms of antenna shapes are currently under investigation in several laboratories.

This style of radiation coupler has allowed a considerable increase in the effective frequency range of the SIS devices. Figure 8 shows a recent compilation of reported receiver noise temperature.

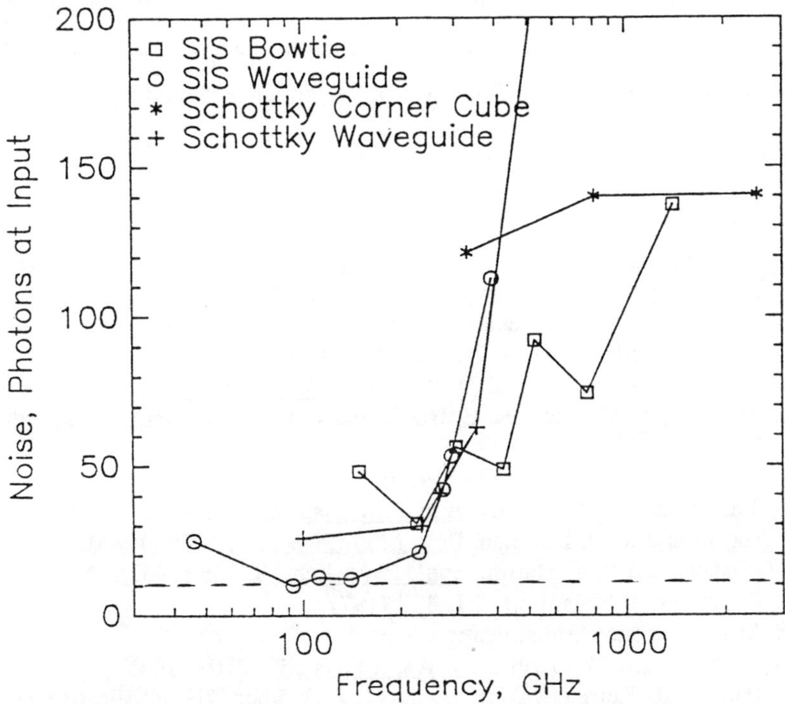

FIGURE 8 Comparison of actual receiver performances in terms of number of noise photons at the input. The dashed curve represents the best achieved result of a 10% quantum efficiency, or 10 photons of noise.

If local oscillator power is available and junction areas can be controlled to the $<$ 1 μm^2 region, it seems that SIS receivers should be able to operate successfully to more than 1,000 GHz.

Finally, the original hopes that microcircuit elements could be incorporated into the lithographic structure of the SIS junctions now seem to be realized. Broad band tuning out of the junction capacities has been achieved in SIS series arrays of junctions by means of inductive tuning stubs for the individual junctions[18].

CONCLUSIONS

In summary it seems that heterodyne techniques, and in particular SIS receivers, have advanced in capabilities to cover the needs of astronomical high resolution spectroscopy from the millimeter to the submillimeter band. It is now possible to construct a receiver in the millimeter band which approaches to within a factor of 10 of the fundamental quantum limit for such systems. This level of performance should soon be achieved in submillimeterwave receivers.

REFERENCES

1. T. G. Phillips and D. M. Watson, The Large Deployable Reflector : Instruments and Technology. NASA, JPL report D-2214 (1984).
2. P. L. Richards and L. T. Greenberg, Infrared and Millimeter Waves, edited by K. J. Button (Academic Press, N. Y., 1982), 6, 150.
3. G. J. Dolan, T. G. Phillips and D. P. Woody, Appl. Phys. Lett., 34, 347 (1979).
4. P. L. Richards, T. M. Shen, R. E. Harris and F. L. Lloyd, Appl. Phys. Lett., 34, 345 (1979).
5. J. R. Tucker, IEEE QE, 15, 1234 (1979).
6. T. G. Phillips and D. P. Woody, Ann. Rev. Astron. Ap., 20, 285 (1982).
7. J. R. Tucker and M. J. Feldman, Rev. Mod. Phys., 57, 1055 (1985).
8. A. M. Dayem and R. J. Martin, Appl. Phys. Lett., 8, 246 (1962).
9. G. J. Dolan, Appl. Phys. Lett., 31, 337 (1977).
10. R. E. Miller, private communication.
11. H. A. Huggins and M. Gervitch, J. Appl. Phys., 57, 2103 (1985).
12. S.-K. Pan, A. R. Kerr, J. W. Lamb and M. J. Feldmen, NRAO Electronics Division Internal Report #268 (1987).
13. A. Shoji, M. Aoyagi, S. Kosaka, F. Shinoki and H. Hayakawa, Appl. Phys. Lett., 46, 1098 (1985).
14. H. G. LeDuc, J. A. Stern, S. Thakoor and S. K. Khanna, IEEE. Trans. Mag., in press (1987).
15. S. Rudner, M. J. Feldman, E. Kollberg and T. Claeson, Squid '80, edited by H, -D. Hahlbohm and H. Lubbig, 901 (1980).
16. D. P. Neikerk, P. P. Tong, D. B. Rutledge, H. Park and P. E. Young, Appl. Phys. Lett., 41, 329 (1982)
17. M. J. Wengler, D. P. Woody, R. E. Miller and T. G. Phillips, Int. J. Infrared and MM Waves, 6, 697 (1985)
18. A. R. Kerr, S.-K. Pan and M. J. Feldman, International Superconductivity Electronics Conference, in press (1987).

LARGE MILLIMETRE AND SUBMILLIMETRE TELESCOPES

RICHARD HILLS
Mullard Radio Astronomy Observatory, Cambridge, England.

Abstract A new generation of single-dish radio telescopes, able to provide both good aperture efficiency at short wavelengths and angular resolution substantially better than one arc minute, are now coming into operation. These employ a wide range of innovative techniques to achieve highly accurate surfaces and to prevent deformations caused by gravity, temperature changes and wind forces. Examples include the use of new materials for the structures and for the panels, of motorized adjusters for setting the surfaces, and of protective enclosures to house the telescopes.

Recent progress is reviewed, with particular emphasis on the recently completed James Clerk Maxwell Telescope on Mauna Kea. In many cases the main limitation on the accuracy that can be achieved is in the measurement of the surface. Many different methods have been tried over the years, most of which have produced disappointing results. Recent progress with "radio holography" is however very encouraging and this is likely to become the standard method for setting large dishes.

INTRODUCTION

Anybody who has visited the Haystack antenna, as we did yesterday, would agree that it well deserves to be described as "large". The importance of Haystack is of course that it is both large and accurate enough to operate at high frequencies. The limits on the performance of such an antenna are largely determined by the ratio, R, of the diameter, D, to the rms surface error, σ. For Haystack $D = 36.6m$, $\sigma = 0.7mm$, so R is a little over 50,000. Operating a telescope at high frequencies increases both its gain and its angular resolution. The gain determines how much signal (antenna temperature) we receive from a source of given strength (brightness temperature) and size so long as the source is small compared to the beamwidth, while the angular resolution determines how much structure one can see in more extended sources. Eventually however, when the wavelength becomes too short, the irregularities in the surface cause coherence to be lost so the gain starts to fall and the beamshape to deteriorate. The maximum gain occurs when the wavelength is about 12 σ. The gain at this frequency has a value of 0.015 R^2, while the beamwidth is about 15/R. For Haystack, with R = 50,000, the peak gain is then 4 x 10^7 or 76 dB, and the beamwidth is 1 arc minute. This paper describes some of the advances which have led to the development of radio telescopes with values of R approaching 500,000, providing an improvement of a factor of 100 in gain and beamwidths reaching down to 6 arcseconds.

Before discussing the technical problems, I should like to comment on the siting and funding of these instruments. In order to obtain sufficient atmospheric transparency to do first-rate astronomy, particularly in the sub-millimetre waveband, it has become necessary to locate the telescopes on remote mountain sites. The cost of operating technically-advanced telescopes on such sites tends to be high, so that the funding only makes sense if one makes a substantial capital investment and builds a major new instrument. This means that in this area of astronomy, as in others, we are moving away from the pattern of telescopes operated by single institutions to that of facilities run on a national or even international basis. This trend has both positive and negative implications, but I fear that astronomers getting involved in such projects nowadays should anticipate longer timescales, more paperwork and a great many more committee meetings than in the days when Haystack was built. It is my impression that astronomers do not yet work together as effectively in these inter-institutional collaborations as do some of our colleagues, in high-energy physics for example. I believe that we must learn to do so if we are to make best use of the opportunities which are being opened up by technical advances such as those described here.

PROBLEMS AND SOLUTIONS

Optics

One of the first questions to be settled in the design of a large single-dish telescope is the optical layout. Here the solution chosen for Haystack - a classical Cassegrain with parabolic primary and hyperbolic secondary - has been widely adopted. The advantages are well known: (i) the secondary focus can be near the centre of gravity where a number of receivers can be accommodated and access to them made easy; (ii) the spillover goes onto the sky and therefore adds little to the system noise; and (iii) one can move the feed a long way off-axis without reducing the gain, so that imaging with arrays of detectors is feasible. A number of alternatives have been tried but all have limitations. The off-axis arrangement, used very successfully on the Bell Labs 7m antenna, has no blocking and very low sidelobes, but it would be difficult to scale-up the mechanical design to larger sizes. A spherical primary would be easier to manufacture and to measure than a paraboloid, but the primary f-ratio has to be large and the secondary would be difficult to make and to support. "Shaping" is often applied to centimetre-wave antennas to increase their illumination efficiency. It would also work at shorter wavelengths, but the shaping destroys the good off-axis properties of the classical Cassegrain, so that the field of view for imaging would be very limited. At millimetre and submillimetre wavelengths the same result of high efficiency can be achieved by modifying the quasi-optics that couple the signal from the telescope into the detector.

Construction

The basic mechanical scheme of Haystack, in which the reflecting surface is made up from a number of accurate panels supported by a space-frame backing structure, has also been widely adopted. A whole range of techniques have been used to make the panels themselves, including machining, stretch-forming and gluing. Many of the most accurate dishes however use aluminium honeycomb "sandwich" construction and the panels are formed on numerically machined molds. Only a small number of different molds are needed so that a lot of care can be taken to provide them with accurate surfaces, which can then be replicated onto the panels themselves. In the case of the JCMT the molds had an accuracy of 7 μm rms and the overall average of the rms error of the panels, as manufactured, was less than 12 μm. In the case of telescopes which have to operate in sunlight, it is worthwhile to use carbon-fibre reinforced-plastic for the skins of the panels in place of aluminium, because of its much lower thermal expansion. This was first done on the inner part of the Japanese 45 m dish, and excellent accuracy has recently been achieved on the IRAM and SEST 15 m telescopes, which have completely CFRP surfaces.

There has also been a lot of progress in designing backing structures which overcome the effects of gravitational deflections. Von Hoerner demonstrated that it is possible to design "homologous" structures in which the surface remains a paraboloid when it is tilted to any elevation angle: only the position of the secondary mirror has to be adjusted to compensate for the shift in the focus of the paraboloid. The Bonn 100 m telescope showed how well this can be made to work in practice. Von Hoerner used designs with a fixed geometry and varied the cross-sections of the beams to minimize the residual errors. On JCMT we used struts of fixed size and altered the geometry - the positions of the nodes in the structure - to obtain the same result. This is more difficult to treat mathematically, and so a less automated procedure than Von Hoerner's was used, but it results in a structure that is easier to manufacture. With patience the predicted residual errors can be reduced to an arbitrarily low level. However, real structures will not behave in exactly the way that the numerical models predict, so it is essential to make the structures stiff enough to keep down the total deflections as well as the residuals. On JCMT the overall deflections of the dish structure due to gravity are about 200 μm rms, while the predicted residuals after subtracting the best-fitting paraboloid are a little over 10 μm.

Protection

One of the most striking aspects of the Haystack antenna is of course the great radome. This provides protection from storms, improves the pointing (because there is no wind loading), and reduces the thermal distortions that would occur if the antenna were exposed to the sun and the cold sky. This approach has been followed by a number of more recent projects, particularly the very successful series of ESSCO antennas. There is of course some penalty in the form of blocking due to the framework of the radome but this is quite small and easily compensated for, in terms of collecting area per unit cost, by building a slightly larger dish. However there are also losses in the fabric of the radome and these rise steeply at the very short wavelengths now being exploited. Two other avenues have therefore been opened up.

For the very large antennas, such as the Japanese 45m and the MPIfR/IRAM 30m, it makes sense to leave the dishes unprotected. The structures are already strong enough to withstand storms and stiff enough to keep down the deformation of the surface due to wind forces, and it has proved possible (although certainly not easy) to find ways of getting ice off the antennas and of pointing them accurately even in quite strong winds. By enclosing the backing structures and providing forced air circulation it has proved possible to keep temperature gradients below 1 degree, which produces an acceptable level of thermal distortion in a steel structure. The alternative solution to the thermal problems of open-air antennas is to use low-expansion materials. This was first done on the Texas dish, which has an Invar structure. More recently it has been adopted for the IRAM and SEST 15m dishes and the MPIfR 10m which all have backing structures composed of carbon-fibre epoxy tubes.

For the smaller antennas, with even more precise surfaces, it becomes economically feasible to use astrodomes - similar in function to the domes of optical telescopes. The advantages, such as the good access to the telescope and the benign environment for working on it even when there is bad weather outside, were demonstrated by the Kitt Peak 36 foot and the Texas 16 foot telescopes. Both of the new antennas on Mauna Kea use astrodomes. Leighton's design for the CSO 10m has what in optical terms would be an up-and-over shutter and fits closely round the antenna, while still providing a surprising amount of space for accommodation. The JCMT design has vertical sides and a flat roof so that standard structural steel techniques can be employed, which cuts costs, and makes it possible for the roof and doors, which are quite massive for this size of telescope, to move horizontally instead of having to be lifted. The outside of the enclosure is painted with titanium-dioxide-loaded paint which is white in the visible but black in the thermal infrared so it keeps cool in sunlight, while both the inside of the enclosure and the telescope stucture have metal surfaces with low thermal emissivity so that there is very little radiative coupling between them. Vents at the top and bottom can be opened to remove temperature gradients in the inside air. With such a large "slit" when the "dome" is open the telescope can still suffer some ill effects from wind forces and solar heat. We have therefore made it possible to install a large "membrane" across the aperture. This is only 0.3mm thick and made of woven teflon (PTFE) so its loss is small even in the submillimetre waveband, but it is strong enough to span the 16m aperture without support and allows us to operate in winds of up to at least 50 mph. Only about 20% of the solar radiation comes through the membrane and this is sufficiently well diffused that one can stand at the focus with the antenna pointed straight at the sun without danger! It may be possible to use this material to construct radomes with very good high frequency properties.

Surface Setting

This is the aspect of the art of building large antennas that I feel is in the least satisfactory state. A great deal of ingenuity has been expended over the years in building special-purpose devices to measure the shape of dish surfaces. In almost every case the results have failed to come up to expectations. The exceptions are probably the ones that are simplest in concept, such as the template arrangement used by Findlay on the NRAO 12m, which has demonstrated repeatability, even over long timescales, of around 20 microns. Leighton has also demonstrated that it is possible to measure a paraboloid very accurately with a laser interferometer in the workshop, but the problems of using such devices with the telescope on site are serious. In the case of JCMT we have constructed a "measuring machine" which uses a combination of an alignment laser and a two-axis interferometer to measure the position of a small probe that can be driven to any point on the surface. The system gives good results in the laboratory and after a big struggle it has been made to work on the telescope. However the process of installing it and setting it up takes at least a week and the accuracy so far (about 90 microns) is disappointing.

The approach which I believe holds the most promise is to use signals at similar wavelengths to the ones that are going to be observed. This is afterall the way optical mirrors are tested. The most direct way of doing this is to measure the phase and amplitude of the far-field pattern of the antenna and then determine the phase errors in the aperture, which reflect the irregularities in the surface, by means of a Fourier transform. This has become known as "radio holography", although "millimetrewave metrology" would perhaps be more appropriate. The technique was first applied to dishes in radio-astronomical interferometers, as the equipment for measuring phase was already available and a strong reference signal could be obtained from the other dishes. It since been used successfully on a number of single dishes, most often with an artificial source on a satellite so that only a small ancilliary antenna is needed to provide a phase reference. Accurate tracking of such a satellite is however difficult and, provided one is satisfied with measurements at a single low elevation, it is easier to use a source on the ground. In the case of JCMT we decided to put the source at the UKIRT, which is nearby - a range of only 700m. The short distance to the transmitter means that the effects of atmospheric fluctuations are kept small. The beam pattern is then being measured well inside the Rayleigh distance, but it is not difficult to make the necessary corrections using the Fresnel approximation. Making the measurement in the Fresnel region also has the advantage that a much smaller dynamic range is needed than in the far field. Dave Morris has shown that it is possible to determine both the phase and amplitude in the aperture of the dish without measuring the phase of the beam pattern, provided two sets of measurements of the amplitude are taken at different focus settings. The hardware for this "phase-recovery" method is of course much simpler than when the phase has to be measured, and I believe that it may well be capable of more precise results at high frequencies because one can measure amplitudes more accurately than phase. The penalty is of course in the more elaborate data reduction needed. Morris suggested the use of an iterative technique due to Misell. Working on the JCMT data, Anthony Lasenby and I have found that this algorithm does find the detailed structure of the errors rather well, but we found it necessary to add a fitting procedure to deal with the large scale errors and to get the image properly centred. This combination of working in the Fresnel region and measuring only amplitude seems to be a good one: two rounds of measuring (at 94 GHz) and adjusting brought the rms error (as deduced from the aperture efficiency) down from about 140 microns to 90 microns to 65 microns. The repeatablity of the measurements is about 30 microns rms, but it is not yet clear what limits this or what systematic errors may be present.

Until recently the actual adjustment of dish surfaces after they have been measured has been carried out manually. There seems to have been some hesitation in adding motorised adjusters to telescopes, possibly because of problems with "active" control systems in the past. However the 45m has motors on each of the structural "nodes" where panels are supported and both the JCMT and IRAM 15m dishes have all the adjustment points motorised. Obviously this makes the process of adjustment a lot easier and more precise, which is especially valuable for the iterative approach that seems to be necessary in practice. On JCMT the whole surface can be adjusted in a few minutes and the resolution of the stepper motors is 3 microns. In practice it has to be admitted that a great deal more work has gone into making the adjusters system work properly (it has 828 stepper motors controlled by a microprocessor) than it would have taken to set the dish by hand. However now that it does work it is very convenient and in the long term it may be possible to use this system to correct residual thermal or gravitaional errors (provided of course one has some way of knowing what they are!).

CONCLUSION

It appears that most of the major problems of building large telescopes for the millimetre and submillimetre bands have been solved. No single "overall best" solution has emerged but a variety of approaches have been shown to work. No doubt astronomers will continue to find reasons why even more challenging specifications have to be met and designers will find ways of meeting them. Just keeping the costs of future projects to manageable levels will be hard enough.

I am grateful to my colleagues, too numerous to name, for the free interchange of ideas and information about the antenna building game without which it would not have been possible to prepare this talk or, indeed, to build the JCMT.

Lew Snyder　　Roger Hildebrand　　Frank Shu

Alan Rogers　　Bill Wilson　　Richard Hills

Jim Moran Jack Welch Tor Hagfors

Phil Schwartz Gisbert Winnewisser Tom Pauls

Tom Armstrong Barry Turner Tom Bania

Jackie Hewitt Pat Thaddeus

V. Contributed Papers

[1]RADIATIVE TRANSFER THROUGH MOLECULAR CLOUDS: P-CYGNI PROFILES

LAWRENCE H. AUER
Los Alamos National Laboratory, MS F-665, Los Alamos, NM

HÉLÈNE R. DICKEL
University of Illinois, 349 Astronomy Bldg., 1011 West Springfield Avenue, Urbana, IL

Abstract A code has been developed to solve the transfer of radiation through molecular clouds with non-monotonic velocity fields and embedded HII regions. P-Cygni-like profiles are common for models with an HII region.

High-resolution maps of the cores of giant molecular clouds reveal considerable structure in both space and velocity, including molecular absorption of the continuum emission from embedded, compact HII regions. To analyze such data, a radiative transfer code has been developed which allows for arbitrary variations of the cloud-parameters with radius, such as the density of the molecular hydrogen and the kinetic temperature. The program treats non-monotonic velocity laws and situations where the turbulent and systematic velocities are of comparable magnitude and it allows for a central HII region.

It is fortunate there exist molecules with radio, rotational transitions that are optically thin enough ($\tau < 10$ in all lines) that the simplest possible approach, Λ-iteration, suffices to solve the multi-level, non-LTE problem. This is the case for the molecules H_2CO, HCO^+, H_2S, CS, and SO but not CO and HCN. In the Λ-iteration method, the formal integral solution for each transition is calculated with a "known" source function. The initial value of the source function is determined from the statistical equilibrium equations wherein the mean intensity is set equal to the cosmic 3 K background radiation. The formal solution - source function - formal solution is iterated to convergence.

A sketch of an expanding cloud with an embedded HII region and representative profiles from this model are presented in Figure 1 for several transitions of CS. Note the P-Cygni-like appearance of the upper two profiles, i.e. there is absorption of the continuum at negative velocities and excess emission at positive velocities. Because of the radial symmetry of the models, a contracting cloud yields an inverse P-Cygni profile. Their exis-

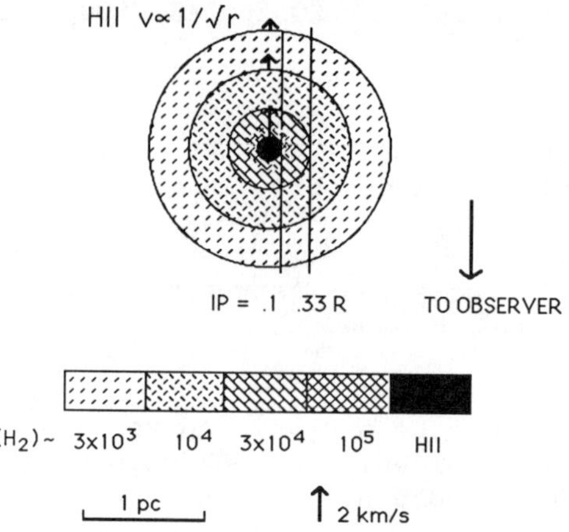

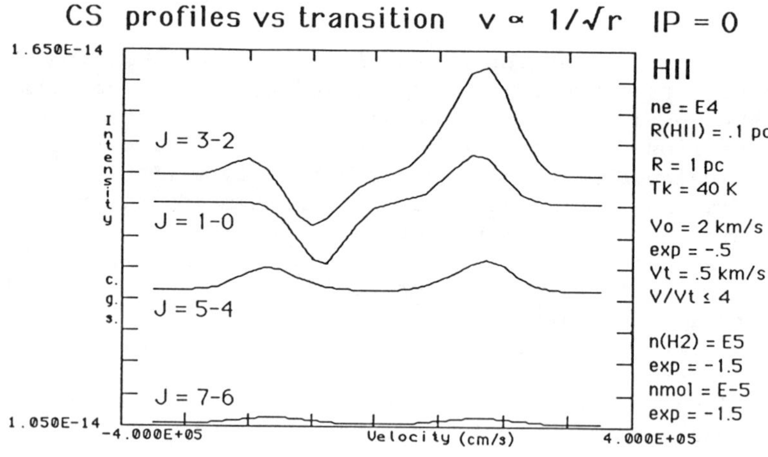

FIGURE 1. Sketch of a cloud model and sample CS profiles. Cloud parameters are given along the right-side of the plot.

tence and strength depends on the continuum-opacity and brightness of the HII region as well as on the velocity field and the opacity of the molecular transition. P-Cygni profiles are common for models with a central HII region and should provide a useful diagnostic of the excitation conditions in a molecular core. Further calculations are planned for a variety of molecular species and models of molecular clouds.

[1]Publication No. LAUR 871510: this research was performed in 1986 when HRD was a resident collaborator at the Los Alamos Lab.

GRAVITATIONAL COLLAPSE OF CLOUD CORES: 2 CANDIDATES

ERIC R. KETO, PAUL T. P. HO, AND MARK J. REID
Harvard–Smithsonian Center for Astrophysics, Cambridge, MA

We report on 10″ resolution observations of NH_3(1,1) in the molecular cores surrounding two ultracompact HII regions, W3(OH) and G34.3+0.2. Both sources show extended emission components with velocity gradients consistent with rotation, and absorption components that are redshifted with respect to the emission suggesting gravitational infall. The velocity structure in both sources, and the density and temperature structure in W3(OH), appear similar to that around the previously observed HII region, G10.6–0.4, which is well modelled by a centrally condensed, and centrally heated, molecular core gravitationally collapsing onto the HII region (Ho and Haschick 1986; Keto, Ho, and Haschick 1987, 1988). We suggest that velocity and temperature structures similar to those seen in W3(OH), G34.3+0.2, and G10.6–0.4 may be a useful signature of gravitational collapse.

The observations were made with the Very Large Array of the National Radio Astronomy Observatory. The approximate integration time on each source was 3 hr with 18 antennas, and 32 channels of 1.2 km s^{-1} resolution were used for each source. The rms noise of the channel maps is 0.034 Jy/beam for W3(OH) and 0.024 Jy/beam for G34.3+0.2.

Figure 1 shows a position-velocity plot of W3(OH)in the NH_3(1,1) transition made along a cut at a position angle 60° east of north. The plot is similar to that seen at the same angular resolution in G10.6–0.4 (Figure 11 of Keto, Ho, and Haschick 1987). Note the velocity gradient seen in emission across the source and the redshifted absorption. In G10.6–0.4 we interpreted this structure as indicating a rotating and infalling dense core of molecular gas. The velocity gradient in W3(OH) is 9.8 ± 4.1 km $s^{-1}pc^{-1}$ over a diameter of 0.27 pc, and the absorption is redshifted from the center velocity of this gradient by an infall velocity of 3.6 ± 1.6 km s^{-1}. Position-velocity plots made along cuts at 10° intervals indicate that the rotation axis lies roughly between 20° and 70° west of north. As for G10.6–0.4, the redshifted gas in W3(OH) has a higher excitation temperature, 70 K, than the more extended gas seen in emission, 30 K. This is expected for a centrally condensed cloud in gravitational collapse: the gas with the highest collapse velocity is closest to the central source and therefore hotter. A central mass of approximately 100 $M_\odot$ would be required to bind both the rotational and collapse motions at a radius of 0.14 pc. The mass estimated from the column density ranges from several tens to several hundreds of solar masses depending on whether $N(H_2)/N(NH_3)$ is assumed to be 10^6 or 10^7. In addition, the ionizing flux implies the presence of $\geq 30\ M_\odot$ in the central star or stars. If the contraction occurs immediately outside of the HII region as in the case of G10.6–0.4, the relevant sizescale is 0.01 pc, and the required binding mass is 30 $M_\odot$.

In Figure 2 we show a position-velocity plot of G34.3+0.2 made along a cut oriented 60° west of north. The velocity structure is similar to that seen in W3(OH) and G10.6–0.4. The velocity gradient seen in emission is 15 ± 6 km s^{-1}pc^{-1} over a diameter of 0.3 pc, and the absorption is redshifted with respect to the center velocity of this gradient by 3.5 ± 1.6 km s^{-1}. Infall may also be indicated by the asymmetric hyperfine satellite lines. The difference between the absorption strength of the blue and red inner satellite lines is about equal to the observed brightness of the emission elsewhere, suggesting blending of the blue inner satellite absorption with emission from the back side of the collapse zone. A central mass of approximately 100 $M_\odot$ would be required to bind these motions at a distance of 0.14 pc. The mass within this radius estimated from the column densities is on the order of 100 $M_\odot$.

PTPH and ERK are supported in part by NSF Grant AST85-09907.

REFERENCES

Ho, P. T. P., and Haschick, A. D. 1986, Astrophys. J., 304, 501.
Keto, E., Ho, P. T. P., and Haschick, A. D. 1987, Astrophys. J., 318, 712.
Keto, E., Ho, P. T. P., and Haschick, A. D. 1988, Astrophys. J., in press.

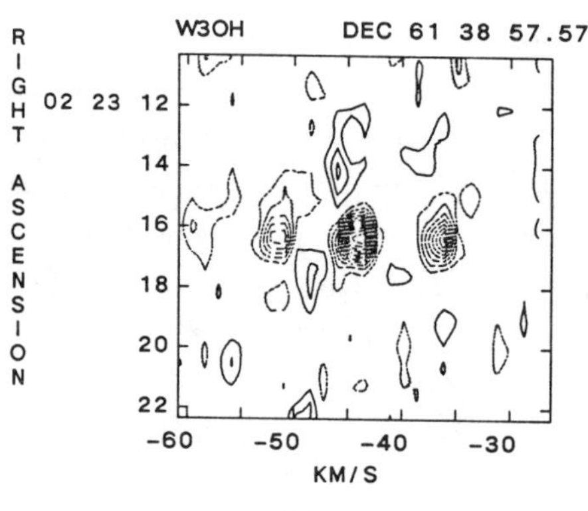

Figure 1

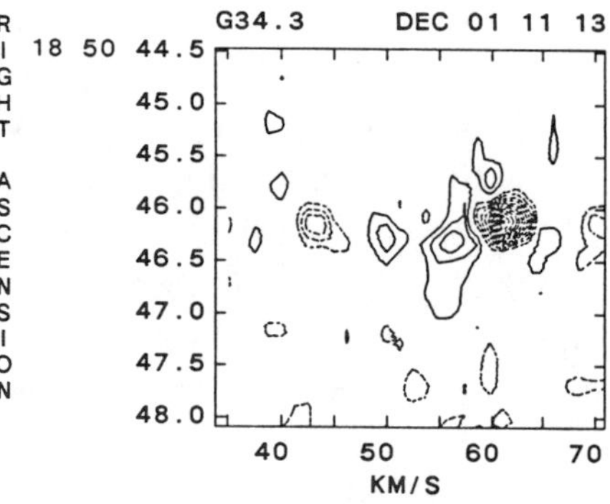

Figure 2

MAGNETIC FIELDS IN MOLECULAR CLOUDS: HOW SUPPORTIVE ARE THEY UNDER PRESSURE?

ALYSSA A. GOODMAN AND PHILIP C. MYERS
Harvard–Smithsonian Center for Astrophysics
Cambridge, Massachusetts, USA

Empirically, correlations between density (n) and radius (R), and between linewidth (σ) and R, have been noticed in spectral line observations of molecular clouds for the past several years. To date, however, there is not a complete or consistent theoretical explanation of these phenomena. The linewidths measured in molecular clouds exceed their thermal value, and the origin of the non-thermal portion of the linewidth (σ_{NT}) has been a subject of much debate. We propose that a relatively weak ($\sim$ 30 μG) magnetic field can account for the supra-thermal linewidths as well as the correlations between n, σ, and R.

From observations of clouds ranging in size from about 0.1 to 100 pc, we have compiled over 120 measurements[1–4] of n, σ, and R, and we have constructed a virial equilibrium model of the correlations n(R) and σ(R). Basically, we assume that the non-thermal motions are magnetic in origin. Then, the gravitational energy of the cloud is balanced by the sum of the cloud's magnetic and thermal energies. The resulting equations allow us to fit the observed correlations with one free parameter–the magnetic field strength, B. We find that a field strength in the relatively small range 15 to 40 μG fits most of the data over the entire size range included in the data. (See Figure 1.)

The few Zeeman splitting field strength measurements available to date indicate that a 30 μG or so field is quite likely to exist in the molecular clouds we have considered in our sample.

The predicted correlations using our virial equilibrium model[5] are: (1) for large clouds ($R \gg 0.1$ pc), $\sigma_{NT} \propto B^{\frac{1}{2}} R^{\frac{1}{2}}$ and $n \propto BR^{-1}$; and (2) for smaller clouds (*i.e.* dense cores, $R \lesssim 0.1$ pc), $\sigma_{NT} \propto BR$ and $n \propto B^0 R^{-2}$, where σ_{NT} is the quadrature difference of σ and the thermal linewidth. The different behavior of the correlations at large and small cloud size demonstrates the dominance of the magnetic support term in the virial equation at large radii, and the thermal support term at small radii.

The relatively weak support in cloud cores may arise from ambipolar diffusion, since low-mass core sizes ($\sim$ 0.1 pc) are close to the magnetic equilibrium size for which ambipolar diffusion is fastest. (See Figure 2, where theoretical curves for ambipolar diffusion time as a function of size[5,6] are shown to have minima near typical low-mass core radii, when the field strength is $\sim 30\ \mu$G.)

In a traditional flux-freezing model, where cloud mass remains constant, one would expect a dense core's B-field to be enhanced by far more than just a factor of 5. Our model, which employs a *constant field strength*, fits the observed correlations, but curves of *constant mass* evolution are orthogonal to the predominant trend (see Figure 1). Therefore, we note that a constant-mass flux-freezing model should not be applied from cloud-to-cloud, but may be relevant in individual (constant mass) stages of a cloud's evolution.

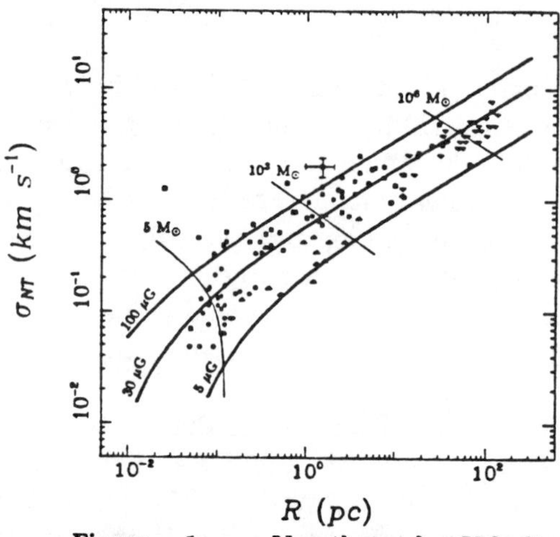

Figure 1: Non-thermal Velocity Dispersion vs. Size Curves show predictions of the virial equilibrium model[5] for $T = 10$ K. *Heavy curves* indicate model predictions for a constant field strength of $B = 5$, 30, and 100 μG. *Light curves* indicate the dependence of non-thermal velocity dispersion on radius if mass is assumed constant at $M = 5$, 1×10^3, and 1×10^6 M$_\odot$. References for symbols: upside-down triangles[1], squares[2], right-side up triangles[3], and circles[4].

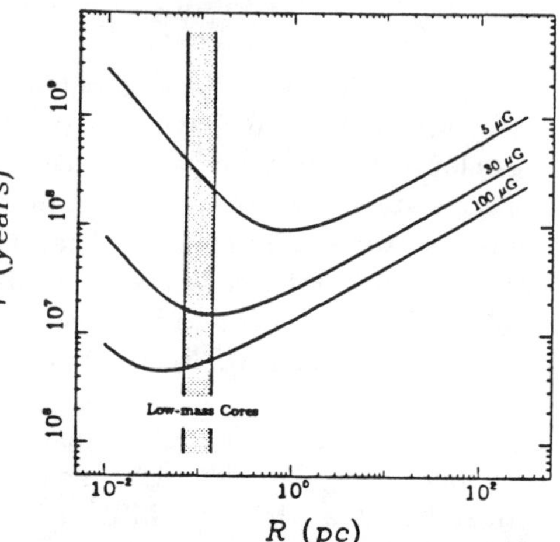

Figure 2: Ambipolar Diffusion Time Scale vs. Size *Heavy lines* indicate the dependence of ambipolar diffusion time scale on size within our virial equilibrium model[5], for $B = 5$, 30, and 100 μG, assuming that the magnetic field size scale is equal to R. *Light lines* enclose the range of cloud radii (*shaded*) occupied by most low-mass dense cores.

REFERENCES

[1]Dame, T., Elmegreen, B., Cohen, R., and Thaddeus, P. 1986, *Ap. J.*, **305**, 892.

[2]Larson, R.B. 1981, *M.N.R.A.S.*, **194**, 809.

[3]Leung, C., Kutner, M., and Mead, K. 1982, *Ap. J.*, **262**, 583.

[4]Myers, P.C. and Benson, P.J. 1983, *Ap. J.*, **266**, 309.

[5]Myers, P.C. and Goodman, A.A. 1987, submitted to *Ap. J.*.

[6]Shu, F., Adams, F.C. and Lizano, S. 1987, *Ann. Rev. Astron. Astrophys.*, **25**, 23.

A REDUCED ROLE OF THE MAGNETIC FIELD IN THE DENSE CORE REGIONS OF THE TAURUS CLOUDS

MARK H. HEYER
Department of Terrestrial Magnetism, Carnegie Institution of Washington
5241 Broad Branch Rd NW, Washington, D.C. 20015

Abstract Rotational axis directions are inferred from velocity gradients of thirteen cloud cores defined by enhanced ^{13}CO J=1-0 emission from the Taurus Molecular clouds. Comparison of these axes with the local magnetic field direction determined from polarization of background stars reveals a uniform distribution. A possible interpretation of this distribution is an ineffective magnetic stress upon the mostly neutral material at densities greater than 10^4 cm^{-3}.

INTRODUCTION

The interstellar magnetic field may provide a significant stress upon the mostly neutral material within cold, molecular clouds. If the field is strongly coupled to the neutral material by collisions between ion and neutral molecules, magnetic torques can redistribute angular momentum of a cloud to the external medium. Mouschovias and Paleologou[1] have shown that the perpendicular component of angular momentum is more efficiently redistributed so that one is more likely to observe magnetically dominated clouds with rotational axes aligned along the magnetic field. Observations of the large scale gas distribution and kinematics and the magnetic field geometry of the Taurus clouds reveal such alignment and suggest a prominent role of the magnetic field in the evolution of the gas at densities $< 5\times10^3$ cm^{-3}.[2,3]

RESULTS

In this study, the kinematics of dense core regions within the Taurus clouds are examined from enhanced ^{13}CO J=1-0 emission. Significant velocity gradients are found to be present within 13 cloud cores. If these motions are interpreted as rotation, then the orientation of the rotational axis θ_{rot}, can be compared to the direction of the local magnetic field θ_B, determined from nearby optical polarization vectors[3] or near infrared polarization measurements.[4] The observed probability of finding a core with a rotational axis offset from the local magnetic field direction by $\Delta\theta_{rot} = |\theta_{rot} - \theta_B|$ is presented in Figure 1. The rotational axes are observed to be uniformly distributed with respect to the local magnetic field direction.

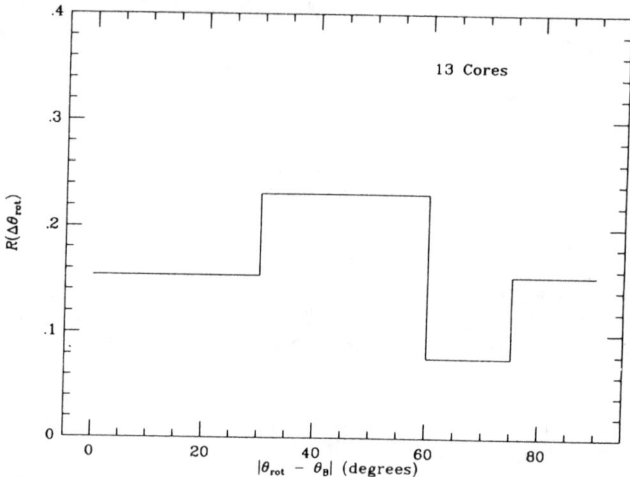

Figure 1 - The observed probability distribution of finding a dense core within the Taurus cloud with a rotational axis offset from the local magnetic field direction by $|\theta_{rot} - \theta_B|$.

Such a uniform distribution may result from the inability of the magnetic field to significantly influence the motions of the mostly neutral material at these densities ($\langle n \rangle \approx 10^4$ cm^{-3}) due to a decrease in the ionization fraction and misaligning interactions with other cores. These neutral gas densities at which the magnetic field's role is reduced are similar to theoretical estimates.[5,6] If the magnetic field is an insignificant stress in regions with large rotational velocities then the angular momentum problem associated with star formation may not be completely resolved by magnetic braking. Rather, fragmentation of the core in the formation of multiple stellar systems or stellar winds, both of which occur at higher densities must ultimately be responsible for the dissipation of angular momentum and the observed rotational velocities of pre-main-sequence stars.

REFERENCES

1. T.Ch. Mouschovias, and E.V. Paleologou, Ap. J., 230, 204 (1979).
2. A. Moneti, J. Pipher, H.L. Helfer, R.S. McMillan, and M.L. Perry, Ap. J., 282, 508 (1984)
3. M.H. Heyer, F.J. Vrba, R.L. Snell, F.P. Schloerb, S.E. Strom, P.F. Goldsmith, and K.M. Strom, Ap. J., 321, (1987).
4. M. Tamura, T. Nagata, S. Sato, and M. Tanaka, MNRAS, 224, 413 (1987).
5. T.Ch. Mouschovias, Ap. J, 211, 147 (1977).
6. T.Ch. Mouschovias, E.V. Paleologou, and R.A. Fiedler, Ap. J., 291, 772 (1985).

Detection of V=0, J=1–0 Silicon Monoxide Absorption in Sgr B2 and Observations of Silicon Monoxide at 43.423 GHz in HII Regions

A. D. HASCHICK
Haystack Observatory, Westford, MA
PAUL T. P. HO
Harvard-Smithsonian Center for Astrophysics, Cambridge, MA

During a survey for further silicon monoxide maser sources in star forming regions, using the 35.5–49 GHz maser amplifier receiver of Haystack Observatory, silicon monoxide (V=0, J=1–0) was detected in absorption against the HII region Sgr B2. The receiver system temperature was ~130K and the aperture efficiency was ~.09 for a 45" beamwidth.

A nine point spectral map of the absorption line at 40" spacings is given in Figure 1. At the central position a broad absorption line at 62 km s^{-1} cuts into a broader emission line centered at roughly 60 km s^{-1}. Emission is detected at all positions on the map and is narrower at positions offset from the center. These profiles agree with NH_3 line observations made by Winnewisser, Churchwell and Walmsley (1979) at the same resolution.

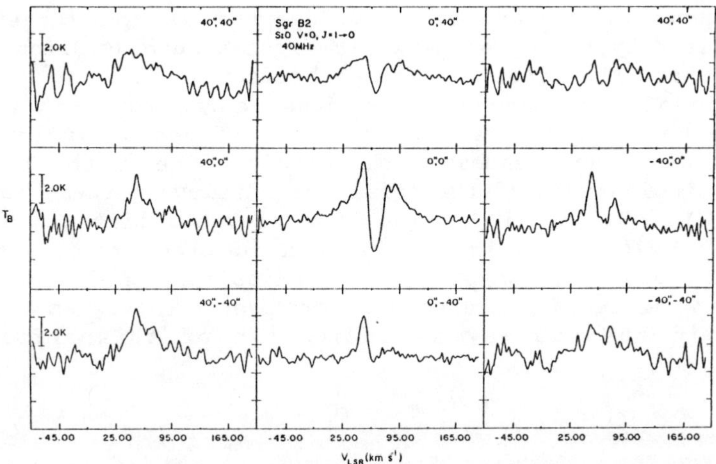

Figure 1. A spectral map of Sgr B2 in the SiO v=0, J=1–0 line.

Model: Many of the weaker lines, particularly ^{13}CO, H_2CO and HNCO show a single peaked profile centered at 62 km s^{-1} and this is taken to be the velocity of the core of Sgr B2. The high abundance molecules such as SiO, NH_3 and CS have absorption dips at this same velocity. These dips would represent absorption by cold gas surrounding Sgr B2. Since the core collapsed from the surrounding material it may be expected that the two regions of the source have the same velocity.

As part of this survey of HII regions for SiO emission we have detected V=0, J=1-0 SiO emission in a number of HII regions. These detections are shown in Figure 2.

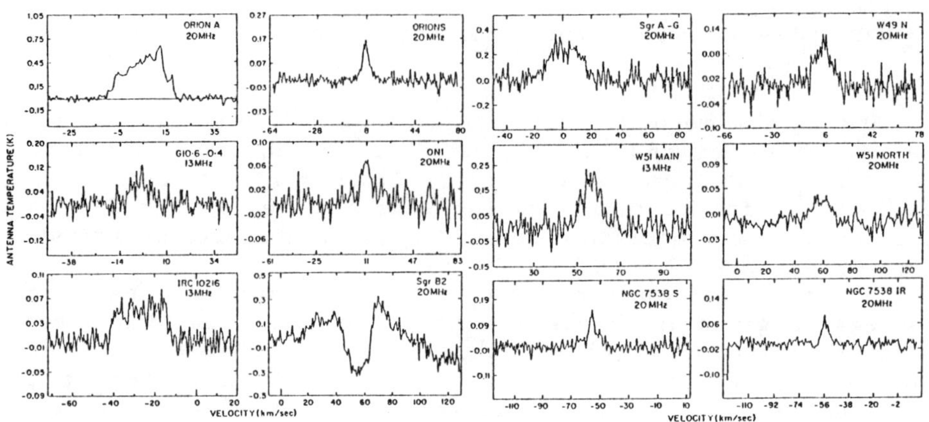

Figure 2. Silicon monoxide (v=0, J=1-0) spectra.

One noteworthy feature of these spectra is the larger line widths exhibited by the more energetic sources such as Orion A, Sgr A and W49N, where the linewidths range between 15-20 km s^{-1} whereas the remainder of the sources' linewidths range from 7-15 km s^{-1}. Orion A is a source having a large outflow of gas as indicated by H_2O maser studies and presumably the broader line widths in Sgr A and W49N are indications of the infall or outflow in these sources. W51M, ON1, G10.6-0.4 and W51N also show elevated line widths over their nominal value of ~7 km s^{-1} as seen in other molecules. Ho and Haschick (1986) have mapped the NH_3 emission from G10.6-0.4 and show a clear example of collapse and rotation. The 12 km s^{-1} line width seen in the SiO may be indicative of this infall and rotation.

REFERENCES:

D. Downes, R. Genzel, A. Hialmarson, L. A. Nyman, and B. Ronnang, Ap. J., **252**, **L29**, (1982).

T. Hasegawa, et al., Masers, Molecules and Mass Outflows in Star Forming Regions, ed., A. D. Haschick, Westford, MA (Haystack Observatory), 275, (1986).

P. T. P. Ho and A. D. Haschick, Ap. J., **304**, 501 (1986).

G. Winnewisser, E. Churchwell, and C. M. Walmsley, Astr. and Ap., 72, **215**, (1979).

SiO J = 1 → 0 EMISSION NEAR THE GALACTIC CENTER: TIDAL STRIPPING AND SHOCKS?

BRUCE L. JACOB,[1] JOHN C. SZCZEPANSKI,[1,2] AND PAUL T. P. HO[1]
[1] Harvard–Smithsonian Center for Astrophysics, Cambridge, MA
[2] Massachusetts Institute of Technology, Cambridge, MA

OBJECTIVE

We studied with 40″ resolution the $J = 1 \rightarrow 0$ SiO emission in a 5′ field south of our Galactic center. We looked for evidence of motions toward the center. We are interested in whether the giant molecular cloud M–0.13–0.18 can resupply the central neutral ring (cf., Gusten *et al.* 1987, *Ap. J.*, **318**, 124).

OBSERVATIONS

During March and April of 1987, we mapped two declination strips south of the Galactic center. We used the 37-m Haystack telescope with the cooled Q-band maser. We operated in the total power mode, beam-switching every minute, and sampling a total of 40 MHz (276 km s^{-1}) bandwidth. Pointing was accurate to $\sim$ 10″ by measuring the nearby SiO maser source W Hydra. The data presented are corrected for atmospheric extinction and elevation-dependent telescope gain variations.

EVIDENCE FOR SHOCKS

To display the observed kinematics, we show in Fig. 1 an average declination-velocity map, along the two declination strips. In this plot, we see emission dominating at $\sim$ 20 km s^{-1}. In addition, there is an extended red-wing that appears to be well localized spatially. The wing emission is associated with two continuum features attributed to nonthermal synchrotron emission from a possible supernova remnant (Ho *et al.* 1985, *Ap. J.*, **288**, 575). The excess high velocity wing emission may be due to the impact of the SNR and the resulting shock of the ambient molecular material. We suggest that the spatial correlation may be evidence of this active interaction.

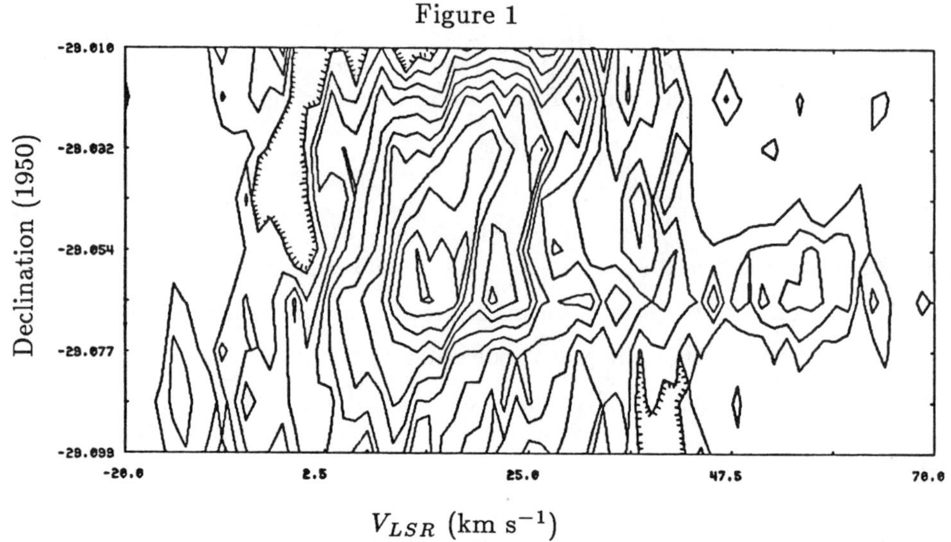

Figure 1

EVIDENCE FOR INFALL?

Within the 20 km s^{-1} cloud, we also detected relative motions. There appears a gradual shift, from 10 to 25 km s^{-1} in velocity, beginning at the central position with the enhanced red emission wing toward the Galactic center. This shift in velocity may be due to an infall toward the Galactic center, or internal rotation within the molecular cloud. If this motion is gravitationally bound, we estimate a required mass of $\geq 10^6\ M_{\odot}$. There is a suggestion that the gradient steepens toward the Galactic center. This would support an infall model.

We suggest that material may be stripped off the molecular cloud M–0.13–0.18 either through tidal forces or impact of a supernova remnant. This material may then approach the central potential well.

This research is supported in part by NSF Grant AST85–09907.

AMMONIA OBSERVATIONS IN THE REGION NGC7023

CECILIA P. BENTON
Wellesley College Dep't. of Astronomy, Wellesley, MA 02181

Abstract The dark cloud star forming region surrounding the reflection nebula NGC7023 was surveyed for emission of the (J,K) = (1,1) inversion line of ammonia.

NGC7023 is a bright reflection nebula in the northern sky, and a site of active star formation. The dark cloud is 30' x 60' with the nebula at the northern end. A Be type star, HD200775, 440 pc from the sun[1] illuminates the nebula and is the primary source for dust heating in the nebula.[2] Observations in infrared wavelengths and detections of ^{12}CO, ^{13}CO, CS line source[3,1], and H_2CO maser[4] support the conclusion that this is a star forming region.

Observations of NH_3 (J,K) = (1,1) and (2,2) lines at a wavelength of 1.3 cm in the region were done in the fall of 1986 and spring of 1987 at Haystack Observatory. The investigation searched nine IRAS sources, four T Tauri stars and ten visually opaque cores. Three separate sets of observations were made with system temperatures ranging from 80-200K. Each source was observed initially at the center position then at surrounding offsets. Three detections were mapped in the (J,K) = (1,1) line. These spectra were fit using Benson's[5] program to determine optical depth, excitation temperature, velocity and intrinsic line width.

The position one arcminute north of 20597+6800 had a strong enough (J,K) = (1,1) line that the (J,K) = (2,2) line was also observed. Twenty-eight runs were completed with an RMS of 0.056 after accumulation and Hanning smoothing. A program using two two-level models written by Benson[5] was applied to this source to determine a kinetic temperature of 10.5K and log column density 14.7 cm^{-3} in the cloud.

Of the 25 positions observed, it is somewhat surprising that only three exhibited positive detections for ammonia. Several possibilities could explain the low detection rate. Three strong ammonia sources have previously been observed and mapped by Benson[5]. So the observed detection rate is not truly indicative of a complete survey. Many ammonia cores are small so zero detection could result from poor correlation between ammonia source and telescope position, or the source could easily be beam diluted.

A comparison with statistical results from Benson's[6] ammonia survey of the Taurus dark cloud complex show that line widths in NGC7023 are larger (~0.6 km/s) and optical depths smaller (~2.2 nepers). The spectra which were fit were taken near the nebula and are probably near more massive stars than the Benson survey. Dense cores in the Taurus region will typically form solar type stars. Massive regions appear to be forming both intermediate mass and low mass stars. The dark cloud region surrounding NGC7023 has larger clumps than Taurus from comparisons with the Palomar Prints. Since NGC7023 is more that twice as distant (440 pc compared with 160 pc) the region is much larger in extent and therefore more massive.

REFERENCES

1. D.M. Elmegreen and B.G. Elmegreen, Ap. J., 220, 510 (1978).
2. S.E. Whitcomb et al., Ap. J., 246, 416 (1981).
3. S.U. Choe, Jour. Korean E.S.E.S., 6, No.2, 41 (1985)
4. V. Pankonin and C.M. Walmsley, Astron. Astrophys., 67, 129 (1978).
5. P.J. Benson, Ph.D. thesis, MIT, Department of Physics (1983).
6. P.J. Benson, in Masers, Molecules and Mass Outflows in Star Forming Regions, edited by A.D. Haschick (Haystack Observatory) p.55 (1985).

DENSE CORES ASSOCIATED WITH HERBIG AE/BE STARS

J. GREGORY STACY[1], PRISCILLA J. BENSON[2],
PHILIP C. MYERS[1], AND ALYSSA A. GOODMAN[1]
[1]Harvard-Smithsonian Center for Astrophysics, Cambridge, MA, USA
[2]Astronomy Department, Wellesley College, Wellesley, MA, USA

Abstract Preliminary results of a search for dense ammonia cores associated with Herbig Ae/Be stars indicate that Ae/Be cores are intermediate in mass (M_{core}~10-400 $M_{\odot}$) compared to the more-studied T Tauri (M_{core}~4 $M_{\odot}$) and OB (M_{core}~10^2-10^4 $M_{\odot}$) cores.

OBSERVATIONS AND ANALYSIS

A catalog of approximately 50 Herbig Ae/Be and similar stars[1] has been observed for evidence of ammonia emission. All observations of the NH_3 (J,K) = (1,1) and (2,2) lines at 1.3 cm wavelength were conducted with the 37 m radio telescope of the Haystack Observatory located in Westford, Massachusetts. Typical observing parameters for the Haystack telescope can be found in Myers and Benson.[2] Our observing program consisted of a systematic search for NH_3 (1,1) emission at candidate stellar positions, along with any nearby opaque regions determined by visual inspection of the appropriate Palomar survey prints. Source detections were mapped in the (1,1) line of NH_3 to the half-maximum intensity contour, and map peak positions were reobserved in the (2,2) line of ammonia in order to permit eventual determination of source temperatures and densities.

The main hyperfine component of the NH_3 (1,1) and (2,2) lines were fit with gaussian profiles to determine the spectral properties of the detected ammonia cores. Estimates of core size were obtained from inspection of contour maps. Preliminary mass estimates were computed assuming a sphere of radius R and uniform number density n. For the purposes of this initial report we have taken a particle density n$\sim 10^4$ cm^{-3}, and a mean particle mass of $\overline{m} \sim 2.33\ m_H$ (*cf.* Myers and Benson[2]). All distances were taken from Finkenzeller and Mundt.[1] More refined mass estimates require a rigorous fitting of the data to statistical equilibrium excitation models[3]; this effort is currently in progress. Our current mass estimates can be considered "lower limits" to the true core masses, since the actual number densities are expected to increase over the value assumed here. We have also estimated rotational temperatures T_R at map peak positions. These values were derived from the ratio of the intensities observed in the main hyperfine component of the (1,1) and (2,2) lines, assuming an optical depth $\tau_m \sim 1$ (see Ho and Townes[4]). The derived physical parameters of selected ammonia cores are listed in Table I.

TABLE I Derived Parameters of Ae/Be Dense Cores

Source	T_R (K)	d (pc)	R (pc)	M ($M_\odot$)
V380 Ori	15	460	0.33	90
(NGC1999)				
LkHα 215	13	800	0.33	85
(NGC2245)				
BD+41°3737	14	1100	0.45	215
BD+41°3731	13	1100	0.54	385
(NGC6914)				
HD 200775	12	600	0.17	12
(NGC7023)				
LkHα 234	14	1000	0.32	75
BD+65°1637	14	1000	0.38	130
(NGC7129)				

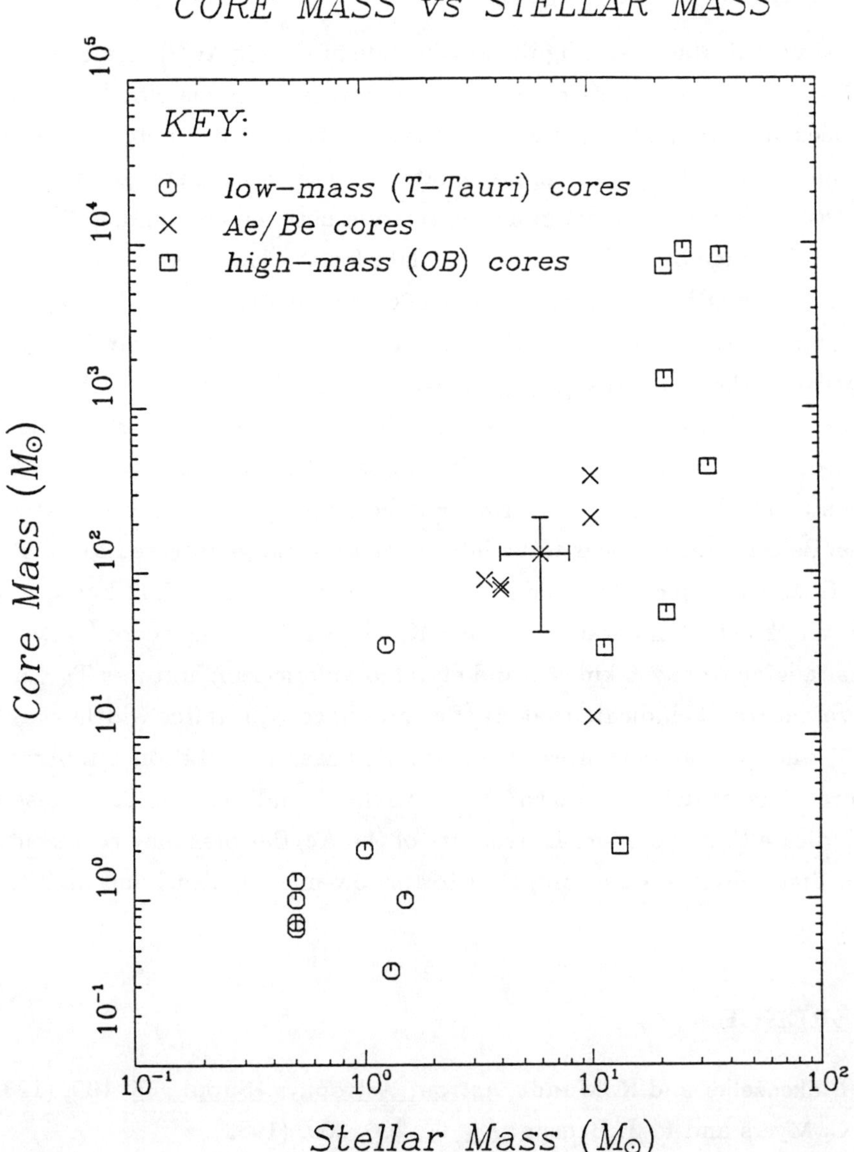

FIGURE 1. A plot of dense core mass versus stellar mass. The low-mass T-Tauri data is taken from Myers and Benson[2], and the high-mass OB data from Ho *et al.*[5] The Ae/Be points represent our current work. Sample error bars are shown.

RESULTS

Our survey indicates that a significant fraction of Herbig Ae/Be stars are associated with dense ammonia cores. Seven separate ammonia cloud complexes have been detected, comprising ten distinct cores. In our detected sample, all stellar objects lie within approximately one core diameter of map peak positions. In Figure 1 we examine the relation between core and stellar mass over a wide range of stellar mass. The data for the low-mass T Tauri cores[2] and high-mass OB cores[5] are compared to the Ae/Be cores which represent our current work. It is evident that the Ae/Be objects effectively bridge the gap between the low-mass (M_{star}~0.5-1 $M_{\odot}$) T Tauri stars and high-mass (M_{star}~10-30 $M_{\odot}$) OB stars. These results lend strong support to the hypothesis that the mass of a protostellar condensation increases with the mass of the star that it is likely to form. Further, the spectral characteristics of the Ae/Be cores also appear to be intermediate to those observed for T Tauri and OB cores. Typical low-mass cores[2] are seen to have ~0.32 km s^{-1} line widths and kinetic temperatures of ~11 K; whereas high-mass cores[5] have line widths ranging from 2-6 km s^{-1} and elevated kinetic temperatures $T_k \gtrsim 20$ K. Our present results indicate that Ae/Be cores have typical line widths of ~1-2 km s^{-1} and gas temperatures of ~13-15 K, based on rotation temperature measurements, which are known[4] to be a reliable indicator of T_k. These results indicate that the internal structure of the Ae/Be cores may represent an intermediate physical case compared to the low-mass T Tauri and high-mass OB cores.

REFERENCES

1. U. Finkenzeller and R. Mundt, Astron. Astrophys. Suppl., 55, 109, (1984).
2. P. C. Myers and P. J. Benson, Ap. J., 266, 309, (1983).
3. P. J. Benson and P. C. Myers, Ap. J. (Letters), 242, L87, (1980).
4. P. T. P. Ho and C. H. Townes, Ann.Rev.Astron.Astrophys., 21, 239, (1983).
5. P. T. P. Ho, R. N. Martin and A. H. Barrett, Ap. J., 246, 761, (1981).

A SOURCE MODEL FOR THE MOLECULAR CORE OF L134N

DARYL A. SWADE AND F. PETER SCHLOERB
FCRAO / University of Massachusetts, Amherst, MA 01003

Abstract A source model based on maps of molecular emission from the dense core of L134N is presented. In this model oxygen is removed from the gas phase chemistry by condensation of water onto dust grains in the densest region of the core.

Maps of the molecular emission from the core of the dark cloud L134N in $C^{18}O$, CS, $H^{13}CO^{+}$, SO, NH_3, and C_3H_2 show different morphologies. These maps are presented in Figure 1. $C^{18}O$ J=1-0 and CS J=2-1 emission emanate from a low density, $n(H_2)=10^{3.5}$ cm^{-3}, envelope and differences in their distributions can be explained by excitation and radiative transfer effects. However, emission from the other molecules mapped probes a higher density core, and these molecules can be divided into two groups whose distribution differences are attributed to a chemical abundance gradient within the core region.

The higher density core, $n(H_2)=10^{4.5}$ cm^{-3}, is delineated by the NH_3 (1,1) map and the C_3H_2 1_{10}-1_{01} map. Other molecular lines surveyed at the peak NH_3 and C_3H_2 position indicate that this density peak has a high abundance of hydrocarbon and cyanopolyyne molecules. Southwest of the high density NH_3 and C_3H_2 region is an intermediate density region traced by the SO N,J=2,3-1,2 map. The $H^{13}CO^{+}$ J=1-0 map shows a slight peak at the high density peak of NH_3 and C_3H_2, but also extends into the region of strong SO emission.

The differences between the NH_3 and C_3H_2 distributions and the SO map may be explained by a gradient in the gas phase C/O ratio in the L134N core region. This C/O gradient could be produced by an oxygen depletion in the dense core due to condensation of water ice onto dust grain mantles in regions of highest density. Hence, molecular species which are readily oxidized, such as hydrocarbons and derivatives of hydrocarbon chemistry, are free to form in the oxygen depleted dense core, while SO forms in the less dense, oxygen-rich environment. The $H^{13}CO^+$ map probes both regions since enhanced $H^{13}CO^+$ emission may occur at the high density peak due to $H^{13}CO^+$ J=1-0 excitation requirements combined with a possible higher abundance from oxidation of hydrocarbons.

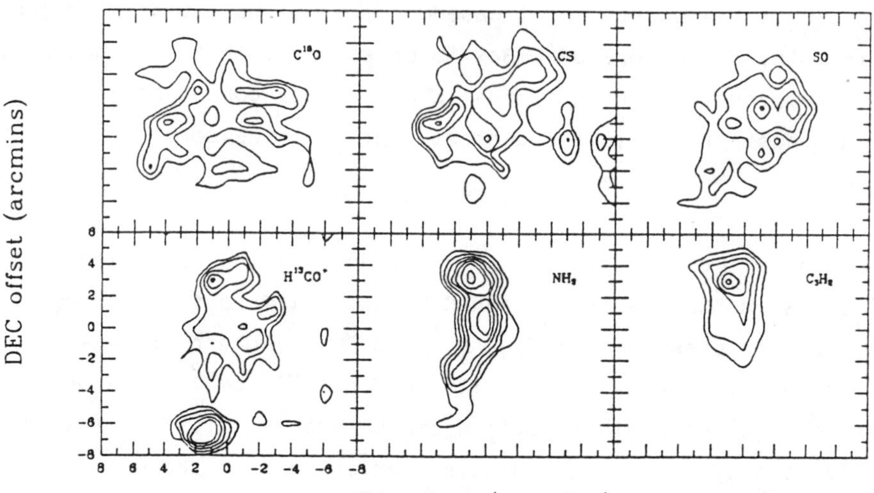

FIGURE 1. Maps of the molecular emission from the core of the dark cloud L134N in $C^{18}O$, CS, SO, $H^{13}CO^+$, NH_3, and C_3H_2. The contour levels representing integrated intensity across the emission line in units of K km s^{-1} are: $C^{18}O$ J=1-0, 1.1-2.0 by 0.3; CS J=2-1, 0.5-1.3 by 0.2; SO N,J=2,3-1,2, 0.8-2.0 by 0.3; $H^{13}CO^+$ J=1-0, 0.3-0.7 by 0.1; NH_3 $(1,1)_M$, 1.0-3.4 by 0.4; and C_3H_2 $1_{10}-1_{01}$, 0.55-1.05 by 0.10. The $C^{18}O$, CS, SO, and $H^{13}CO^+$ maps were constructed from spectra obtained with the Five College Radio Astronomy Observatory 14 m telescope. The NH_3 map was obtained with the Haystack Observatory 37 m radio telescope, and the C_3H_2 map was obtained with the National Radio Astronomy Observatory 43 m radio telescope. All offsets are referenced to the position: RA = $15^h51^m30^s$, DEC = $-2^o43'31''$ (1950.0).

WATER EMISSION IN ORION AT 183 AND 380 GHZ

M. A. FRERKING and T. B. H. KUIPER
Jet Propulsion Laboratory, California Institute of Technology, Pasadena, CA 91109

Abstract The $3_{13} \rightarrow 2_{20}$ and the $4_{14} \rightarrow 3_{21}$ transitions of water at 183 and 380 GHz have been mapped toward the Kleinmann-Low Nebula in Orion. The observations were taken with a dual frequency heterodyne radiometer from the Kuiper Airborne Observatory in the spring of 1985.

Molecular line emission from the Orion molecular cloud arises from several components. A ridge of relatively cool quiescent gas, extending over several minutes of arc in the N-S direction, has been observed in many molecular species. In addition there are also other more turbulent, compact regions of the gas associated with IRc 2. A region of high velocity outflow, commonly called the plateau region because of its spectral shape, surrounds IRc 2. Located inside the region swept out by the high velocity winds are several clumps, the hottest of which is called the hot core while less intense regions have been identified as the compact ridge. These have been observed in high resolution maps of HDO, NH_3, and CS[1].

Each of these regions are characterized by different temperatures, densities, and velocities as summarized in the following table[2].

Feature	Quiescent Ridge	Plateau	Compact Ridge	Hot Source
Temperature (K)	55-60	95-150	80-140	150-300
V_{lsr} (km s^{-1})	9.6	7.5	8	6
ΔV (km s^{-1})	4	22	3	10
$n(H_2)$ (cm^{-3})	10^5	$> 10^6$	10^6	$> 10^8$
$N(H_2)$ (cm^{-2})	3 x 10^{23}	< 5 x 10^{22}	-	1 x 10^{24}
Size	7′ x 2′	20″ x 10″	16″ x 13″	6″ x 4″

We have observed the $3_{13} \rightarrow 2_{20}$ and the $4_{14} \rightarrow 3_{21}$ transitions of water at 183 and 380 GHz toward the Kleinmann-Low Nebula in Orion. The emission was not resolved by the antenna FWHM beamwidth of 6.′9 and 3.′3 at 183 and 380 GHz respectively.

A composite line made up of the line shapes for the quiescent ridge, the plateau, and the hot source have been fit to the lines observed at the nominal center position. The only free parameter in the fits was the relative amplitude of the various components. The results are summarized in the table below and in figure 1. Antenna temperatures corrected for beam dilution are also listed in the table.

Feature		Quiescent Ridge	Plateau	Compact Ridge	Hot Source
T_A^* (K)	(183 GHz)	7	4	-	2
T_b (K)	(183 GHz)	27	3900	-	16,000
T_A^* (K)	(380 GHz)	3	11	-	5
T_b (K)	(380 GHz)	5	2500	-	9300

In all the compact regions the observed brightness temperature is larger than the characteristic temperature determined from other molecules. This indicates that the emission arises from non-thermal processes.

In the quiescent ridge the observations were analyzed assuming thermodynamic equilibrium, no trapping, and an ortho - para ratio of 1. The following parameters were determined; excitation temperature = 40K, 183 GHz line optical depth = 1.4, 380 GHz line optical depth = 0.17, $N(H_2O) = 7 \times 10^{17}$ cm^{-2}, and $\chi(H_2O) = 2 \times 10^{-6}$. No consistent solution was found with an ortho - para ratio of 3.

This work was performed at the Jet Propulsion Laboratory, California Institute of Technology, under contract with the National Aeronautics and Space Administration.

REFERENCES

1. Mundy, L.G., et.al., 1986, Ap. J., 304, L51; and references cited therein.
2. Blake, G. A., 1985, Ph.D. Thesis, California Institute; and references cited therein.

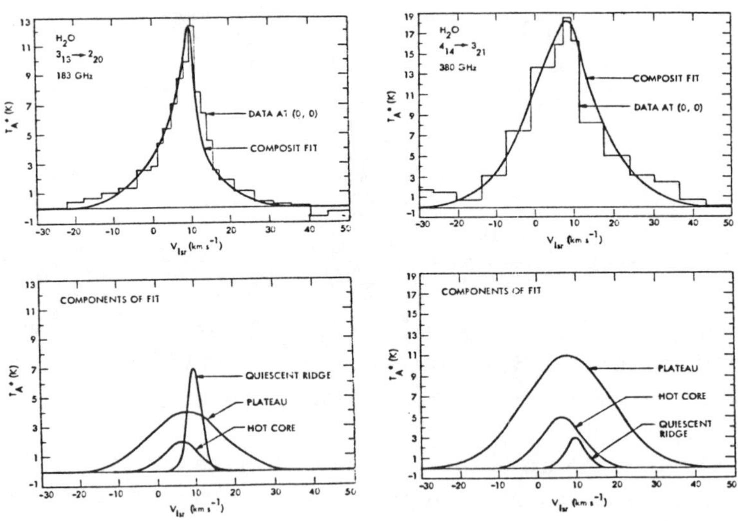

Figure 1. Observed 183 and 380 GHz lines and composite fits.

FAR-INFRARED, ROTATIONALLY EXCITED OH EMISSION IN ORION-KL

GARY J. MELNICK
Harvard–Smithsonian Center for Astrophysics, Cambridge, MA

REINHARD GENZEL, ALBRECHT POGLITSCH
Max-Planck-Institut für Physik und Astrophysik, Garching, F.R. Germany

GORDON J. STACEY AND JOHN B. LUGTEN
University of California, Berkeley, CA

INTRODUCTION

We have undertaken a study of Orion-KL in a number of ^{16}OH far-infrared rotation lines and a ground-state rotational transition of ^{18}OH. The four ^{16}OH and one ^{18}OH transitions detected thus far greatly constrain the range of physical conditions in the OH emitting region. If the OH emission arises in the post shocked gas, as the large velocity widths ($\geq$ 50 km s^{-1}) measured for the OH lines suggest, then these transitions provide valuable information about the cooler, denser portion of the post shocked region not sampled by the higher excitation lines of CO (J > 15) or H_2.

OBSERVATIONS AND RESULTS

The OH far-infrared lines were observed from the Kuiper Airborne Observatory using the U. C. Berkeley tandem Fabry-Perot spectrometer[1]. The most recent measurements include the observations of 3 lines: the ^{16}OH $^2\Pi_{1/2}$ $J = 3/2^- \rightarrow$ $^2\Pi_{3/2}$ $J = 3/2^+$ transition at 53.351 μm with a 40″ beam and 38 km s^{-1} resolution, the ^{16}OH $^2\Pi_{3/2}$ $J = 5/2^- \rightarrow 3/2^+$ transition at 119.234 μm with a 45″ beam and 24 km s^{-1} resolution, and the ^{18}OH $^2\Pi_{3/2}$ $J = 5/2^+ \rightarrow 3/2^-$ transition at 120.172 μm with a 45″ beam and 55 km s^{-1} resolution (see Fig. 1).

The 163.121, 119.234, and 84.597 μm lines are all seen in emission with intensities of 1.6, 1.9, and 2.1 erg s^{-1} cm^{-2} sr^{-1}, respectively. Both the ^{16}OH 53.351 μm and the ^{18}OH 120.172 μm lines display a P-Cygni profile: the 53.351 μm line has an intensity in absorption and emission of 7.3 and 1.6 erg s^{-1} cm^{-2} sr^{-1}, respectively, and the 120.172 μm line has an intensity in absorption and emission of 0.11 and 0.05 erg s^{-1} cm^{-2} sr^{-1}, respectively. In both P-Cygni spectra, the peak in the absorption is blueshifted with respect to the peak in the emission.

CONCLUSIONS

Earlier modeling of the OH far-infrared rotational emission[2], based on the 163.121, 119.234, and 84.597 μm line intensities, led to a number of conclusions about the OH emitting region:

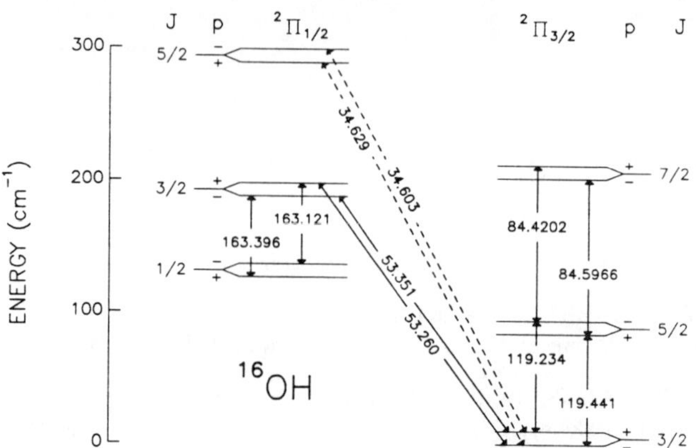

Fig. 1. Part of the rotational energy level diagram of OH. The transitions which have been detected are shown with solid lines along with the wavelength in microns.

1) The near equality of the 84.597 and 119.234 μm line intensities is within a factor of two to three of the intensity ratio in a fully thermalized, optically thick gas at infinite temperature. This requires large OH column densities ($\geq 10^{15}$ cm^{-2}) and either large volume densities ($n_{H_2} > 10^8$ cm^{-3}) or a combination of moderately large densities (10^6 or 10^7 cm^{-3}) and large OH optical depth, which can trap OH line radiation and increase the population in the upper levels.

2) The 163.121 : 84.587 μm line intensity ratio implies that the gas temperature in the emitting region is less than 200 K.

3) The 163.121 : 119.234 μm line intensity ratio can only marginally be reproduced in a region which experiences only collisional excitations. More likely, radiative excitations, along with collisional excitations, must be considered.

The main results of the most recent observations are as follows:

1) That the ^{16}OH 53.350 μm line is seen largely in absorption confirms that the gas temperature in the OH emitting region is less than 200 K.

2) The strength of the ^{16}OH 53.350 μm absorption feature implies that 50% or more of the 163.121 and 163.396 μm emission we observe must be due to radiative pumping.

3) The P-Cygni profile exhibited by both the ^{16}OH 53.350 μm and the ^{18}OH 120.2 μm transitions is consistent with an expanding shell of cool gas in which the blueshifted material absorbs background continuum radiation, presumably due to IRc2, while the redshifted gas, behind the continuum source, is seen in emission.

4) The strength of the ^{18}OH 120 μm emission implies an optical depth in the ^{16}OH 119 μm lines of ~ 10 (assuming ^{18}OH : ^{16}OH $= 1:500$).

REFERENCES

1. J. B. Lugten 1987, Ph.D. Thesis, University of California, Berkeley.
2. G. J. Melnick, R. Genzel, and J. B. Lugten 1987, *Ap. J*, **321**, 530.

40″-RESOLUTION CS MAPS OF THE CORE OF THE ORION MOLECULAR CLOUD

JOHN C. SZCZEPANSKI,[1,2] AND PAUL T. P. HO[1]
[1] Harvard–Smithsonian Center for Astrophysics,Cambridge,MA
[2] Massachusetts Institute of Technology,Cambridge,MA

INTRODUCTION The 120″ x 120″ region centered on Orion-KL has been fully-sampled with a resolution of $\sim 40''$ using the CS $J = 1 \rightarrow 0$ line. The goals are twofold: (1) Calibrate the telescope at 49 GHz, and (2) Investigate the spatial extent and structure of two possible clouds toward the center of OMC1. Interferometric studies[1] of HCN and Nobeyama $J = 1 \rightarrow 0$ CS studies[2] suggest that the larger line width toward the center of OMC1 may be due to rotation. This is contrary to extended NH_3 mapping results[3] where two distinct cloud components were identified and the larger line width in the center was suggested to be due to the overlap of the two cloud components.

OBSERVATIONS The data were obtained during January 1987 with the 36.6-m radio telescope and the Q-band maser receiver of the Haystack Observatory. Typical system temperature was about 350 K . Our velocity resolution was 0.06 km/s. Continuum drift scans of the 43 GHz SiO maser in Orion were performed in order to reduce pointing errors to within $\sim 10''$. Observations were made in the total power mode. A reference position was repeated after every five spectra as an additional check on calibration and applied corrections. Positional offsets were taken in units of half-beam spacing from the KL nebula: RA(1950) = $5^h32^m46\overset{s}{.}7$, DEC(1950) = $-5°24'20\overset{''}{.}0$.

RESULTS Comparison with Nobeyama data[2] of integrated intensity, peak velocity distributions, and spectra shows good agreement. Minor differences are probably due to our larger beamwidth and the incomplete sampling in the Nobeyama data near the central peak. The velocity gradient along a NE-SW line through the center ($\pm 28''$) is in excellent agreement. The position-velocity map in Figure 1 is especially interesting since the data seem to indicate a rotation model as proposed by Hasegawa[2].

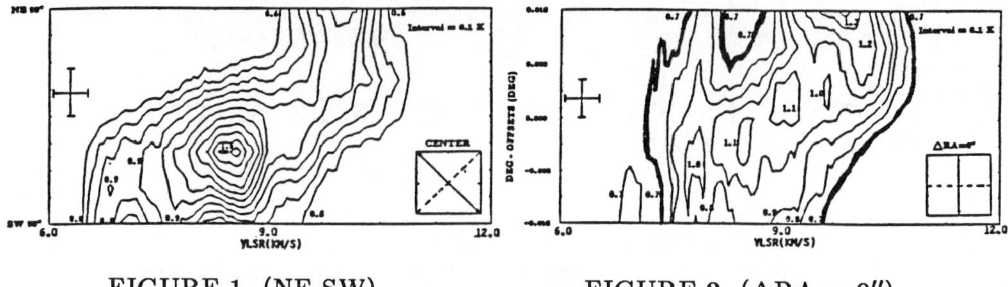

FIGURE 1 (NE-SW) FIGURE 2 (ΔRA $= 0''$)

However, in the position-velocity map at $\Delta\mathrm{RA} = 0''$ of Figure 2, note that the 0.7-K contour (bold) shows a low velocity component at positive DEC offsets in addition to the high velocity component. This feature is also indicated in the spectra of Figure 3. A simple rotation model with major axis in the NE-SW direction cannot account for this feature.

Figure 3 shows preliminary results of double-Gaussian fitting for several spectra. Averaging the center velocities yields components at 7.6 ± 0.3 km/s and 9.9 ± 0.2 km/s. These results are in excellent agreement with earlier NH_3 observations.

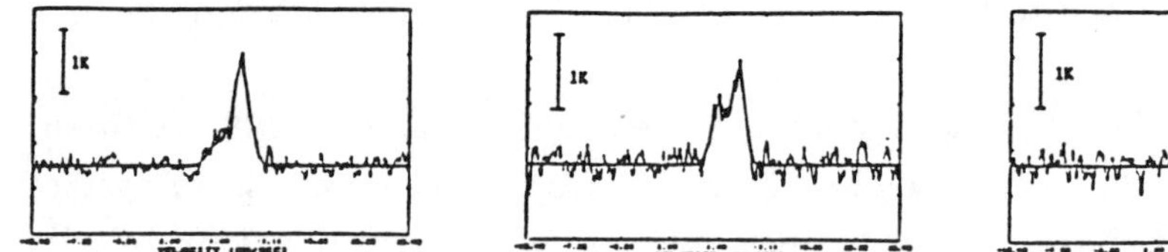

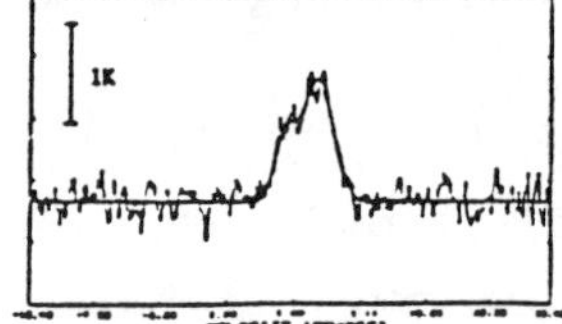

FIGURE 3 Double-Gaussian fits to selected spectra.

We conclude: (1)Telescope operation at 49 GHz is reliable, and (2) A simple rotation model has difficulty accounting for the observed two-component spectra.

This research is supported in part by NSF Grant AST85-09907.

REFERENCES

1. S. N. Vogel, J. H. Beiging, R. L. Plambeck, W. J. Welch and M. C. H. Wright, Ap. J., 296 ,600-605 (1985).
2. T. Hasegawa, N. Kaifu, J. Inatani, M. Morimoto, Y. Chikada, H. Hirabayashi, H. Iwashita, K. Morita, A. Tojo and K. Akabane, Ap. J., 283, 117-122 (1984).
3. P. T. P. Ho and A. H. Barrett, Ap. J.(Letters), 224 , L23-L26 (1978).

THE DUPLICITY OF THE 'PROTOSTAR' IRAS16293-2422

ALWYN WOOTTEN
NRAO†, Edgemont Road, Charlottesville, VA 22903-2475

Abstract Continuum maps at λ 6cm and 2 cm of the region of the infrared object IRAS16293-2422 reveal two sources separated by ~5"(750 AU), presenting a rare opportunity to observe a binary system in its earliest stages of formation.

DISCUSSION

The remarkable protostellar source IRAS16293-2422 may be the youngest star yet located in the ρ Ophiuchi star formation complex[1]. It lies within the complex of cloud cores described by Wootten and Loren[2], and probably harbors the source of a bipolar flow apparently centered on the warm core containing the IRAS source.

Observations of L1689IR were made with the NRAO Very Large Array (VLA) near Socorro, N.M. in the A/B configuration at λ6 cm on 1986 July 3 and in the B/C configuration at λ6 cm and 2 cm on 1986 Oct 5. A detail of the final natural weighted maps near IRAS16293-2422 is shown in Figure 1a for 6cm, and Figure 1b for 2cm.

Three sources were detected in the vicinity of IRAS16293-2422 at 6cm, of which two lie sufficiently close to the infrared source that they might

† The National Radio Astronomy Observatory (NRAO) is operated by Associated Universities, Inc., under contract with the National Science Foundation.

be identified with it. These sources will be called IRAS16293-2422A and IRAS16293-2422B, or simply sources A and B for reference, with A being the stronger source. Another source was found nearby, which we call 16293-2423. Its steep spectrum suggests this source may be a background extragalactic source. Another IRAS source, IRAS16293-2424, lies within the field at 6cm. No source stronger than 0.1 mJy (4 σ) was detected at the accurate position provided by Wilking (private communication). No other sources were detected within the 9' primary beam at 6cm. In the less sensitive 2cm map, only source A was detected. The positions and fluxes of the associated sources are: IR1629A, $\alpha(1950)$ 16^h 29^m $21^s.025$, $\delta(1950)$-24° 22' 15".6, S_{6cm}=2.68 $\pm$ 0.02 mJy, S_{2cm}=3.8 $\pm$ 0.2 mJy and IR1629B, $\alpha(1950)$ 16^h 29^m $20^s.770$, $\delta(1950)$-24° 22' 11".7, S_{6cm}=0.25 $\pm$ 0.025 mJy, $S_{2cm} \lesssim 0.5$ mJy (3 σ).

At their distance of 160 pc (Whittet 1974), the continuum sources extend less than 120 AU and are separated by about 750 AU (5") along the plane of the disk-like structure mapped by Mundy *et al.*[3]. Both lie well within the 6500 AU region of accretion[2].

Comparison of the map in Figure 1 to the 110 GHz maps[3] reveals good positional coincidence between the stronger 6cm source A and an extended component in the high frequency map. The weaker 6cm source B corresponds to the brightest part of the millimeter emission. As Mundy *et al.*[3] note, an appealing interpretation of the 110GHz maps is emission from a rotating disk. As they point out, however, such cold massive disks are unstable (Larson 1984). The evidence from the 6cm maps is that the structure observed at 3mm has fragmented into at least two components, one associated with each of the 6cm sources. A simple model fitted to the 3mm map suggests that 6cm source B can be identified with an unresolved point source with an integrated flux of 209 mJy at 3mm, and that 6cm source A can be identified with a source of 9" x 4" in position angle 165° which has an integrated flux of 325 mJy at 3mm. Such a component may represent a disc surrounding IRAS1629A.

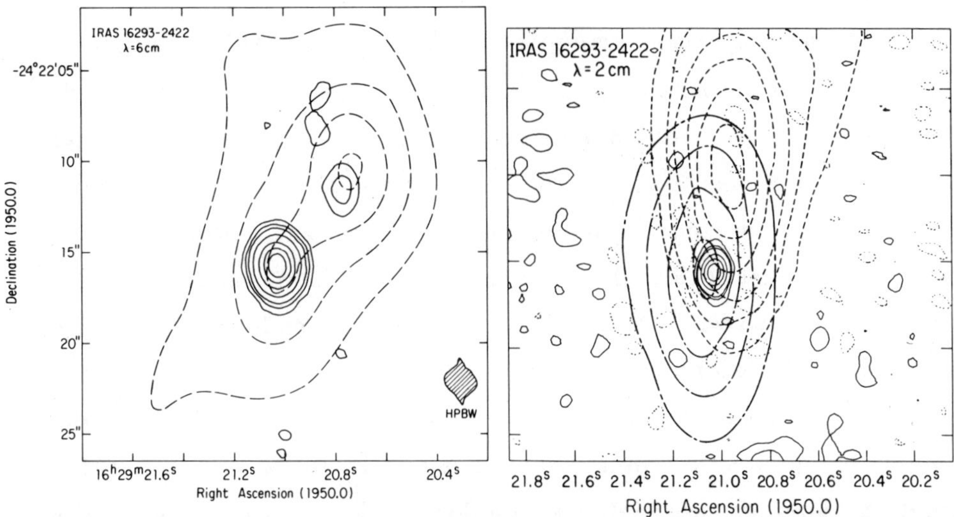

Figure 1. a). Map of the 6cm continuum emission (left panel, solid contours) from the region of IRAS16293-2422. Contours are drawn at intervals of 3, 6, 10, 20, 40, 60, and 80 times the rms noise level of the map (26 μJy). Representative contours (dashed lines) from the 3mm continuum map[3] (beamsize 3.9 x 6.0 ") are superimposed. The 2cm map (right panel, solid contours) is contoured at 3, 6, 9, 18, 25 and 35 times the rms noise level (100 μJy). Representative contours (dashed lines) from the ^{13}CO map[3] (beamsize 5.3 x 16.2").

The lower resolution ^{13}CO map presented[3] also shows two features which lie approximately at the position of the 6cm sources A and B. The velocity difference ($\sim$3 km s^{-1}) between the line features might be interpreted as the radial component of the velocity of the sources ultimately responsible for the emission.

The flat spectral index observed for source A (α_{6-2}=0.30 $\pm$ 0.12) and the lack of polarization, suggest that it arises in thermal emission from ionized gas associated with a star. A similar origin for source B seems likely. Localized energy sources, possibly stars or protostars, for the ionization of the gas must be present.

For several reasons, components A and B probably originate in two distinct objects, rather than, for example, lobes of a continuum manifestation of a flow. The most compelling reason is the coincidence of the centimeter and millimeter components. Since the centimeter continuum traces ionized gas and the millimeter continuum dust mass, and since the centimeter sources lie embedded in dust surface density peaks, a simple and consistent model suggests that the two centimeter sources lie in gravitationally distinct regions. A second reason is morphological: the millimeter structure of object A suggests gravitational evolution has produced a flattened region of enhanced density parallel to the position angle between A and B. It seems unlikely that expanding ionized gas associated with a flow would follow the ridge of greatest density. These arguments suggest a model in which components A and B originate in two distinct stellar objects. IRAS16293-2422 appears to be an extremely young binary system.

REFERENCES

1. C. K. Walker, C. J. Lada, E. T. Young, P. R. Maloney, and B. A. Wilking, Ap.J.Letters, **309**, L47 (1986).

2. A. Wootten and R. B. Loren, Ap.J. **317**, 220 (1987).

3. Mundy, L. G., B. A. Wilking, and S. T. Myers, Ap. J. Letters, **311**, L75 (1986).

A COMPOSITE CO SURVEY OF THE ENTIRE MILKY WAY

T. M. DAME, H. UNGERECHTS, AND P. THADDEUS
Harvard-Smithsonian Center for Astrophysics, 60 Garden St.
Cambridge, MA 02138

Abstract Two nearly identical 1.2 meter telescopes were used to survey the entire Milky Way in the J=1→0 line of CO at 115 GHz. The survey reveals the large-scale structure of the molecular Galaxy with unprecedented clarity and permits a fairly complete inventory of molecular clouds within 1 kpc of the Sun.

Five large unbiased CO surveys of segments of the Galactic plane have been combined with eleven surveys of particular local clouds to produce a composite survey 10°-30° wide in latitude, covering the entire Galactic plane at an angular resolution of 0.5°. The full survey contains more than 31,000 spectra and fully samples ~7700 deg^2, nearly a fifth of the entire sky. Its clear from a comparison with dark nebulae, and from the IRAS far-infrared and COS-B gamma ray surveys, that little molecular gas lies beyond the boundary of the composite survey.

The complete longitude and wide latitude coverage of the survey permits, for the first time, a detailed study of molecular clouds near the Sun (Figure 1). The molecular mass within 1 kpc of the Sun is found to be four times greater in the northern Milky Way (ℓ=0°-180°) than in the southern (ℓ=180°-360°). Furthermore, nearly all the clouds within 1 kpc in the first and fourth quadrants apparently lie on a fairly straight ridge more than 1 kpc long which may trace the inner edge of the Local spiral arm. Within 1 kpc of the Sun, the z dispersion of molecular gas is 74 pc, corresponding to a Gaussian FWHM thickness of 87 pc, and the mean mass surface density is 1.3 $M_\odot$ pc^{-2}. The mean midplane density of molecular gas at the solar circle is estimated to be 0.10 H_2 cm^{-3}.

Large-scale spatial and longitude-velocity maps produced from the survey will be published elsewhere[1]; false-color versions of these maps in poster form are available from the authors.

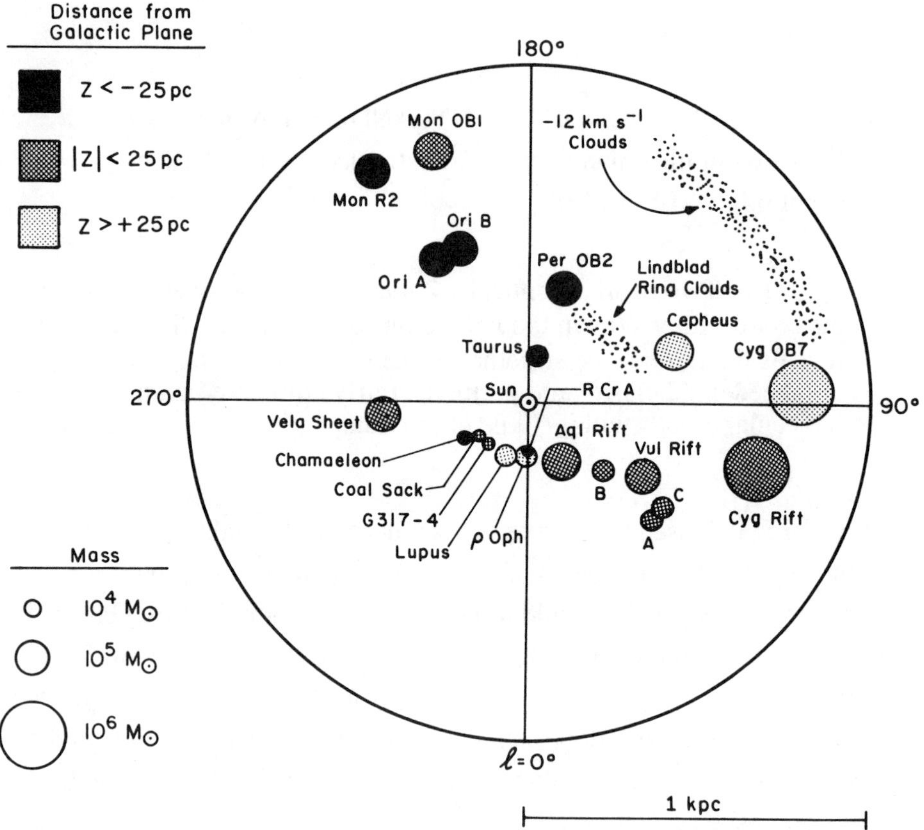

Figure 1. The distribution in the Galactic plane of molecular clouds within 1 kpc of the Sun. Shading indicates distance from the plane.

REFERENCES

1. T. M. Dame *et al.* 1987, *Ap. J.*, **322**, 706.

APERTURE SYNTHESIS OBSERVATIONS OF CO IN THE STARBURST GALAXY NGC 2146

JAMES M. JACKSON
Radio Astronomy Laboratory, Univ. of California, Berkeley, CA 94720

CO MAP

CO $J = 1 \rightarrow 0$ emission from the starburst galaxy NGC2146 was mapped with the Hat Creek Millimeter Wave Interferometer (Figure 1). The CO lies in a long, narrow structure that is not coincident with optical dust lanes. This structure is either a bar or a disk viewed edge-on. There are two peaks in the CO distribution separated by about 700 pc. These peaks flank the kinematic center of the galaxy. The central CO minimum may be an actual absence of molecular gas or a depopulation of the lower levels due to heating.

The CO map matches the distribution of Hα[1] and 20 cm continuum[2] emission. If the CO luminosity traces the mass of molecular gas and the radio continuum the distribution of supernovae, then their close correspondence suggests that the supernova rate and the star formation rate are proportional to the molecular mass. Alternatively, the supernovae may heat the molecular gas and increase the CO brightness. The CO $J = 2 \rightarrow 1$ emission[3] mapped at $30''$ resolution is consistent with optically thick emission.

The CO rotation curve is in fair agreement with Hα.[4] The CO shows a larger velocity gradient than the Hα and is probably closer to the nucleus. There is no evidence for large noncircular motions. If the starburst in NGC2146 was triggered by a collision with a companion galaxy, then the collision has not perturbed the CO gas, or else the gas has had time to relax into a stable configuration.

COMPARISON WITH M82

If the standard [CO]/[H_2] conversion factor[5] holds for NGC2146, then the mass of molecular gas deduced is $4.1 \times 10^9\ M_\odot$. This mass 9–23% of the total dynamical mass[4,6] and comparable to the total HI mass[7] of $4 \times 10^9\ M_\odot$. The ratio of the total infrared luminosity to the total H_2 mass is then $16L_\odot/M_\odot$. This is similar to the value for M82 if the same [CO]/[H_2] conversion is used. The efficiency of star formation is therefore equally high in the case of NGC2146. The higher luminosity of NGC2146 as compared to M82 can be explained by a somewhat larger region undergoing an equally intense burst of star formation.

REFERENCES

[1]Keel, W. C. 1984, *Ap. J.*, **282**, 75.
[2]Kronberg, P. P., and Biermann, P. 1981, *Ap. J.*, **243**, 89.
[3]Jackson, J. M., and Ho, P. T. P., in preparation.
[4]Benvenuti, P., Capaccioli, M., and D'Odorico, S. 1975, *Astr. Ap.*, **41**, 91.
[5]Young, J.S., and Scoville, N.Z. 1982, *Ap. J.*, **258**, 467.
[6]Burbidge, E.M., Burbidge, G.R., and Prendergast, K.H. 1959, *Ap. J.*, **130**, 739.
[7]Fisher, J.R., and Tully, R.B. 1976, *Astr. Ap.*, **53**, 397.

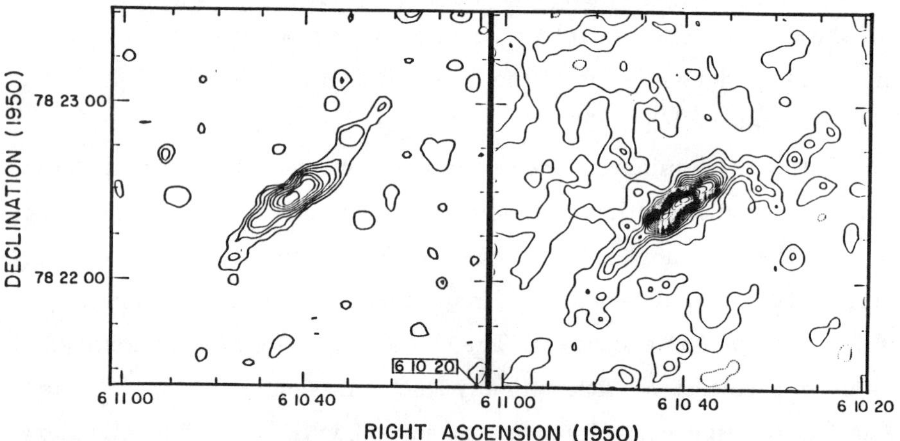

Figure 1. The integrated CO emission from NGC2146 is on the left. The contours are in steps of 6.2 K km s^{-1} brightness temperature in an $8'' \times 8''$ beam. The 20 cm VLA continuum map from Kronberg and Biermann (1981) is on the right.

MOLECULAR OUTFLOWS IN GALACTIC NUCLEI

Willem A. Baan
Arecibo Observatory, P.O.Box 995, Arecibo, PR 00604

Aubrey D. Haschick
Haystack Observatory, Westford, MA 01886

Megamaser galaxies form a small subgroup of active galaxies representing a very early stage in the evolution of a Seyfert or HII nucleus[1]. During this stage large amounts of molecular gas and dust must exist close to the active regions in the nucleus. The model for these sources involves low-gain amplification of the nuclear radio continuum by foreground molecular material[2,3]. The OH in these large scale molecular regions is pumped by the FIR radiation field combined with the radio continuum at 6 cm[2,4]. For three OH megamasers molecular outflows have been seen, originating close to the nucleus. The data of two of these OH megamasers, IC4553 and IIIZW35, will be discussed here.

IC4553=Arp220 IC4553 is a highly peculiar and very luminous FIR source. Using the results of studies of the HI absorption[5] and the OH emission[6,7], it is possible to recognize three components in the 18-cm OH spectra[8]: a) the dominant emission region I at the systemic velocity represents the rotating molecular disk, b) region II is an infalling molecular complex at a velocity 200 km s^{-1} above the systemic velocity, and c) region W is a molecular outflow producing an extended low velocity wing in the 1720 MHz line and weak absorption in the 1667 MHz line. The wing extends to 700-800 km s^{-1} below the systemic velocity. The line ratios of the three regions are distinctly different, requiring different excitation processes.

IIIZW35 This galaxy has an OH line and FIR luminosity very similar to IC4553[9]. Two distinct components are recognized in the main OH lines[10]: a) a main emission region I being a molecular disk centered on the nucleus and b) an outflow region W giving a blue wing to the 1667 MHz line extending 530 km s^{-1} below the systemic velocity.

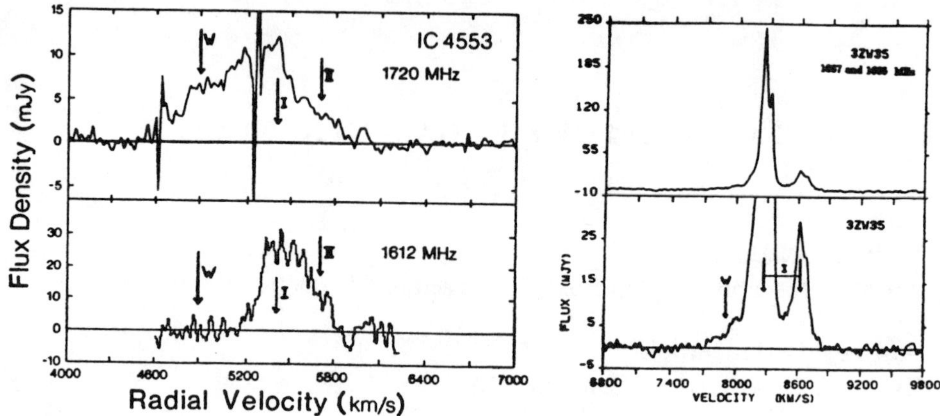

Figure 1. The 18 cm OH spectra of IC4553[8] and IIIZW35[10].

Discussion The low velocity wings in the OH megamaser lines represent symmetric molecular outflow from the nuclear region. If radiation pressure is driving the outflow, the gas must be relatively close to the nucleus (1-10 parsecs) in order to attain such velocities. Shock fronts passing through such spherical shells may facilitate collisional pumping of the OH and further enhance the OH abundances. A sherical outflow in a low-gain amplification model should only result in a blueshifted wing on the emission lines. The 1720 MHz line of IC4553 shows a red and a blue wing, but the red wing is fully accounted for by the infalling region II. For the new megamaser, IR12112+0305[10], there is clear evidence for only a blueshifted wing, since there is little confusion with the 1665 MHz line in the spectrum.

References

1. W. Baan, C. Henkel, and A. Haschick, Ap.J.,220,in press(1987).
2. W. Baan, Nature,315,26(1985)
3. A. Haschick and W. Baan, Nature,314,144(1985).
4. C. Henkel, R. Güsten, and W. Baan,A.A.,in press(1987).
5. W. Baan, J. van Gorkom, J. Schmelz, and I. Mirabel, Ap.J.,313,102(1986).
6. W. Baan and A. Haschick, Ap.J.,279,541(1984)
7. R. Norris et al., M.N.R.A.S.,213,821(1985).
8. W. Baan and A. Haschick, Ap.J.,318,139(1987).
9. L. Staveley-Smith et al., M.N.R.A.S.,in press(1987).
10. W. Baan, A. Haschick, and C. Henkel, in preparation(1987).

THE CEPHEUS-A AND MON-R2 OUTFLOWS; 15" RESOLUTION MAPS

JOHN BALLY
AT&T Bell Laboratories, Holmdel, N.J.

M. HAYASHI, S. HAYASHI, T. TAKANO, and M. TANAKA
Nobeyama Radio Observatory, Nagano, Japan

OBSERVATIONS AND RESULTS

Ceph-A and Mon-R2 are among the first bipolar outflows to be discovered since they are relatively close (about 1 kpc and 750 pc respectively), have large angular extent, and prominent line wings in ^{12}CO. Both outflows appear to be at a late stage in their evolution; the stars located at their centers have produced HII regions and the total mass of accelerated gas is large. We have produced large scale maps of the ^{12}CO distribution in the Cepheus-A and Mon-R2 outflow regions using the 45-meter N.R.O. mm-wavelength radio telescope. The full extent of the Cepheus A flow was mapped with 15" resolution along with the red-shifted lobe of the Mon-R2 outflow. The blue shifted part of the Mon-R2 flow was observed in strip maps. The distribution of J=2-1 CS emission was mapped in Mon-R2, also on a 15" grid.

We find evidence for "oscillations" in the velocity field and structure of these flows. In Ceph A the highest velocity gas alternates between red and blue shifted components along the major axis of the flow (Figure 1). Although the core of the flow appears bipolar about the central IR source, a western extension contains 3 distinct blue-shifted peaks and 2 red ones. These peaks define a highly collimated structure over 500" long and less than 50" wide. The intermediate velocity gas exhibits a less collimated, more amorphous morphology, with a hint of shell structure near the central IR source. The most likely interpretation is that there is a single outflow source near the IR source; the flow is re-directed by dense structures in the host cloud. The first such re-direction in the western part of the flow coincides in position with the GGD 37 group of Herbig-Haro Objects. In Mon-R2, the highest velocity gas describes a meandering pattern (Figure 2), reminiscent of a river. Again it is likely that this morphology is produced by the interaction of an outflow from the MonR2 core with density structures in the ambient cloud.

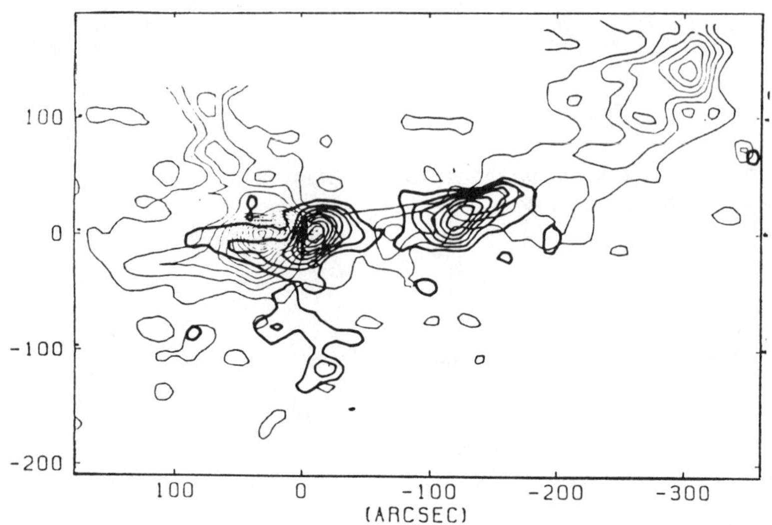

FIGURE 1 The Ceph-A outflow in ^{12}CO. Heavy lines correspond to the red-shifted wings (V_{lsr} = -3 to 3 km s^{-1}) while the light lines corespond to the blue shifted gas (V_{lsr} = -23 to -17 km s^{-1}).

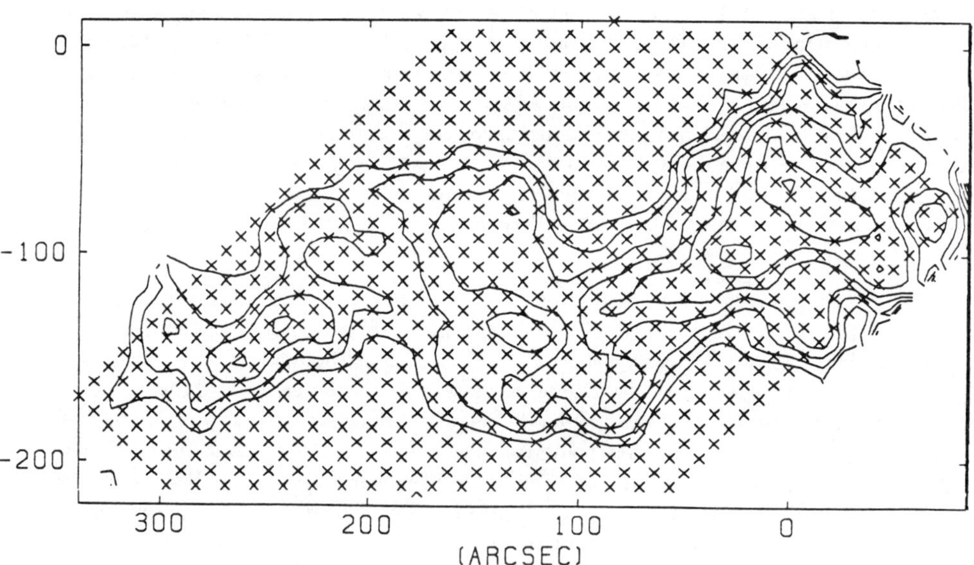

FIGURE 2 A ^{12}CO maps showing the red-shifted lobe of Mon-R2.

L43: An Example of the Interaction of Molecular Outflows and Dense Cores

ROBERT D. MATHIEU[1], PRISCILLA J. BENSON[2], GARY A. FULLER[1,3], PHILIP C. MYERS[1], RUDOLPH E. SCHILD[1]

[1] Harvard-Smithsonian Center for Astrophysics
[2] Department of Astronomy, Wellesley College
[3] Astronomy Department, University of California, Berkeley

Abstract Images of the dark cloud L43 and the associated young star RNO 91 in both optical continuum and millimeter wavelengths lines reveal a dense core in the process of disruption by molecular outflow.

L43 is a dark cloud associated with the Sco-Cen OB2 complex. The region includes two T Tauri stars (RNO 90 and 91), a dense molecular core, a molecular outflow and reflection nebulosity. I-band CCD images of the RNO 91 field dramatically reveal the presence of a large (~0.2 pc) bay cut out of the dark cloud surrounding RNO 91. Three sides of the bay are outlined by high extinction, but to the south the bay is unbounded and the surface brightness within the bay extends from the mouth of the bay southward beyond the boundaries of the immediately adjacent cloud.

The coincidence of the CO blue-shifted line-wing emission with the bay is striking; the outflow emission is, in projection, entirely confined within the bay. The only extension of the blue-shifted emission is out the mouth of the bay, coincident with the extended surface brightness. The red-shifted outflow emission also projects beyond the border of the dark cloud to the north. NH_3 emission is observed to be closely coincident with the regions of high extinction to the east and west of the bay. The NH_3 emission decreases rapidly at the bay boundary and little emission is detected within the bay; indeed the NH_3 and ^{12}CO blue-wing emission roughly anti-correlate. RNO 91 is offset by 0.07 pc in projection from the peak of the NH_3 emission.

In contrast, CS emission peaks at RNO 91 and is, in projection, present throughout the bay. The CS emission is markedly more extended than the NH_3 emission, which is difficult to reconcile with its higher critical density. A suggested explanation is that due to scattering the CS emission actually traces gas at lower densities than its critical density. Thus the CS emission may reflect a more extended envelope about the dense core.

The kinematic data also reveal outflow-core interaction. Both the NH_3 and the CS line widths increase significantly within the boundaries of both the red- and blue-shifted CO outflow. The maximum CS line widths are associated with two regions of the outflow well removed from RNO 91 itself; indeed the substantial increases in line width in these two relatively small areas suggest that at these positions the outflow is plowing directly into the surrounding gas.

Both the NH_3 and the CS velocity data reveal an extended gradient in the velocity field lying along a northwest/southeast line, with the maximum gradient being 4 km s^{-1} pc^{-1}. If interpreted as evidence for rotation, the rotation axis is roughly perpendicular to the outflow axis. This orientation does not therefore suggest that this large scale rotation is associated with the collimation of the outflow. On the other hand, the axis passes in projection very near to RNO 91. Furthermore, the rotation is a relatively small-scale phenomenon in the vicinity of RNO 91; the spatial limits of the velocity gradient are only 0.1 pc distant from the star. This suggests an association between the rotation and formation of RNO 91, and perhaps is evidence that the core spun up during the process of star formation.

This ensemble of observations blend together into one coherent picture in which the outflow, originating at RNO 91, has blown through the surrounding dense core into the lower density environs, in the process revealing the star. The data clearly show that the increased NH_3 line widths observed in many cores with outflows are intimately associated with the CO outflow itself. Outside of the projected boundaries of the L43 outflow the NH_3 line widths are $\sim$0.3 km sec^{-1}; within the boundaries they increase to $\sim$0.5 km sec^{-1}. Similarly, there is a close association of increased CS line widths with both the red- and blue-shifted wings of the CO outflow over distances up to 0.15 pc from the star in projection. Furthermore, the near perfect coincidence in projection of the blue-shifted lobe of the outflow and the bay cut into the middle of the L43 core leaves little doubt regarding whether outflow momentum can be significantly coupled with a surrounding core and the capability of outflows to modify the structure of a core.

However, the data also show that both the increased internal motions and the alteration of the core structure occur in a limited volume of the L43 core. Much of the mass in the L43 core is relatively undisturbed; at present the outflow energy is coupled with only about half of the core mass. Apparently the mechanical action of the outflow has been spatially restricted by its collimation and/or possibly its youth.

HIGH RESOLUTION RADIO OBSERVATIONS OF THE REGION NEAR HH 7-11 - OUTFLOWS AND DISKS

ALEXANDER RUDOLPH
Radio Astronomy Laboratory, University of California, Berkeley, CA
and Department of Physics, University of Chicago, Chicago, IL

ABSTRACT

Aperture synthesis maps of the region near the objects HH 7-11 were made with the Hat Creek Millimeter Interferometer. The $HCO^+(J = 1 \rightarrow 0)$ and the CS $(J = 2 \rightarrow 1)$ transitions were mapped with $8'' \times 7''$ and $8'' \times 6''$ resolutions respectively. The HCO^+ observations reveal a striking correlation of the peaks of line intensity with the positions of the optical HH objects. The CS maps show two emission peaks, one coincident with SVS13 the other $30''$ to the southwest. The main peak shows extension approximately perpendicular to the CO flow and the line made by HH 7-11.

INTRODUCTION

HH 7-11 are a string of almost equally spaced Herbig-Haro objects associated with the young stellar object SVS13 (Strom, Vrba and Strom 1976) in the molecular cloud NGC1333. The five HH's are observed to be surrounded by the blue lobe of a bipolar CO outflow (Snell and Edwards 1981). Herbig and Jones (1983) have measured proper motions for HH 7-11 and find only HH 11 has a measurable motion of 58 km s^{-1} at a distance of 350 pc. The optical spectra of HH 7 and HH 11 have been modelled as due to low-excitation shocks $v \lesssim 40$ km s^{-1} (Böhm, Brugel and Olmsted 1983). Shock-excited $2\,\mu$ H_2 emission has been detected from the region immediately surrounding HH 7-11 as well as to the northwest of SVS13 in the direction of the red lobe of the CO flow (Lightfoot and Glencross 1986). A 21 cm spectrum towards SVS13 shows broad wings of decreasing intensity out to $v = 170$ km s^{-1} suggestive of a decelerated neutral wind (Lizano, *et.al.* 1987).

HCO^+ OBSERVATIONS

An overlay of an integrated HCO^+ map with an optical photograph (Herbig 1974) shows a correlation of the emission peaks with HH 8-11. The HCO^+ peaks are *downwind* from HH 8,9,10 and to one side of HH 11. This difference is probably due to the proper motion of HH 11. Spectra of the HCO^+ emission peaks show that they have velocities very close to the ambient cloud velocity as determined from earlier molecular studies ($|v - 8.2$ km $s^{-1}| \lesssim 1$ km s^{-1}) (Lada, *et.al.* 1974, Ho and Barrett 1980). The spectra also show very narrow linewidths ($\Delta v \lesssim 0.5$ km s^{-1}). Thus the peaks of emission are almost at rest with respect to the cloud and are internally quiescent.

These facts all support a picture in which the HH objects are the shocks forming in the wind when it impinges on the dense ambient clumps observed in HCO^+ (Schwartz 1978). The positioning of the dense cloudlets *downwind* is as expected in this picture as is the fact that the clumps are stationary. In the bullet model (Norman and Silk 1979) the *dense* projectile (i.e. the HCO^+ peak) would be *upwind* from the shocked ambient material (the HH object). Also, in the bullet model the dense object is moving at high velocity contrary to these observations.

Density and Mass

Collisional excitation of HCO^+ requires an H_2 density greater than the critical density n_{crit} = *few* $\times 10^5$ cm^{-3} (Vogel and Welch 1983). Thus these clumps are dense. This density implies, for clumps 7″ across, a mass of .002 $M_\odot$.

One can also estimate the mass from the integrated emission (see for example Blake, *et.al.* 1987). Assuming the emission is optically thin in LTE at T_{kin} = 20 K (Lada, *et.al* 1974), and with an HCO^+ abundance of $X_{HCO^+} = 2 \times 10^{-9}$ (Blake, *et.al.* 1987) one gets masses of $1 - 2 \times 10^{-3} M_\odot$ in good agreement with the crude estimate above.

CS OBSERVATIONS

The CS maps show two peaks of emission which differ by ≈ 1 km s^{-1}. The main peak is centered $\approx 5''$ northwest of SVS13 and is slightly elongated perpendicular to the CO outflow and the line made by HH 7-11. A position-velocity cut along the axis of this ridge of material shows a velocity gradient ≈ 12 km s^{-1} pc^{-1} possibly due to rotation. If this material is in centrifugal equilibrium then this implies a mass of $M \approx 0.1 M_\odot / \sin^2 i$. Using the ratio of proper to radial motion of HH 11 (Herbig and Jones 1983) as an estimate of sin i gives a mass of $M \approx 0.75$ $M_\odot$.

An estimates from the integrated line intensity assuming the gas is optically thin and in LTE at T_{kin} = 20 K gives masses of 0.12 and 0.08 $M_\odot$ for the two peaks. This is assuming a CS abundance of $X_{CS} = 2.5 \times 10^{-9}$ (Blake, *et.al.* 1987). This estimate is lower than the dynamical estimate above; however this is a lower estimate and could be higher if the gas is optically thick.

The line profile is wider at the position of SVS13 ($\Delta v \approx 2$ km s^{-1}) consistent with infall towards a $0.5 - 1.0$ $M_\odot$ star.

The conclusions are:

1) The main peak is a flattened rotating ridge of dense material at or around SVS13. This probably represents the infall region, the volume from which mass is falling onto a much smaller thiner accretion disk. The ridge is flattened due to the original rotation of the cloud.

2) The two peaks may represent two pieces of the same core and therefore might be

rotating about each other. If the smaller piece has no embedded IR source then it either does not have sufficient mass to produce a star or it may represent a protostar at an earlier stage in its evolution.

References

Blake, G. A., Sutton, E. C., Masson, C. R., and Phillips, T. G. 1987, *Ap. J.*, **315**, 621.

Böhm, K. H., Brugel, E. W., and Olmsted, E. 1983, *Astr. Ap.*, **125**, 23.

Herbig, G. H. 1974, Lick Observatory Bulletin # 658.

Herbig, G. H., and Jones, B. F. 1983, *A. J.*, **88**, 1040.

Ho, P. T. P., and Barrett, A. H. 1980, *Ap. J.*, **237** ,38.

Lada, C. J., Gottlieb, C. A.,Litvak, M. M., and Lilley, A. E. 1974, *Ap. J.*, **194**, 609.

Lightfoot, J. F., and Glencross, W. M. 1986, *MNRAS*, **221**, 993.

Lizano, S., Heiles, C., Rodriguez, L. F., Koo, B., Shu, F. H., Hasegawa, T., Hayashi, S., and Mirabel, I. F. 1987, *Ap. J.*, submitted.

Norman, C., and Silk, J. 1979, *Ap. J.*, **228**, 197.

Schwartz, R. D. 1978, *Ap. J.*, **223**, 884.

Snell, R. L., and Edwards, S. 1981, *Ap. J.*, **251**, 103.

Strom, S. E., Vrba, F. J., and Strom, K. M. 1976, *Astron. J.*, **81**, 314.

Vogel, S. N., and Welch, W. J. 1983, *Ap. J.*, **269**, 568.

PRECISE OBSERVATIONS OF HH7-11

TOSHIAKI TAKANO
Nobeyama Radio Observatory, Nobeyama, Minamisaku, Nagano 384-13, Japan
Present address: Radio Astronomy, Toyokawa Observatory, Nagoya University, Toyokawa 442, Japan

Abstract Precise maps of HH7-11 region are presented in CO, NH_3, and CS lines. There is a large void of the ambient clouds between HH7-11 and HH12, in which the red CO flow seems to fill. This suggests strong interaction between CO outflow and ambient clouds.

Using the Nobeyama 45m millimater-wave telescope, we made precise observations of HH7-11 with an angular resolution of 17 arcsec in 115GHz CO line[1]. Figure 1 shows distribution of the blue (-10~0 km s^{-1}) and the red (16~26 km s^{-1}) shifted outflow. With its fine angular resolution we have found following results:

i) The blue shifted flow is compressed in a small area and does not extend in the direction of the Herbig-Haro chain but has dual peaks which line up in the orthogonal direction of the chain.

ii) The red shifted flow widely extends in the north-west direction with a shell-like structure.

These results imply that the bipolar molecular flow is not the same out-flowing material that expels the Herbig-Haro objects.

The NH_3[2] and CS[3] (Fig.2) maps obtained with the Effelsberg 100m and the Onsala 20m telescope allow us to see large scale distributions of ambient molecular clouds. One of the most conspicuous structures in these maps is that there is a large void of the ambient clouds between HH7-11 and HH12. The red CO flow shown in an extant map[4] seems to fill in the void (Fig. 2). These facts suggest that CO flow interacts strongly against the ambient clouds.

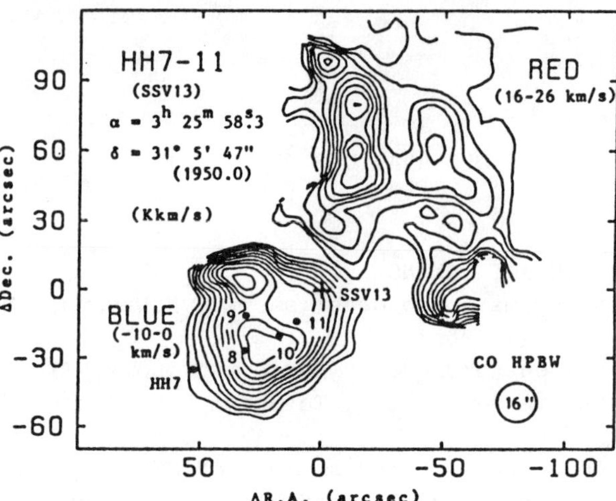

FIGURE 1 Map of CO bipolar flow in HH7-11 with the Nobeyama 45m telescope[1].

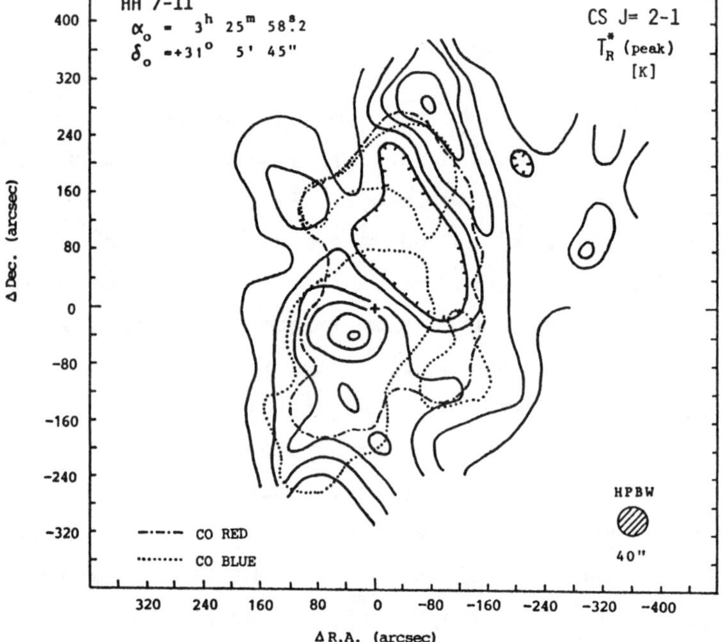

FIGURE 2 CS map of NGC1333 (HH7-11)[3] with an extent map of CO flow[4].

REFERENCES

1. T. Takano, Y. Fukui, and I. Gatley, in preparation, (1987).
2. T. Takano, Y. Fukui, and G. Winnewisser, in preparation, (1987).
3. T. Takano and G. Sandell, in preparation, (1987).
4. L.B.G. Knee, R. Liseau, G. Sandell, and W.J. Zealey, preprint, (1986).

AN UNBIASED SURVEY OF MOLECULAR OUTFLOWS ASSOCIATED WITH BRIGHT FAR-INFRARED SOURCES

RONALD SNELL, Y.-L. HUANG, ROBERT DICKMAN, AND M.J. CLAUSSEN
Five College Radio Astronomy Observatory and Department of Physics and Astronomy, the University of Massachusetts.

Abstract CO observations of 38 bright 100μm sources have revealed 11 new molecular outflows. Because our survey was unbiased by any *a priori* search criteria, we are able to show that the lifetime of young, luminous stars as far-infrared sources must be less than about 4x10% years.

We have carried out a systematic search for high velocity ^{12}CO emission associated with bright 100 μm sources from the IRAS Point Source Catalog (1985), in an effort to better understand the frequency with which outflows occur and the nature of the stars responsible for them. The observations were obtained using the 14 m telescope of the Five College Radio Astronomy Observatory. A complete sample of bright far-infrared sources with 100 μm flux densities >500 Jy, that lie between 0^h and 12^h RA and with Dec. > 0° were chosen. The latter criteria were imposed to avoid lines of sight through the inner Galaxy, where it is difficult to separate the complicated Galactic emission from high velocity outflows, and to insure that the sources transited at reasonably high elevations. A total of 50 sources (excluding IRC+10216 and M82) met these criteria, of which 38 were observed; the other 12 were already known to be outflow sources (Lada 1985). Each IRAS source was found to lie close to a peak in the CO emission, thus suggesting a physical association of the source with an ambient molecular cloud. This relationship, along with the infrared fluxes and colors of the objects in our survey, indicates that all sources studied are intrinsically luminous (10^4 $L_\odot$), young stellar objects still embedded in their parent cloud.

Eleven of the far-infrared sources in our survey show clear evidence of high velocity molecular emission and six of these have definite bipolar morphologies. The eleven newly detected outflows are the following: IRAS 00338+6312, 00494+5617 (S184), 02575+6017 (IC1848A), 05274+3345 (AFGL5142), 05345+3157 (AFGL5157), 05358+3543, 05490+2658, 05553+1631 (AFGL5173), 06056+2131 (AFGL6366S), 06058+2138 (AFGL5180), and 06308+0402. Full velocity widths at zero intensity for the CO emission in these flows vary from 12 to 30 km s^{-1}. The eleven newly-detected outflows have average outflow masses of 6 $M_\odot$, outflow energies of 10^{45} ergs, and ages of 2×10^5 years.

The additional 12 known outflow sources that met our selection criteria are S187, W3, W3OH, AFGL437, AFGL490, HH12, LkHα101, S235B, NGC2071, S255, AFGL961, and NGC2264. Thus, 23 of the 50 bright far-infrared sources in our sample have associated molecular outflows. Selection of these objects was not biased by prior information on molecular emission from these regions; therefore, the high detection rate of outflows must reflect a high occurrence of the outflow phenomenon among luminous, young stars. In addition, the nearly 50% detection rate and the relatively short outflow lifetimes suggest that the lifetime of high-luminosity stars as bright far-infrared sources is at most 4×10^5 years. After this time, most of the surrounding material must be dispersed and the visual extinction greatly reduced, probably by the action of the outflow activity.

REFERENCES

IRAS Point Source Catalog 1985, Joint IRAS Science Working Group, (Washington D.C.: U.S. Government Printing Office).

Lada, C.J. 1985, *Ann. Rev. Astron. Astrophy.*, **23**, 267.

HOT HIGH-VELOCITY OUTFLOW IN NGC2071

TOSHIAKI TAKANO
Radio Astronomy, Toyokawa Observatory, Nagoya University, Toyokawa 442, Japan

YASUO FUKUI
Department of Astrophysics, Nagoya University, Chikusa, Nagoya 464, Japan

G. WINNEWISSER
I. Physikalisches Institut, Universität zu Köln, Zülpicher Str.77, D-5000 Köln 41, FRG

Abstract From NH_3 observations of three transitions, kinetic temperature for high-velocity outflowing gas in the bipolar flow source NGC2071 has been derived to be $\gtrsim$80 K, which is much hotter than that for ambient molecular clouds.

We have observed the NH_3 (J,K)=(3,3) lines in NGC2071 with the Effelsberg 100m telescope[1]. As shown in Figure 1, high-velocity wing emission was detected as well as strong, narrow emission. The velocity extent of the high-velocity wing emission is 20 km s^{-1}, which is similar to those of the (1,1) and (2,2) lines[2].

In order to derive kinetic temperature, NH_3 populations on the three (J,K) levels are estimated using the obtained profiles. Level populations plotted in Figure 2 clearly shows that the rotation temperature for the high-velocity wing emission is higher than that for the narrow emission. The kinetic temperature was derived using a T_{kin}-T_{Rot} relation[3] and summarized in Table I together with that from CO observations[4].

The kinetic temperature for the high-velocity emission in NH_3 is to be $\gtrsim$80 K which is much higher than 30 K for the ambient clouds and than 12 K for the high-velocity emission in CO. This fact limits models for acceleration mechanisms and/or shock heating mechanisms of high-velocity outflow.

TABLE I Kinetic temperatures from NH_3 and CO observations

Emission component	NH_3	CO
High-velocity wing	$\gtrsim$80 K	12 K
Narrow emission	30 K	25 K

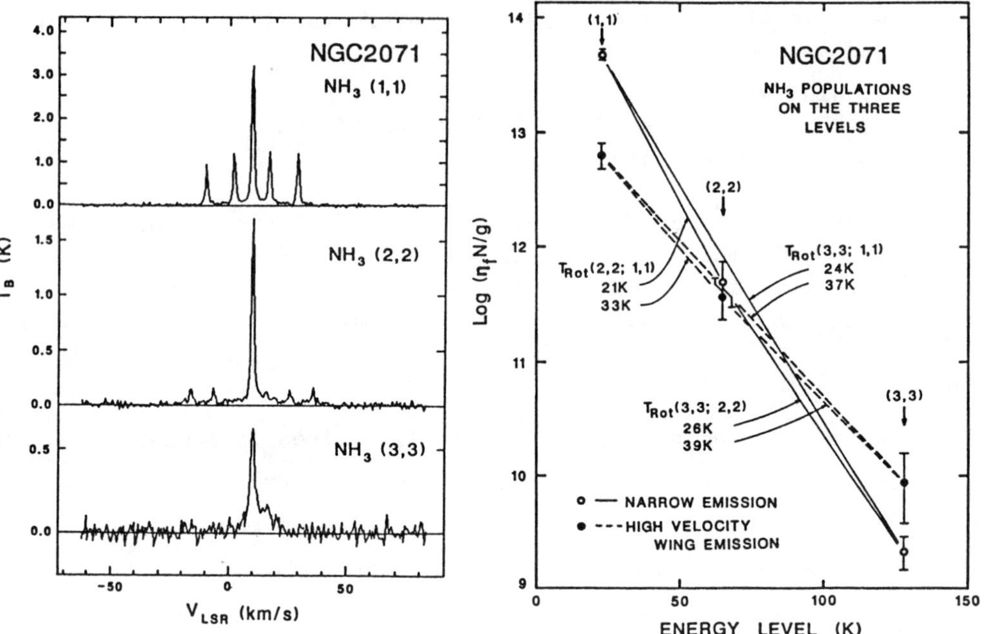

FIGURE 1 Observed NH_3 spectra of three (J,K) transitions in NGC2071.

FIGURE 2 Populations of NH_3 molecules on the three (J,K) levels.

REFERENCES

1. T. Takano, Y. Fukui, M. Miller, and G. Winnewisser, submitted to Astrophys.J.Letters, (1987).
2. T. Takano, J. Stutzki, G. Winnewisser, and Y. Fukui, Astron.Ap., 144, L20, (1985).
3. C.M. Walmsley and H. Ungerechts, Astron.Ap., 122, 164, (1983).
4. R.L. Snell, N.Z. Scoville, D.B. Sanders, and N.R. Erickson, Astrophys.J., 284, 176, (1984).

EXTENDED OUTFLOW SOURCES IN NEARBY MOLECULAR CLOUDS

J. T. ARMSTRONG, L. HAIKALA, G. WINNEWISSER
University of Cologne

Abstract We report a molecular outflow in L1228, and a probable outflow in L673. Their angular extents, 10′ to 20′, are among the largest yet found. The L1228 outflow contains an IRAS point source, but the L673 outflow apparently does not. The fact that most known outflows are of smaller angular size may be a selection effect.

We observed CO (1-0) emission with the University of Cologne 3 m telescope[1] (3.′8 beam). Further CO and ^{13}CO data for L1228 were taken with the 14 m Metsähovi telescope of the University of Technology, Espoo, Finland (45" beam). We used IRAS data with the zodiacal[2] and Galactic background models subtracted.

In both L1228 and L673, both the CO line shapes and the bipolar distribution of wing emission indicate outflows. (See Winnewisser[1], Fig. 4, for L1228.) The ^{13}CO data for L1228 show that the wings have much smaller optical depth than the line core, making it unlikely that the wing emission comes from a separate component or another cloud along the line of sight. As yet, we have no ^{13}CO data for L673, so the outflow there is still tentative. However, we assume here that it too is real.

Table I shows outflow positions and estimates of distance, size, mass, dynamical time scale, kinetic energy, mechanical luminosity, momentum input, and IR point source luminosity. We assume optically thin emission, unity filling factor, no emission from outflow gas in the line cores, and T(ex) = 5 K and 10 K for L1228 and L673 respectively; these all lead to underestimates of the outflow masses. For L1228, we use the kinematic distance, 650 pc. The distance to the nearby outflow in Cep A has been estimated as 700 pc[3]. For L673 we use the near kinematic distance, 400 pc. We reject the further distance, 11 kpc, since L673 is clearly visible on the POSS plates.

The IRAS point source 20582+7724 is at the center of the L1228 outflow. Its IR colors are typical of sources embedded in molecular clouds. However, there is no IRAS point source at the center of the L673 outflow, although there are three between 3′ and 4′ away (Table II). If they are not associated with the outflow, L673 is an unusual outflow source. The only one with two high-quality fluxes has a 12 μm/25 μm color typical of a visible star; but the upper limits for these sources at 60 μm and 100 μm are high, so it is possible that some have embedded-source colors. The extended IR emission in this area is quite confused, but there is none clearly correlated with the outflow; if anything, the outflow lies in a local minimum in all four IRAS bands.

These sources are at the upper limit of angular size of previously known outflows. This may reflect a selection effect, i.e., that it is time-consuming to map large areas with large telescopes. Fukui et al.[4] found four new outflows, two of 10′ to 15′ extent, in an unbiased survey of the southern Orion molecular cloud with the Nagoya 4 m telescope. They estimate that the number of known outflows could increase by 50% to 90% with complete surveys. If this is true, outflows could be the primary contributors to the mechanical energy of molecular clouds.

TABLE I Outflow Parameters

	L1228	L673
Position	$20^h58^m11^s$, 77°24′00″ 111°.7, 20°.2	$19^h18^m15^s$, 11°15′30″ 46°.3, -1°.2
Distance	650 pc	400 pc
Size	5 pc × 3 pc	2 pc × 1 pc
Mass	8 to 16 M(sol)	20 M(sol)
t(dyn)	1.2×10^5 yr	2×10^5 yr
E(kin)	4.2×10^{45} erg	1×10^{46} erg
L(mech)	0.3 L(sol)	0.8 L(sol)
dP/dt	2×10^{-4} M(sol) km/s/yr	5×10^{-4} M(sol) km/s/yr
L(IR)	13 L(sol)	---

TABLE II IRAS Point Sources near the L673 Outflow

	Fluxes (Jy)				Dist. to Outflow Center	Luminosity (L(sol))
	12 μm	25 μm	60 μm	100 μm		
19180+1114	<0.41	1.60	<2.5	<104	3′.0	0.4
19181+1112	<0.39	0.43	<5.5	<105	3′.6	0.1
19184+1118	2.96	1.08	<5.4	< 71	3′.4	0.4

REFERENCES

1. G. Winnewisser, this volume.
2. P. Cox and A. Leene, Astr. Ap., in press (1987).
3. J. Bally and C. J. Lada, Ap. J., 265, 824 (1983).
4. Y. Fukui, K. Sugitani, H. Takaba, T. Iwata, A. Mizuno, H. Ogawa, and K. Kawabata, Ap. J. (Letters), 311, L85 (1986).

AIRBORNE INFRARED SPECTROSCOPY OF EVOLVED STARS

S.C. BECK, S.V.BECKWITH, J.WYANT

Department of Astronomy, Cornell University, Ithaca N.Y.

L. ALLAMANDOLA

NASA Ames Research Center

N.J.EVANS II, J.CARR

Department of Astronomy, University of Texas at Austin

A. NATTA

Centro per l'Astronomia, CNR

Abstract We present observations of five late-type stars performed with spectral resolving power of greater than 10^3 at wavelengths between 5 and 8 μm. IRC+10216 and CIT 6 were found to be essentially featureless. The fundamental vibration-rotation bands of SiO were clearly visible in α Ori and U Ori. The spectrum of U Ori also shows a series of absorption features around 7.4 μm which have not been identified.

OBSERVATIONS

Evolved stars play very important but not well-understood roles in the processing of stellar material and the evolution of the interstellar medium. They have been difficult to observe thoroughly as the shells these stars generate and the more obscured stellar photospheres can only be studied in the infrared. We have used the Kuiper Airborne Observatory to observe five evolved stars with circumstellar shells at wavelengths between 5 and 8.2 μm. Until now, most work in the middle-infrared has used either resolving power too low ($R=\lambda/\delta\lambda \approx 100$) to distinguish among bands of different molecules or so high that only a very small fraction of the spectrum could be measured. In addition, the high-R work has been done from ground-based telescopes and hence could not explore much of this wavelength region. The R used here was high enough ($\approx 10^3$) to resolve molecular bands yet low enough that large portions of the spectrum could be scanned in the fairly short time available on KAO flights.

The cooled grating spectrometer described by Beckwith et al 1983 was used with beam size 24" and resolving power R=1300. Atmospheric absorption was measured and removed by comparision to a lunar spectrum and the flux was calibrated using spectra of the Moon and α Ori. Five sources were observed : IRC+10216 and CIT 6, heavily obscured carbon-rich objects, and U Ori, VYCMa, and α Ori which are less heavily obscured and oxygen-rich.

The observations found the 8μm fundamental vibrational bands of SiO in α Ori and U Ori. This molecule had previously been detected in these objects via the overtone bands. Several molecules which have been observed in these sources were not detected in our spectra or were found to be much weaker that expected. There are several possible explanations including P Cygni profiles and veiling by dust emission. We suggest that higher spectral resolution will be needed to detect molecular features from the shells of very highly obscured objects such as IRC+10216 and CIT 6. U Ori and possibly VY CMa have in their spectra features that we have not suceeded in identifying with any gas-phase molecule or dust feature. The VY CMa spectrum may resemble features seen in ices such as CH_3OH/H_2O mixture, and the U Ori spectrum resembles patterns of absorptions seen in hydrocarbon ices such as hexane. Such ices have been observed in the spectra of protostars and molecular cloud sources (Tielens et al 1984), but U Ori is evolved, oxygen-rich, and at least a factor 10 warmer than any of the young objects where the ices have been detected. If it can be confirmed that such ices exist in U Ori it may have important implications for the evolution of grains and mantles.

REFERENCES

1) Beckwith, S.V., *et al.* 1983, *Ap. J.*, **264**, 152.
2) Tielens, A.G.G.M., *et al.* 1984, *Ap. J.*, **287**, 697.

THE PHOTODISSOCIATION OF CO IN CIRCUMSTELLAR ENVELOPES

A.E. GLASSGOLD, G.A. MAMON, AND P.J. HUGGINS
New York University, New York, NY

The basis for the photodissociation of CO occurring through lines rather than the continuum (Glassgold *et al.* 1985) has now been firmly established by the recent laboratory experiments of Letzelter *et al.* (1987). We have calculated the rate of CO photodissociation in the unshielded interstellar medium using the far UV radiation field given by Jura (1974), and obtain a value of 2.0×10^{-10} s^{-1}—a factor of 10 larger than usually assumed. As one example of the numerous applications of this new result, we have investigated the variation of the CO photodissociation rate in models of the circumstellar envelopes of cool giant stars undergoing substantial mass loss. Self shielding by CO and mutual shielding by dust and by H_2 molecules have been included. The model calculations have important implications for circumstellar chemistry.

REFERENCES

1. A.E. Glassgold, P.J. Huggins and W.D. Langer, Ap. J., 290, 615 (1985).
2. M. Jura, Ap. J., 191, 375 (1974).
3. C. Letzelter, M. Eidelsberg, F. Rostas, J. Breton, and B. Thieblemont, Chem. Phys., in press (1987).

THE DISTRIBUTION OF CYANOACETYLENE IN IRC +10216

A.E. GLASSGOLD and G.A. MAMON
New York University, New York, NY 10003, USA

According to the photochemical model for C-rich circumstellar envelopes around cool evolved stars[1], the fragments of molecular dissociation generate an active chemistry at intermediate distances - driven in large part by ion-molecule reactions[2]. The primary reaction for producing cyanoacetylene (HC_3N) is the measured reaction[3]

$$C_2H_2^+ + HCN \longrightarrow H_2C_3N^+ + H \qquad (A)$$

followed by dissociative recombination. Nejad and Millar[4] have proposed other mechanisms, notably

$$H_2CN^+ + C_2H_2 \longrightarrow H_2C_3N^+ + H_2 \qquad (B)$$

We have calculated the distribution of HC_3N for our standard model of IRC +10216. The peak abundance of HC_3N (few times 10^{-4} relative to CO) is peaked at about 3×10^{16} cm (or 10"). If only (A) is operative, HC_3N occurs in a relatively thin shell. Reaction (B) leads to a substantial abundance in the inside of the envelope. Existing interferometers can distinguish between these two cases[5]. Substantial amounts of isocyanoacetylene (C_3N) are also produced by photodissociation of HC_3N, and measurement of its spatial distribution would provide another important test of the photochemical model.

REFERENCES

1. P.J. Huggins and A.E. Glassgold, Ap. J. 252, 201 (1982).
2. A.E. Glassgold, A. Omont, and R. Lucas, Astr. Ap. 157, 35 (1986).
3. A.E. Glassgold, G.A. Mamon, A. Omont, and R. Lucas, Astr. Ap. in press (1987).
4. L.A.M. Nejad and T.J. Millar, Astr. and Ap. in press (1987).
5. J.H. Bieging and A.H. Wootten, private communications.

VIBRATIONALLY EXCITED CS AND SiS IN IRC10216

B. E. TURNER
National Radio Astronomy Observatory, Charlottesville, VA

Abstract: Vibrationally-excited (v=1) CS and SiS have been detected in IRC10216. The excitation and chemical abundances in the inner core are deduced.

Except for the masering v=1 and v=2 rotational transitions of SiO in O-rich CSEs, vibrationally excited molecules have not until recently been detected in CSEs. Several species (HCN, CO, CS, SiO, SiS) have vibrational energies in the 1000-3000 K range, and thus are excited only in hot, dense gas, or by strong IR radiation fields such as occur in CSEs. Here we report the detection in IRC10216 of vibrationally excited (v=1) CS and SiS.

The v=1 and v=0 profiles of CS have the same width (± 14.4 km/s) to zero power as that of other species in IRC10216. Thus the v=1 CS emission arises throughout the entire envelope. A large self-absorption, unlike anything seen before, occurs in the blue wing of the v=1, J=5-4 profile but not in the v=1, J=2-1 profile.

The v=1 SiS profiles (J=14-13, 13-12, 12-11) are centered on the systemic velocity but are distinctly narrower than v=0 and hence must arise within the innermost acceleration zone of the outflow (size < 8 R_* ~ 1.2(15) cm ~ 0.4" at distance 200 pc) where terminal velocities have not yet been attained.

The principle questions posed by these observations are (1) why is SiS v=1 confined to the inner core while CS v=1 is not? (2) Why are the abundant species CO, SiO not seen in their v=1 states?

If SiS v=1 is confined to a region ≦ 0.4" in size, the brightness temperature of the v=1 lines is ≧ 600 K. Since the inner core of IRC10216 consists of a core of opaque dust at 600 K and radius 0.2", the SiS v=1 lines may be "thermally" excited by the IR radiation field of the core provided that $\tau_{rot}(v=1) > 1$ in the rotational transitions of v=1. We calculate the total column density N = N(v=1) + N(v=0) for each species assuming that τ_{rot} (v=1) = 1, that the rotational temperature is 600 K in the inner core, and using $N(v=1)/N(v=0) = R_{IR}/(R_{IR} + A_{IR}) = (1+\zeta)^{-1}$ with $\zeta \equiv \exp(\theta T^*)^{-1}$. The vibrational energy in K is θ = 1080, 1774, 1835, 3131 for SiS, SiO, CS, CO respectively. The total column density through the core is ~1(24) cm^{-2}. The fractional abundances X required to satisfy $\tau_{rot}(v=1) = 1$ are found to be higher than observed for CO, and higher than expected for CS and

SiO, but actually lower than expected for SiS (on the basis of v=0 studies in the outer envelope). Thus we conclude that "thermal" radiative excitation is consistent with observations of SiS v=1 and with negative results for CO and SiO v=1 if these species are all confined to the inner core. CS v=1 is not confined to the core, and the contribution from thermal radiative excitation in the core is negligible compared with the contribution from the overall envelope.

Collisional excitation of v=1 states within the inner core is negligible compared with radiative excitation for densities < 3(13) cm^{-2} for SiS and higher for the other species.

v=1 excitation of a given species will be confined to the inner core if the IR opacity in the v=1 transition is large through the core. For $\tau_{rot}(v=1) = 1$ we find τ_{IR} = 0.4, 2.5, 7.5, 9390 for SiS, SiO, CS, CO respectively. With these results we explain the v=1 observations as follows:

CO: If N(CO) = 1(20) cm^{-2} as observed in the IR, then $\tau_{IR}(CO) = 760$ and $\tau_{rot}(v=1) \approx 0.08$ through the inner core. Thus the v=1 intensities are 12 times below our detection limit, while the v=1 excitation is confined to the center of the inner core.

SiO: $\tau_{rot}(v=1) \sim 1$ implies X(SiO) ~ 7(-7) in the inner core, or X(SiO)/X(SiS) ~ 6. If X(SiO)/X(SiS) ~ 0.2 as deduced for the outer envelope from v=0 studies, then $\tau_{rot}(v=1) \sim 0.04$, 25 times below the detection limit for the inner core. Then $\tau_{IR} \sim 0.1$ so SiO(v=1) should not be confined to the inner core. However, T_R^*(SiO v=1)/T_R^*(CS v=1) = 0.35* X(SiO)/X(CS) if both species are optically thin in v=1; hence T_R^*(SiO v=1) is 12 times smaller than CS v=1 in the outer envelope, and thus should be undetectable despite availability of IR flux, consistent with observations.

CS: Since v=1 is seen throughout the outer envelope, we require $\tau_{IR} < 1$ or X(CS) < 3(-8) for the inner core, ~10 times less than deduced from v=0 studies for the outer envelope. For optically thin v=1 emission, T_R^*(CS v=1)/T_R^*(SiS v=1) = 0.51*X(CS)/X(SiS), so either X(CS)/X(SiS) must exceed ~6 in the outer envelope, or SiS v=1 must fail to be excited there and instead be confined to the inner core. The latter alternative seems necessary, since X(CS)/X(SiS) ~ 1 in the outer envelope from v=0 studies.

SiS: Since SiS v=1 is confined to the inner core, either $\tau_{IR} \gg 1$ or X(SiS) must decrease sharply outside the inner core. $\tau_{IR} \gg 1$ is necessary because: (1) the value of X(SiS) ~ 2.4 (-7) found from v=0 studies for the outer envelope already implies $\tau_{IR} \gg 1$ in the inner core if the abundance is the same there; (ii) if the IR is not confined, then X(SiS) ~ 2.4 (-7) implies stronger SiS v=1 emission than CS v=1 emission from the outer envelope, contrary to observations. $\tau_{IR} \gg 1$ implies X(SiS) >> 3(-7) in the inner core. We find that X(SiS) is at least 3 times higher and X(CS) at least 10 times smaller in the inner core than in the outer envelope. SiS is likely lost to grains, while the augmentation of CS outside the core is a separate chemical phenomenon, as yet unexplained.

LINEAR POLARIZATION OF HCN MASER EMISSION IN CIT6

P.F. GOLDSMITH, D.C. LIS
FCRAO, University of Massachusetts, Amherst
S. GUILLOTEAU, R. LUCAS, and A. OMONT
Groupe d'Astrophysique, Observatoire de Grenoble

ABSTRACT We have measured the linear polarization of the intense emission from vibrationally excited HCN in the circumstellar envelope of the carbon star CIT6 detected by Guiloteau, Omont, and Lucas (1987). We found the emission to have a linear polarization of 15 percent at a position angle of -17°, with no significant variation in percentage or angle across the line profile. The ground vibrational state line was found to be unpolarized, with an upper limit of 3 percent.

INTRODUCTION

Maser emission from OH, H_2O, and SiO molecules (*cf.* Reid and Moran 1981) is common in the atmospheres of evolved stars. SiS (Nquyen-Q-Rieu *et. al.* 1984) and possibly CO (Zuckerman and Dyck 1986) heve been detected masing in at least one object. The detection of polarized emission in a masing transition can give information about the magnetic field as well as other parameters in the region with gain (*cf.* Goldreich, Keeley, and Kwan 1973; Western and Watson 1984).

We report measurements of the linear polarization of the intense radiation from the HCN transition in the (0,2°,0) vibrationally excited state. While there is no theory available specifically for this transition, the high degree of linear polarization detected supports the presence of maser amplification.

OBSERVATIONS

The observations were carried out between 20 March and 24 April 1987 using the 14m FCRAO radome enclosed telescope. A Schottky mixer receiver provided SSB system temperatures typically 400 K referred to above the earth's atmosphere. Pointing and focus were checked by continuum observations of 3C84 or by HCN J=1-0 emission from IRC+10216. The beamwidth and aperture efficiency were not measured during this program, but were found to be 58 arcsec FWHM and 0.49, respectively in a study of HCO^+ polarization (Lis *et al.* 1987). The resulting conversion factor from antenna temperature T_A^* to unpolarized flux density is 38.4 Jy/K.

Linear polarization was obtained using a rotating half wave plate. The instrumentation and data taking procedures are described by Barvainis and Predmore (1985). A reference position offset by 0.5 degree in azimuth was used to determine baselines for the spectra. Spectral resolution was obtained with a 100 kHz filterbank, with corresponding velocity width 0.34 km s^{-1} per channel. A spectrum expander operating at a factor of 4 expansion was employed for additional observations of the vibrationally excited line.

The peak antenna temperature of the vibrationally excited line as observed with 100 kHz resolution spectrometer is 0.92 K (average of two orthogonal polarizations). The FWHM linewidth is 1.1 km s^{-1}, and the mean velocity of emission is -3.8 km s^{-1}. Corresponding parameters of the ground state line are 0.42 K, 25 km s^{-1} and 4.6 km s^{-1}, respectively. The linear polarization referred to a 1 km s^{-1} velocity interval centered on the line peak velocity is (15.1±1.1) percent and the position angle is (-17±3)°. The upper limit for the polarization of the ground state line in a 1 km s^{-1} interval is 3 percent. The Stokes parameters of the vibrationally excited line are shown in Fig. 1.

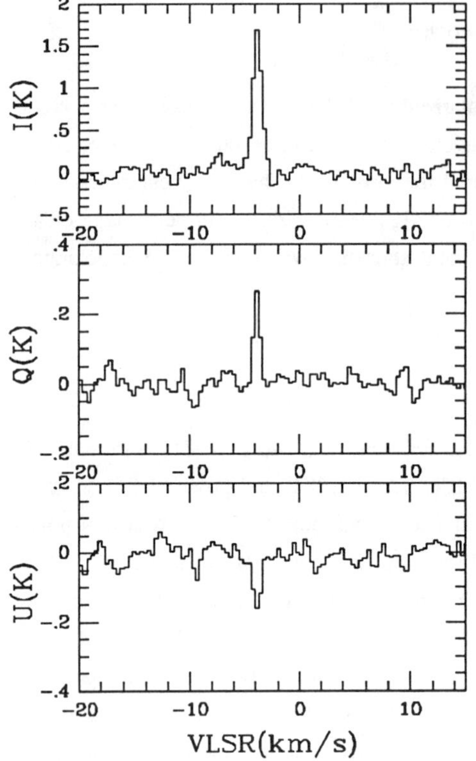

FIGURE 1 Observed Stokes parameters of HCN maser line in CIT6.

REFERENCES

1. R. Barvainis and C.R. Predmore, Ap. J., 288, 694 (1985).
2. P. Goldreich, D.A. Keeley and J. Kwan, Ap. J., 179, 111 (1973).
3. S. Guiloteau, A. Omont and R. Lucas, Astr. Ap., 176, L24 (1987).
4. D.C. Lis, P.F. Goldsmith, R.L. Dickman, C.R. Predmore, J. Cernicharo and A. Omont, in preparation (1987).
5. Nguyen-Q-Rieu, V. Bujarrabal, H. Oloffson, L.E.B. Johansson and B.E. Turner, Ap. J., 286, 276 (1984).
6. M.J. Reid, and J.M. Moran, Ann. Rev. Astr. Ap., 19, 231 (1981).
7. L.R. Western and W.D. Watson, Ap. J., 285, 158 (1984).
8. B. Zuckerman and M. Dyck, Ap. J., 311, 345 (1986).

WATER MASERS AROUND TWO SHORT PERIOD MIRAS: R CETI and RZ SCORPII

IRENE R. LITTLE-MARENIN
AFGL/OPC, Hanscom AFB, MA 01731

PRISCILLA J. BENSON
Whitin Observatory, Wellesley College, Wellesley, MA 02181

DALE F. DICKINSON
Lockheed Palo Alto Research Lab, Palo Alto, CA 94304

ABSTRACT The 22 GHz water maser line of R Cet varies in phase with the infrared lightcurve but with a phase lag of 0.1 phase. The water maser emission of RZ Sco also appears to vary in phase.

We report the first detections of the water maser emission line at 22,235.080 MHz from two short period M star Mira variables: R Cet (M4e-M9, $P=166^d$) and RZ Sco (M3e-M4e, $P=160^d$) with the 37m Haystack Observatory[1] radio telescope. The intensity of the R Cet maser line has varied by about a factor of 10 from 4.2 Jy (1/23/85; phase=0.01) to 40.8 Jy (6/9/86; phase=0.21; $T_A^*=3.40$ (0.09) K, $V_{LSR}=+31.20$ km s^{-1} and $\Delta V=0.64$ km s^{-1}) in phase with the visual lightcurve but with a phase lag of about 0.28 phase (Fig. 1a) corresponding to a phase lag of about 0.1 phase with respect to the K magnitude lightcurve. No other water maser lines are visible with flux > 1 Jy (3 σ). The phase lag of 0.1 phase (about 16 days) with respect to the infrared suggests a collisionally pumped maser as proposed by Cooke and Elitzur[1]. A maser pumped by the stellar radiation field should experience a negligible phase lag since the stellar radiation should transit

---

[1]Radio astronomy at Haystack Observatory of the Northeast Radio Observatory Corporation is supported by the National Science Foundation under grant AST78-18227.

through the water masing region (less than 70 AU) in less than 1 light day.

The intensity of the water maser line from RZ Sco also appears to vary in phase being below the detection limit of 1 Jy at phase=0.98 (5/19/87) and having an intensity of 7.2 Jy at phase=0.62 (3/23/87) (Fig. 1b). The line observed at phase=0.62 (3/23/1987) shows three components separated by 1.5 km s^{-1}. The strongest component is well fitted with a Gaussian: $T_A^* = 0.60$ (0.05) K, $V_{LSR}=-163.87$ km s^{-1} and $\Delta V=0.56$ km s^{-1}.

Despite the suggestion by Feast et al.[2] that water masers are preferentially detected from apparent bolometrically bright (i.e. close-by) Miras, we find that neither R Cet nor RZ Sco are exceptionally bright; $m_{bol}=5.6$ mag (R Cet, d=800pc) and $m_{bol}=7$ mag (RZ Sco, d=1600pc). The infrared colors of both stars are typical of other maser stars. However, both Miras have asymmetric lightcurves and R Cet shows strong 10 and 18 μm silicate emission features (IRAS low resolution spectrum) indicative of a fairly optically thick circumstellar shell and a large J, H and K amplitude. These are characteristics of longer period Miras.

REFERENCES

1. Cooke, B., and Elitzur, M. 1985, Ap.J., 295, 175
2. Feast, M. W. et al. 1982, Mon. Not., 201, 439.

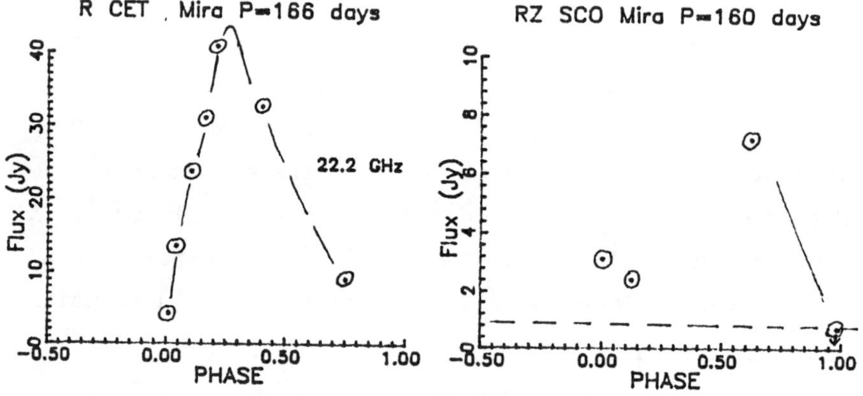

Figure 1a,b. The variation of water maser flux with phase

CIRCULAR POLARIMETRY OF SiO MASERS

GORDON C. MCINTOSH, C. READ PREDMORE
University of Massachusetts, Amherst, MA,
and RICHARD E. BARVAINIS
National Radio Astronomy Observatory, Charlottesville, VA,

Abstract Stokes polarimetry, including the measurement of circular polarization (Stokes V), has been performed on the SiO (v=1, J=1-0) masers around several late type stars. The circular polarization, detected in VYCMa, RLeo, WHya, VX Sgr, and RCas, indicates large magnetic fields (8-80 gauss) are present in the near circumstellar environment.

INTRODUCTION

Linear Stokes polarimetry of SiO masers has been studied by several investigators, but no confirmed observations of circular polarization in circumstellar masers have been reported previously. We detected the circular polarization of these masers on 2-5 Nov 1986 at the Five College Radio Astronomy Observatory.

ANALYSIS

No unambiguous Zeeman pairs exist in the measured spectra. However, possible Zeeman pairs exist for WHya at several velocities. (See Figure 1.) If these features are Zeeman pairs the longitudinal magnetic field can be calculated from

$$B = 510 \quad v \, V_{pp}/I \text{ gauss}$$

where v is the velocity separation of the left and right maxima in km/s, V_{pp} is the peak to peak flux density of the V spectrum, and I is the total flux density for the feature. This equation yields magnetic fields of 10 to 20 gauss for the features present in WHya.

Deguchi and Watson[1] have proposed a mechanism to determine the magnetic field present when the Stokes V spectra do not exhibit Zeeman pairs. Assuming the Zeeman splitting is small compared to the linewidth this mechanism yields magnetic fields of 8 to 35 gauss for the features in WHya.

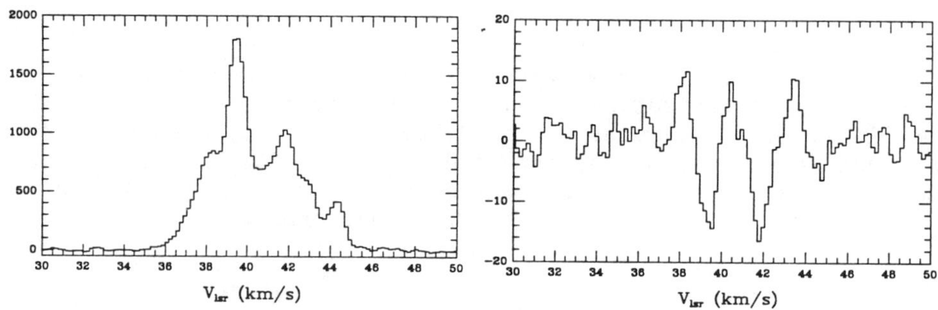

FIGURE 1 WHya, I and V in Janskys.

IMPLICATIONS FOR MASERS AND CIRCUMSTELLAR PHYSICS

The magnetic fields derived from the circular polarization measurements are large enough to explain the observed linear polarization of SiO masers in accord with the theory of Goldreich, Keeley, and Kwan[2]. The existence of these fields also affects the dynamics and geometry of the circumstellar region and may explain the lack of any discernible pattern in the SiO masers as observed with VLBI techniques[3-4].

REFERENCES

1. Deguchi, S., and Watson, W.D. 1986, Ap.J., 289, 621.
2. Goldreich, P., Keeley, D.R., and Kwan, J.H. 1974, Ap.J., 179, 111.
3. Lane, A.P. 1982, Ph.D. Dissertation, Univ. of Mass.
4. McIntosh, G.C. 1987, Ph.D. Dissertation, Univ. of Mass.

REFLECTION NEBULAE OF MON R1 IN THE FAR-INFRARED AND SUBMILLIMETER

S.C. CASEY, D.A. HARPER
Yerkes Observatory, University of Chicago, Williams Bay, WI

Abstract We present 100 μm, 160 μm, and 350 μm ($8' \times 8'$) maps (45″/beam) of reflection nebulae in the Mon R1 region (IC446 and NGC2245). The far-infrared and submillimeter observations employ 32-beam bolometer arrays (6×6 square) designed for the NASA Kuiper Airborne Observatory and the Infrared Telescope Facility.

INTRODUCTION

We are currently engaged in an evaluation of the far-infrared radiative emissivity of interstellar dust grains (radius a=0.1 μm) through a careful determination of the heating and cooling processes within visual reflection nebulae. A complementary program is underway to map the surface brightness of visual reflection nebulae in the far-infrared (50-200 μm) and submillimeter (300-500 μm). The embedded sources of nebular excitation are generally *B* stars not sufficiently energetic to generate observable regions of ionized hydrogen but are capable of producing the non-equilibrium emission characteristic of small dust grains (a=10Å) or polycyclic aromatic hydrocarbons[1]. Non-equilibrium emission from small grains may contribute a significant flux at wavelengths as long as 60 μm[2]. Thus photometric measurements at 100–400 μm may be essential for a proper determination of the large dust grain ($a = 0.1\mu$m) equilibrium temperature.

OBSERVATIONS

We employ two similar photometers with similar pixel diameters and spacings (45″ and 48″). Surface brightness maps are presented in the figure of two reflection nebulae in the region Mon R1 (IC446 and NGC2245). Our 360 μm images were obtained at the Infrared Telescope Facility in February 1987 while our 100 and 160 μm images were obtained aboard the Kuiper Airborne Observatory in March 1987.

PHOTOMETRY

Given the infrared fluxes in several pass bands, $F(\nu)$, we are able to calculate dust color temperatures and optical depths. Thermal radiation from dust grains is modeled by a modified blackbody of the form $Q(\nu)B(\nu,T)$. Current evidence suggests that $Q(\nu) \propto \nu^{\beta}$ [3,4,5] and therefore the color temperature is determined that produces the measured ratio, $F(\nu_1)/F(\nu_2) = \nu_1^{\beta}B(\nu_1,T)/\nu_2^{\beta}(\nu_2,T)$.

The ratio of the peak emission at 100 and 160 μm from IC446 and NGC2245 is best fit by a modified blackbody of average temperature 33 ± 2K with a spectral index of $\beta = 1.0$. Our contour maps are at 10% intervals of the peak flux listed below.

Object	100 μm(Jy)	160 μm(Jy)	360 μm(Jy)
IC446	190 ± 9.5	135 ± 6.7	33 ± 3
NGC2245	126 ± 6.3	97 ± 4.9	30 ± 3

For both NGC 2245 and IC446 the submillimeter peaks corresponds to nebular regions with lower than average color temperatures ($T \lesssim 20\ ^{0}$K).

REFERENCES

1. K. Sellgren Ap. J., 277, 623 (1984).
2. J.L. Weiland, L. Blitz, E. Dwek, M.G. Hauser, L. Magnani, and L.J. Rickard Ap. J. (Lett.), 306 L101 (1986).
3. S.E. Whitcomb, I. Gatley, R.H. Hildebrand, J. Keene, K. Sellgren, and M.W. Werner Ap. J., 246 416 (1981).
4. J. Keene Ap. J., 245 115 (1981).
5. R.H. Hildebrand Quart. J. Royal Astr. Soc., 24 267 (1983).

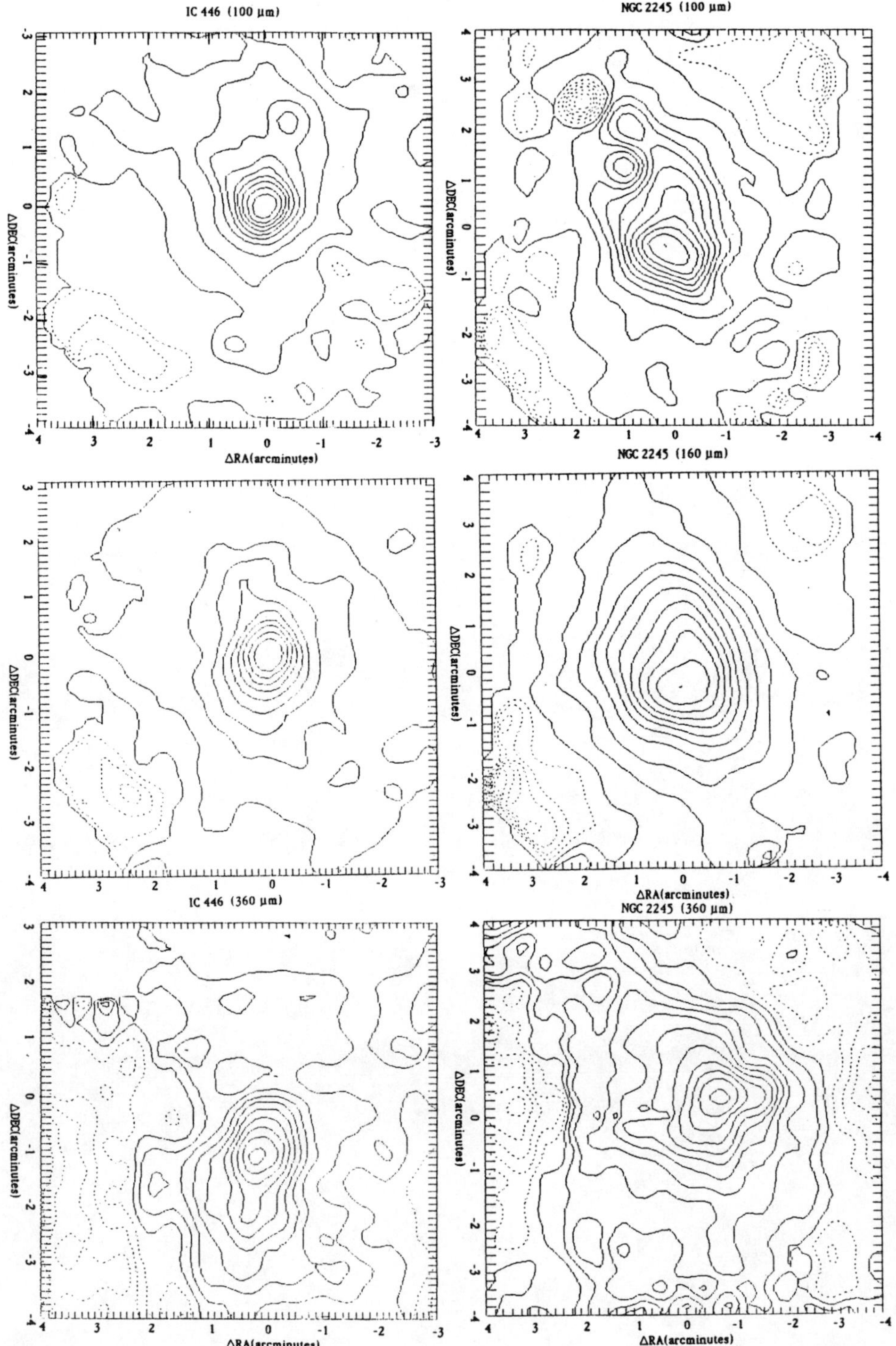
IC 446 (100 μm)
NGC 2245 (100 μm)
NGC 2245 (160 μm)
IC 446 (360 μm)
NGC 2245 (360 μm)
ΔDEC(arcminutes)
ΔRA(arcminutes)

H_2O MASER EMISSION FROM SHOCKED REGIONS

M. ELITZUR[a], D. J. HOLLENBACH[b] and C. F. MCKEE[c]
[a]NASA Goddard Space Flight Center and University of Kentucky
[b]NASA Ames Research Center
[c]University of California, Berkeley

H_2O maser radiation is one of the more important indicators of the star formation process. The powerful H_2O masers in star forming regions appear[1] as individual features, streaming away from some center of activity at velocities that can sometimes approach ~ 200 km s^{-1}. Such high velocities will drive powerful shocks.

Detailed studies[2] show that fast shocks (velocities ~ 100 km s^{-1}) in dense clouds (N ~ $10^6 - 10^7$ cm^{-3}) develop a high temperature (T ~ 600 K) plateau, rich in H_2O. This happens as follows: the molecules are first completely dissociated by the high post-shock temperatures. Later on, when the material cools down, H_2 molecules reform on the dust grains and are ejected to the gas phase with sizeable kinetic energies, providing a source of heating for the gas. The resulting high temperatures enable chemical exchange reactions with neutral H and H_2 to proceed rapidly, and these channel all the available oxygen into H_2O (see e.g. ref. 3). The column density of the heated region can be as large as ~ 10^{22} cm^{-2}. This plateau therefore suggests itself as a natural site for H_2O maser action.

We have performed detailed calculations of the H_2O cooling and maser emission expected from the high temperature plateau of the shocked region. The populations of the first 37 levels of ortho H_2O (all the levels with energies below 1200 cm^{-1}) were solved using the escape probability method, taking into account all the collisional and radiative processes that connect them. The complete calculation is an iterative process because the exact structure of the shock in the relevant region is determined by H_2O cooling — water is the main coolant in the temperature range ~ 400 – 1000 K. The calculations were performed using new H_2O cross sections, generously supplied by S. Green[4]. These rates were computed utilizing the most recent advances in molecular potential theory, and are an improvement over the previously published rates[5]. The exact H_2O cooling function obtained can be successfully approximated with the analytic expressions of Hollenbach and McKee[6] with simple modifications of some of the parameters.

Inversion of the H_2O 22 GHz line was obtained for a wide range of densities (~ 10^6 – 10^{10} cm^{-3}) and temperatures (~ 400 – 800 K). The inversion is due to collisions with the H_2 molecules, and is similar in nature to that obtained[7] for late-type stars. As was the case in that calculation, other lines are also strongly inverted but because of their longer wavelengths the corresponding brightness temperatures are not as high as those for the 22 GHz line. An interesting feature of the results obtained with the new cross sections is that the maser is quenched at lower densities than was the case with the previous cross sections. Another interesting feature is that higher densities require higher temperatures for peak maser emission.

The high brightness temperatures ($\gtrsim 10^{14}$ K) observed in some of the stronger H_2O masers[1] have long been a challenge for theoretical models. The problem can be traced to simple thermodynamic arguments[8] which show that for radiative pumps where the same geometrical factors apply to both pumping radiation and maser photons, the maser brightness temperature cannot exceed ~ $10^9 - 10^{10}$ K. Collisional pumps are usually subject to similar constraints, although the restrictions are not quite as severe. Our present calculations display the same limitations when we use the same escape geometry for both maser photons and the non-inverted transitions. The highest brightness temperatures obtained in this case are only a few 10^9 K. However, detailed mapping of molecular clouds show that their structure is filamentary rather than homogeneous[9]. In the case of Orion, where the linear scale is about three orders of magnitude larger than in the shocks we study here, individual features having length up to 40 times their width are identified. It is therefore reasonable to assume that the material behind a shock front will also break up into filaments and that, due to the smaller linear scale, aspect ratios up to, at least, ~50 can be expected. The maser emission from a filament with an aspect ratio (length/width) a increases over that from an isotropic structure because of two reasons: (1) The photons generated in the non-inverted, pump-cycle transitions can escape sideways, where the escape probability is larger by a factor a. This relaxes the thermodynamic limit on the brightness temperature of the maser emission by a similar factor. (2) The filamentary geometry concentrates the maser emission into a small angle of order a, leading to a further increase of ~ a^2 in the brightness temperature.

Detailed calculations with a filamentary maser geometry confirm that the maser brightness temperature is enhanced by ~ a^3. Our model is capable of explaining the maser emission with parameters consistent with those observed. In particular, we obtain brightness temperatures in excess of ~ 10^{13} K for filaments with density of ~ 10^9 cm^{-3}, diameter ~ 5×10^{12} cm, temperature ~ 700 K and aspect ratio of ~ 20.

REFERENCES

1. Genzel, R. 1986, in *Masers, Molecules and Mass Outflows in Star Forming Regions*, ed. A.D. Haschick (Haystack Observatory), p. 233.
2. Hollenbach,D.J. and McKee, C.F. 1987, in preparation.
3. Elitzur, M. 1979, *Ap. J.* **229**, 560.
4. Green, S. 1987, private communication.
5. Green, S. 1980, *Ap. J. Suppl.* **42**, 103.
6. Hollenbach,D.J. and McKee, C.F. 1979, *Ap. J. Suppl.* **41**, 555.
7 Cooke, B. and Elitzur, M. 1985, *Ap. J.*, **295**, 175.
8 Elitzur, M. 1982, *Rev. Mod. Phys.* **54**, 1225.
9. Bally, J. *et al.* 1986, *Summer School on Interstellar Processes,* NASA TM 88342, eds. D.J. Hollenbach and H.A. Thronson, p. 29.

HOW ABUNDANT ARE COMPLEX INTERSTELLAR MOLECULES?

ERIC HERBST
Department of Physics, Duke University, Durham, North Carolina, U. S. A.

CHUN MING LEUNG
Department of Physics, Rensselaer Polytechnic Institute, Troy, New York, U. S. A.

T. J. MILLAR
Mathematics Department, U. M. I. S. T., Manchester M60 1QD, U.K.

Abstract The differing results of two recent models of the gas phase chemistry of dense interstellar clouds have been reconciled. The consequences for the growth of complex molecules are discussed.

MODEL CALCULATIONS

Two groups of investigators have recently published detailed time-dependent models of dense interstellar cloud chemistry in which moderately complex molecules are included. The models - those of Herbst and Leung[1] (HL) and Millar and Nejad[2] (MN) - differ considerably from one another in their calculated abundances of many complex molecules. In general the MN and HL models show the same time dependence for the calculated abundances of complex molecules; i.e., the abundances are predicted to peak at early times before decreasing to their steady-state values. However, the MN results are often in agreement with observation at steady state whereas the HL results, *which are orders of magnitude lower*, agree with observation only at earlier times. The characteristic time dependence of complex molecules in both models occurs for most complex molecules only when the elemental abundance of oxygen exceeds that of carbon.[1] The assorted gas phase reactions that produce and destroy complex species have been reviewed by Winnewisser and Herbst.[3]

In an attempt to understand the large differences between the calculated abundances of HL and of MN, we have augmented the smaller MN model with a small number of reactions and additional product channels found in HL. (A few other changes have been made in the MN model.[4]) The additional reactions all involve neutral oxygen atoms whereas the added product channels are from dissociative recombination reactions. Neither the oxygen atom reactions nor the product channels have yet been studied in the laboratory. Since the augmented model of MN has been found to be in very good agreement with HL,[4] measurement of these reactions and product channels are urgently needed. The dependence of the abundances of complex molecules on the inclusion or exclusion of a small number of important reactions is surprisingly large.

MORE COMPLEX MOLECULES

Based on the MN and HL results, we conclude that efficient growth of hydrocarbons and cyanoacetylenes larger than currently included in either model can occur for both the MN and HL models at early times but only for the MN model at steady state. If the additional reactions between oxygen atoms and neutral species included in the augmented MN model are excluded but reactions involving oxygen atoms and ionic species retained, we obtain model results that show efficient growth of complex molecules even at steady state.[4]

A more detailed and comprehensive investigation of the growth of molecules as complex as C_9H and HC_9N is currently being undertaken by Leung and Herbst.

This work has been supported in part by the NSF via grant AST-8513151 to E. H.

REFERENCES

1. E. Herbst and C. M. Leung, Mon. Not. Roy. Astr. Soc., 222, 689 (1986).
2. T. J. Millar and L. A. M. Nejad, Mon. Not. Roy. Astr. Soc., 217, 507 (1985).
3. G. Winnewisser and E. Herbst, Topics in Current Chemistry, 139, 119 (1987).
4. T. J. Millar, C. M. Leung, and E. Herbst, Astr. Astrophys., in press.

RADIO OBSERVATIONS OF CARBON MONOXIDE TOWARDS ZETA OPHIUCHI: STRUCTURE, ISOTOPIC ABUNDANCES, AND PHYSICAL PROPERTIES

WILLIAM D. LANGER
Princeton University, Princeton, NJ
ALFRED E. GLASSGOLD
New York University, New York, NY
ROBERT W. WILSON
AT&T Bell Laboratories, Holmdel, NJ

We present a map of the $^{12}CO(1 \rightarrow 0)$ emission from the region around Zeta Oph and report detections of $^{12}CO(2 \rightarrow 1)$ and $^{13}CO(1 \rightarrow 0)$ along the line of sight to the star. The CO emission arises from at least four components with peak emission at -2.0, -0.7, 0.0, and $+0.6$ km/s, each with intrinsic velocity dispersion < 0.2 km/s. About one half of the CO column density is in the main component at -0.7 km/s, and the emission from all the components appears to be optically thin ($\tau < 0.5$). An excitation calculation gives a density of about 400 cm^{-3} for the main component and about 1000 cm^{-3} for the weaker components at 0.0 and 0.6 km/s. The main component emission is uniform over the map, but the other components are variable, which suggests that the cloud is clumpy. The observations provide new insights into the physical and dynamical properties of the cloud. The existence of numerous narrow velocity components can change abundance determinations and influence the radiation transfer in cloud models. Our measurements do not provide any direct evidence for interstellar shocks, although the observations of CH^+ and rotationally excited H_2 suggest that they are present. No CO is detected near +5 km/s, which is usually assumed to be the velocity of the preshock gas. On the basis of the CO observations we suggest that some of the CH^+ may be produced in cloud-cloud collisions which may explain the CH^+ linewidths and the lack of any correlation between its velocity and that of CH in other diffuse clouds. The twelve to thirteen isotope ratio is about 80 with a three sigma range of 50 to 300.

EMISSION FROM INTERSTELLAR HYDROGENATED AMORPHOUS CARBON GRAINS

A. N. WITT
Ritter Astrophysical Research Ctr., The University of Toledo, Toledo, OH, USA

T. A. BOROSON
Dept. of Astronomy, University of Michigan, Ann Arbor, MI, USA

R. E. SCHILD
Harvard-Smithsonian Center for Astrophysics, Cambridge, MA, USA

Abstract We report the detection of a broad emission feature, peaking near 0.67μ wavelength with a FWHM of 0.12μ, in spectra of numerous reflection nebulae. We identify this feature as luminescence from hydrogenated amorphous carbon dust, excited by the short-wavelength radiation field present in the nebulae. There is strong observational evidence for grain processing in the nebular environment, resulting in varying degrees of hydrogenation and of resulting luminescence efficiency. There is a pronounced spatial correlation between the observed amorphous carbon luminescence and near-IR emission from H_2 and the 3.3μ band.

INTRODUCTION

Surface photometry of reflection nebulae has revealed the presence of extended diffuse emission in the R and I photometric bands in excess beyond what is expected from dust scattering alone[1-3]. Here we report on further studies involving spectroscopy and imaging, which have led to an identification of the extended red emission.

OBSERVATIONS

Spectroscopic observations were made with the intensified Reticon scanner (IRS) on the KPNO No. 2-0.9m telescope ($0.62\mu<\lambda<0.81\mu$; $\lambda/\Delta\lambda \simeq 600$) and the Mk III spectrograph on the 1.3m telescope ($0.5\mu<\lambda<0.9\mu$; $\lambda/\Delta\lambda \simeq 120$) of the McGraw-Hill Observatory. A GEC-CCD detector was employed with the latter instrument, which was used in a 7"x300" long-slit mode. The IRS was used in the nebular mode with dual 45".5 circular apertures, separated by 60" center-to-center in E-W direction. With these two instruments we obtained sky-compensated low-resolution spectra of 17 reflection nebulae and their respective illuminating stars.

Imaging observations were carried out with the 0.6-m reflector of the F. L. Whipple Observatory and its thinned (blue-sensitive) RCA-CCD detector.

RESULTS

The presence of a broad emission feature in the nebular spectra becomes apparent when the nebular spectra are digitally divided by the spectra of the associated illuminating stars. Ten out of 17 nebulae show clear evidence of a band peaking near 0.67μ wavelength with FWHM of 0.12μ, notably among them NGC 2023, 7023, 1333, 2247, 2327 and the Red Rectangle, while the band is not present in some other nebulae as well as in some locations in NGC 2023 and NGC 7023. Other examples for non-detection are IC 435 and the Maia and Merope nebulae in the Pleiades.

Comparison with laboratory results[4] as well as abundance arguments suggest strongly that we have observed the luminescence of hydrogenated amorphous carbon in interstellar dust. This interpretation is further strengthened by our observation that the emission feature is notably enhanced in a filamentary shell structure[5] in NGC 2023, which is also seen in near-IR fluorescence of H_2 and in 3.3μ band emission.[6] In nebulae intensively studied (NGC 2023, 7023) we find the band shape and central wavelength to vary, consistent with laboratory results for amorphous carbon samples with varying degrees of hydrogenation.

REFERENCES

1. A. N. Witt, R. E. Schild and J. B. Kraiman, Astroph. J., 281, 708 (1984).
2. A. N. Witt and R. E. Schild, Astroph. J., 294, 225 (1985).
3. A. N. Witt and R. E. Schild, Astroph. J. Suppl., 62, 839 (1986).
4. I. Watanabe, S. Hasegawa and Y. Kurata, Jap. J. Appl. Phys., 21, 856 (1982).
5. A. N. Witt and R. E. Schild, Astroph. J. (in press) (1987).
6. I. Gatley, T. Hasegawa, H. Suzuki, R. Garden, P. Brand, J. Lightfoot, W. Glencross, H. Okuda and T. Nagata, Astroph. J. Letters, 318, L73 (1987).

INFLUENCE OF LARGE MOLECULES ON INTERSTELLAR GAS PHASE CHEMISTRY

V. BUCH, S. LEPP, A. DALGARNO
Harvard Smithsonian Center for Astrophysics, Cambridge, MA, USA

Computer simulations are performed of interstellar gas phase chemistry under various cloud conditions. The chemistry includes Large Molecules (LM) as carriers of negative charge and as ion neutralizing agents. Inclusion of LMs improves significantly the agreement between the gas phase results and the observations. Abundances of all the major species in TMC_1 and in the OMC_1 Ridge are reproducible within an order of magnitude for $x_{LM} \geq 6 10^{-7}$, $x_C = 1.46\ 10^{-4}$, $x_O = (1.76\text{-}2.46)\ 10^{-4}$ and $x_N = (2.15\text{-}3.01)\ 10^{-5}$. Adequate amounts of ammonia are calculated in the cold clouds without including the questionable reaction $N + H_3^+$. At $x_{LM}=2 10^{-6}$, significant steady state abundances of cyano chain molecules can be generated in the gas phase at densities as low as $10^3\ cm^{-3}$. Major contributions to chain lengthening are assumed to come from condensations of large ions with small molecules like C, C_2H_2, CN and HCN, an assumption that requires experimental confirmation. The largest cyano molecules and N_2H^+ may serve as chemical clocks, tracing the conversion of carbon to CO and nitrogen to N_2. In TMC_1 the first appears incomplete and the second well advanced, suggesting an age of $\sim 10^6$ years. In the OMC_1 Hot Core, enhancement of hydrogenated species HCN and NH_3 and decrease in the radicals CN and C_2H are reproduced by a steady state gas phase chemistry, in which LMs play an important role in assisting the hydrogenation. A possibly serious problem with the model is its prediction of an HCO abundance in Orion Ridge which is two orders of magnitude above the observed upper limit.

THE GREGORIAN PROJECT AT ARECIBO

Willem A. Baan
Arecibo Observatory, P.O.Box 995, Arecibo, PR 00613

In its twenty-four years of operation, the 1000-foot diameter reflector antenna at Arecibo Observatory has contributed greatly to widely diverse fields of astronomy. The instrument is unrivalled in radar studies of objects in the solar system and provides the most sensitive incoherent scatter radar in the world for studies of natural and artificially modified ionospheric plasmas and of tropospheric and stratospheric dynamics.

The performance of the instrument with its spherical primary can be improved still further. Engineering studies show that major improvements are feasible by increasing the accuracy of the primary reflector and providing for it efficient low-noise illumination over a wide range of frequencies and zenith angles. This can be achieved by adding a reflecting ground screen around the perimeter of the primary, improving the surface accuracy of the main reflector and of the pointing, replacing one "carriage house" (receiver cabin with "line feeds") with a Gregorian secondary/tertiary feed system, and shielding that feed system with a metal radome. Noise at high zenith angles will be reduced significantly by the ground screen. A shaped secondary/tertiary subreflector system will permit an accurate, wavelength independent correction for the spherical aberration over the frequency range from 300 MHz to 8 GHz. The primary surface has already been reset to an accuracy better than 2.2 mm rms, allowing operation to 8 GHz.

The Gregorian Upgrade program will achieve the following charcteristics for the new Arecibo: a. decreased system noise (20 K), b. improved sensitivity (12 K/Jy), c. extended frequency coverage (to 8 GHz), d. continuous frequency coverage (300 MHz to 8 GHz), e. greatly increased instantaneous bandwidth (>15 percent), f. increased angular resolution (40 arcseconds) and pointing accuracy (5 arcseconds), g. shielding from interference and ground clutter, and h. greatly increased radar sensitivity.

Phase 1 of the project consists of two parts. One is the construction of a smaller version of the Gregorian feed system for testing the design concepts

and the performance characteristics of the full system. This Mini-Gregorian will serve as a high frequency feed until the full Gregorian arrives, illuminate 350 feet of the surface (1.6 K/Jy at 5 GHz), and be fully operational in 1988. The other part is the construction of the 60-foot high ground screen. The latter is the subject of the Phase 1 Upgrading Proposal now pending before the NSF.
Phase 2 incorporates the replacement of a carriage house with the full scale Gregorian subreflector system inside an 80 foot radome. Additional support cables need to be installed to take the increased load. The Gregorian feed system will illuminate a slightly elliptical area of 700 by 850 feet of the primary surface having a 75 percent aperture efficiency for an effective area of 32,500 m^2 (12 K/Jy). The Phase 2 projects are the subject of a proposal to be submitted to the NSF.

The Gregorian Upgrade will have great impact on astronomical research at Arecibo (Davis and Taylor 1987). The HI redshift surveys done at Arecibo are essential for the study of large scale structures in the universe. Its role as a "redshift machine" will become even greater after the Upgrade. Similarly, since the discovery and timing of millisecond pulsars are unique for Arecibo, pulsar studies will benefit greatly from the interference protection, increased sensitivity and frequency agility.

Even greater will be its impact on molecular astronomy. Arecibo has been well instrumented for 21-cm line research but, except for the 18-cm OH line feed, instrumentation for molecular research was limited or absent. The Gregorian will provide the frequency agility and uncompromised performance needed for molecular research. The broad-band capability with the multi-feed carriage house will allow the molecular observer to switch easily from transition to transition, from source to source (in redshift) and from molecule to molecule. The sensitivity will be an order of magnitude greater than that of other telescopes in this frequency regime. Initially most attention will be focussed on the molecules CH, OH and H_2CO but there is no lack of molecular transitions below 10 GHz. Single dish studies in this frequency range have generally been hampered by the inability to obtain arcminute resolution needed for the study of the clumpy molecular structures. The high sensitivity and small beamsize of Arecibo will fill a gap in molecular research in this region of relatively low frequencies.

References

1. "The Scientific Impact of the Gregorian Upgrade", M.M. Davis and J. Taylor, eds., NAIC:Ithaca, in press (1987).

A NEW RADIO TELESCOPE FOR CHINA

Samuel L. Hensel and Albert Cohen
Electronic Space Systems Corporation
Old Powder Mill Road, Concord, MA 01742

Abstract In 1987, installation and check out of a new 45 foot antenna for the Purple Mountain Observatory (PMO) at Deling Ha in Qing Hai Province will be complete. The precision 45 foot radio telescope will be capable of operating at mm-wave and potentially at sub mm-wave frequencies.

The antenna is radome enclosed to provide for precision all-weather operation. It was designed and built under a cooperative agreement between Electronic Space Systems Corporation (ESSCO) and the Nanjing Astronomical Instruments Factory (NAIF). NAIF organizations are located in Nanjing, in the People's Republic of China and are part of Academica Sinica, the Chinese Academy of Sciences, located in Beijing. Under the agreement, the NAIF manufactured major subsystems of an ESSCO designed 45 foot diameter millimeter wave radio telescope for PMO. ESSCO supplied the precision reflector panels, the subreflector and ogival spars, the metal space frame radome and the inductosyn subsystem. The NAIF supplied the pedestal, reflector backstructure and the servo control subsystem. This radio telescope is located at Deling Ha in Qing Hai Province in Western China on a 10,500 foot plateau. Although access to the site takes several days by train, this location is free of almost all types of radio interference. The humidity is very low, even in summer and in general, the observing conditions are excellent, possibly the best in the northern hemisphere.

The telescope was installed in the summer and early fall of 1986 under the guidance of ESSCO field supervisors. Dr. Tong Fu, Director of PMO, and Mr. Wang Yi, who has the responsibility of this project for Academica Sinica, both expressed their personal satisfaction with the high quality of the equipment and especially the ESSCO precision reflector panels, which had an aggregate surface tolerance of 28 μm (0.88 mils) rms. The telescope will initially be utilized for molecular studies and PMO will cooperate with many other observatories around the world including ESSCO installations in Sweden, Finland, Brazil, Spain, the U.S. and potentially, South Korea.

The telescope is in its startup and calibration phase and finishing touches are being completed at the site. Dedication is expected in the summer or fall of 1987. This promises to be one of the finest radio telescopes in the world and the dedication is being planned as an international event.

TABLE I PMO Antenna Performance Parameters

Pointing Accuracy	1-3 arc seconds rms each axis
Tracking Accuracy	1-3 arc seconds rms each axis
Velocity Range	0.001° - 1° Sec. each axis
Acceleration	$1^{\circ}/\text{Sec.}^{2}$ each axis
Reflector Tolerance at 60° Reference Elev. Angle	$\leq$ 4.1 mils rms
Aggregate Surface Tolerance of Panels (as determined by factory tests)	$\leq$ .88 mils rms
Subreflector Tolerance (as determined by in-factory tests)	$\leq$ 1 mils rms
Movable Subreflector Mount two axes	1 inch motion (peak to peak) both axes
Subreflector and Mount Assembly Manual Adjustment and Mounting to Quadrapod Structure	Transverse to focal axis $\pm$ 5/8 inch, along focal axis $\pm$ 2-1/2 inch

THE SUBMILLIMETER TELESCOPE (SMT) PROJECT

ROBERT N. MARTIN
Steward Observatory, Tucson, Az.

Abstract The Steward Observatory and the Max-Planck-Institut für Radioastronomie are building a 10 m diameter dedicated submillimeter telescope to be located at 3260 m elevation (10,700 ft) on Mount Graham (SE Arizona). Reflector accuracy (15 μm) and pointing stability (1″) will provide excellent performance at 350 μm wavelength. Extensive use of carbon fiber reinforced plastic for reflector panels and structural members minimizes thermal deformations during operation while the telescope is exposed to the environment. During adverse weather the telescope will be protected by a co-rotating enclosure of a novel and simple design. The reflector panels are replicated from pyrex glass molds ground to 3 μm accuracy at the University of Arizona. The telescope will be operational in 1989.

Submillimeter astronomy represents, perhaps, the last wavelength frontier of radio astronomy. There have been three main impediments to advances in this field. (1) The atmospheric transmission is poor and is limited to a few wavelength windows even at the best, high, dry sites. (2) Sensitive submillimeter detector systems are difficult and recent technology. And, (3) suitable telescopes are only now being built. Past measurements have had to be made on optical and infrared telescopes.

To exploit the potential of submillimeter astronomy, the Steward Observatory of the University of Arizona and the Max-Planck-Institut für Radioastronomie, Bonn, Germany have collaborated on the construction of a dedicated submillimeter telescope facility. The telescope is 10 m in diameter with a primary F/D = 0.35 and a secondary F/D = 13.8. The AZ-EL mount is made of steel with thermal insulation and will have a pointing accuracy of 1″.

To control the structural deformations caused by thermal gradients as well as by gravity, carbon fiber reinforced plastic (CFRP) is used for the members of the reflector backup structure. The coefficient of thermal expansion of CFRP is about an order of magnitude smaller than that of steel. The reflector panels are a CFRP skin and aluminum honeycomb sandwich. The reflector error budget calls for 6 μm panel accuracy, 5.5 μm reflector panel deformation under load, 8 μm backup structure deformation under load, and 10 μm panel setting accuracy. This results in an overall surface accuracy of 15 μm rss. At the shortest wavelength of operation (350 μm atmospheric window) the surface roughness loss is only 25 percent, and the beam size is 8″.

The telescope has been designed to maintain its accuracy in the open air and at a wind velocity of 12 m/s. To protect the telescope during bad weather conditions, a simple co-rotating enclosure has been designed. The telescope will have two Nasmyth and/or bent Cassegrain focii outside of the elevation bearings and inside enclosed receiver rooms provided by the co-rotating enclosure. The control room and computer room are located in the co-rotating enclosure.

Mt. Graham in southeast Arizona has been chosen as the site because of its high altitude and dry conditions. The telescope will be located at 3260 m elevation near the summit. The precipitable water vapor above the site has been monitored with a 22 GHz radiometer since December 1985. This instrument automatically takes data every four minutes. From these data and radiosonde measurements, we estimate that the precipitable water vapor overhead is less than 1.5 mm for about 30 percent of the time during the non-summer months.

The construction of the telescope structure is underway by Krupp Industrietechnik and MAN Technologie in Germany. The mount will be complete in January 1988, and the reflector in fall 1988. Construction of the co-rotating enclosure on Mt. Graham is scheduled to begin in spring 1988. The telescope will be installed at the site at the end of 1988. The telescope will be operational in 1989.

A SUBMILLIMETER RECEIVER AT 492 GHZ

J.A. TAUBER, N.R. ERICKSON, P.F. GOLDSMITH, R.L. SNELL
Five College Radio Astronomy Observatory, University of Massachusetts, Amherst, MA 01003, U.S.A.

Abstract We describe a receiver that operates at 492 GHz

We have built a heterodyne receiver that operates at frequencies near 492 GHz, which is the frequency corresponding to the $^3P_1$--$^3P_0$ ground state transition of atomic neutral carbon. The mixer used in the receiver is a cryogenically cooled Schottky-barrier diode; it is followed by an FET amplifier which operates at 1.4 GHz. Both the mixer and the FET are placed inside a vacuum dewar which is cooled by a helium-cycle refrigerator to 75 K. The LO chain consists of an 82 GHz Gunn-diode oscillator followed by a frequency sextupler; this is made by cascading a tripler and a doubler. The LO and signal are combined into the same spatial mode by a polarization-rotating dual-beam interferometer. A reflective quarter-wave plate is placed in the signal path to prevent resonances between the receiver and the secondary mirror on the telescope, which could produce large baseline ripples. There is no rejection of the unwanted sideband. The 1.4 GHz IF signal is downconverted to 470 MHz, at which frequency it is sent on to the spectrometer.

The spectrometer used is a 32-channel filterbank operating at frequencies between 440 and 500 MHz, and affords a velocity resolution of 1.2 km/s. This filterbank is of a novel design, featuring sections of coaxial waveguide as the filtering elements. Its main advantage over conventional solid-state filterbanks are: low cost, and the large number of channels which are juxtaposeable in frequency space. This new design is being developed for use in the 15-element array which will operate at a wavelength of 3 mm on the FCRAO 14 meter telescope.

The receiver itself will be installed at the 2.3 meter reflector of the Wyoming Infrared Observatory in December 1987, to attempt observations of neutral carbon in several molecular clouds.

A COOLED 2 MM SCHOTTKY RECEIVER FOR RADIO ASTRONOMY

L.M. ZIURYS, N.R. ERICKSON, and R.M. GROSSLEIN
FCRAO, University of Massachusetts, Amherst

ABSTRACT A low noise heterodyne receiver with a Schottky diode as the mixing element has been built for radio astronomical observations in the 125-170 GHz frequency range, and has been tested on the FCRAO 14 meter telescope. The receiver double sideband noise temperature is 140 K at 150 GHz, and increases up to 200 K at 125 GHz, with comparable performance expected up to 170 GHz.

INTRODUCTION

The 2 mm window, which lies between the atmospheric absorption line of oxygen at 118 GHz, and the atmospheric line of water at 183 GHz, has been relatively unexplored by astronomical observations. In fact, very limited data exists in this band above 150 GHz. Consequently, a heterodyne receiver has been constructed for observations in this frequency window for use on the FCRAO 14 meter telescope.

OPTICS

The signal beam to the mixer passes from the telescope subreflector to a mirror, which reflects the beam through a quartz window and into the receiver dewar (Fig. 1). The beam is then reflected from a second mirror and through a lens, which re-images the beam into a scalar feed-horn attatched to the ring filter block. The second mirror and lens are cooled to 20 K. The mirror external to the dewar will be replaced by a polarizing wire grid, which will allow for simultaneous observations with the 2 and 3 mm receivers. A single sideband filter will replace the mirror internal to the dewar and will be cooled to near 20 K, as well.

RECEIVER

The receiver is a cooled Schottky diode mixer with a tunable backshort. The local oscillator signal is coupled to the mixer using a tunable ring filter. The local oscillator source for the receiver is derived by tripling the output of phase-locked klystrons. Two triplers, which are located outside the dewar, are used to cover the frequency range of the mixer. Tripled power passes into the dewar and to the ring filter via a waveguide input. The 1.4 GHz IF output of the mixer is amplified by a low noise 3-stage FET amplifier.

SYSTEM PERFORMANCE

The double-sideband system noise temperature of the receiver is shown in Fig. 2. At the moment, only the lower frequency tripler is in use, and therefore there are no data points above 150 GHz. As the figure shows, the receiver noise temperature is 140 K (DSB) at 150 GHz, and increases to near 200 K at 125 GHz. Comparable performance is expected up to 170 GHz.

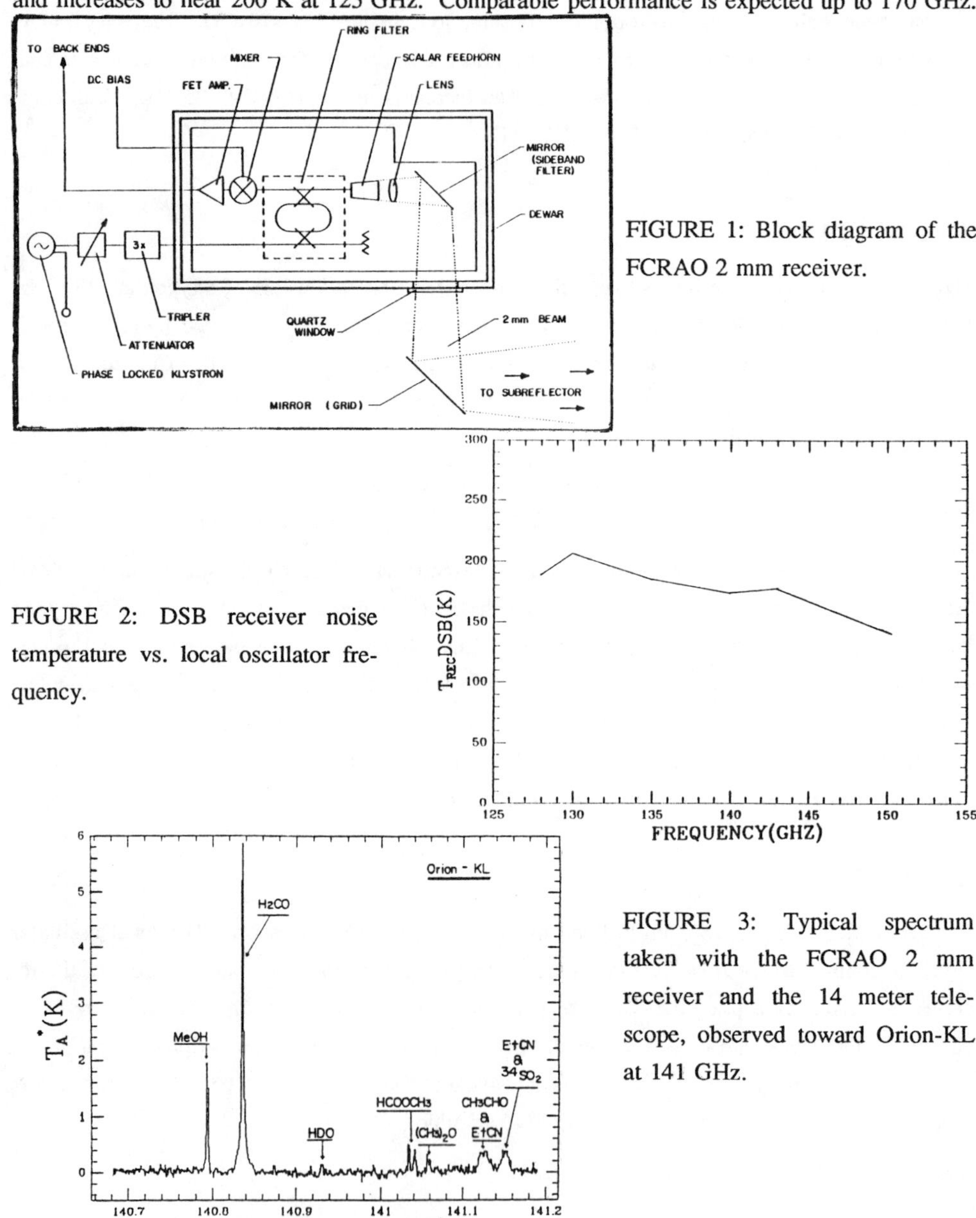

FIGURE 1: Block diagram of the FCRAO 2 mm receiver.

FIGURE 2: DSB receiver noise temperature vs. local oscillator frequency.

FIGURE 3: Typical spectrum taken with the FCRAO 2 mm receiver and the 14 meter telescope, observed toward Orion-KL at 141 GHz.

A FOURIER TRANSFORM SPECTROMETER AND BOLOMETER SYSTEM FOR MM AND SUBMM ASTRONOMY

ROBERT N. MARTIN
Steward Observatory, Tucson, Az.

Abstract A Fourier Transform Spectrometer (FTS) and bolometer system has been built and installed on the IRAM 30m telescope for broadband spectroscopic and continuum measurements. The FTS is of the Martin-Puplett type with free standing wire grids as beam splitters. It is usable over the wavelength range of 4 mm to 0.5 mm with a resolution of 110 MHz. Two ^{3}He cooled (T=0.3 K) bolometers are used as detectors, optimized for 1.2 mm wavelength. Broad passband filters can be inserted to select specific wavelength windows. The system can make coarse resolution spectroscopic measurements over a single or several atmospheric windows. It is useful for spectroscopic studies of atmospheric transmission and calibration, absolute source calibration, and searches for strong broad lines. With the FTS set at the central fringe, the system can be used as a broadband continuum detector and polarimeter.

At radio wavelengths, heterodyne receivers coupled with filterbank, acousto-optical, or autocorrelation spectrometers are the mainstay of spectral line observations. However, virtually all heterodyne systems at 1 mm wavelength are limited to a maximum of 500 MHz (500 km s^{-1}) passband. Most Galactic objects can be studied in detail but broad linewidth external galaxies and planetary atmospheres cannot be adequately studied. For this reason, we have constructed an FTS to undertake broadband, coarse resolution spectroscopy.

Millimeter and submillimeter continuum observations have also proved to be extremely valuable to the study of galactic and extragalactic sources. Bolometer systems have been the mainstay of these observations. For maximum sensitivity it is common to allow the bolometer passband to be as broad as possible (up to a 30 percent at the λ=1.2 mm wavelength atmospheric window). With such broad bandwidths and variable transmission across the atmospheric window, it is difficult to define a mean wavelength for the observation. In addition, without measuring spectral properties of the atmospheric transmission during the astronomical observations, calibration of the results becomes difficult at best. The FTS/bolometer system presented here alleviates these problems. Because of the polarization sensitive nature of this FTS and the dual channel detectors, the system can be used as a polarimeter. Measurement of the Stokes I, Q, and U parameters is possible. Initial tests indicate sensitivity to a few percent polarization.

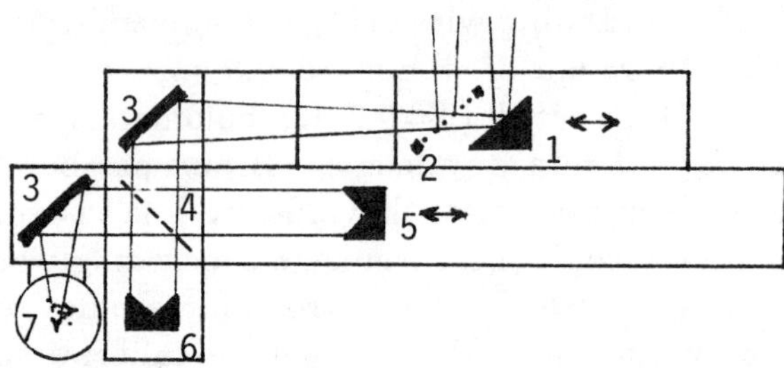

1) flat mirror for main beam.
2) rotating wire grid chopper.
3) offset parabolas.
4) wire grid beam splitter.
5) movable roof mirror.
6) fixed roof mirror
7) dewar with detectors.

For continuum measurements, the movable FTS mirror is put at the zero path difference position and the input wire grid is rotated at about 600 rpm. This provides 20 Hz sine wave modulation between the two input sky positions. The resulting dual beam maps are deconvolved in post-real-time reduction software. The separation of the two input beams is adjustable. An example of the mapping capability is given in figure 1.

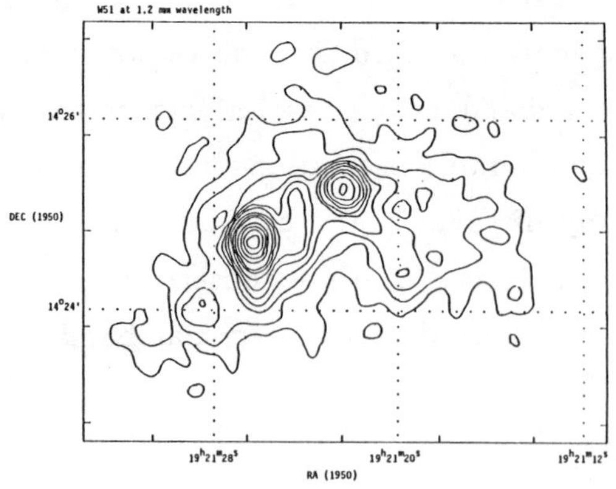

Figure 1
W51 in λ=1.2 mm continuum

Al Barrett　　　　Phil Myers

Al
Barrett

Cosmo Papa Ginny Barrett Stan Zisk

Jack Barrett Cosmo Papa Kathy McCue Henry Zimmermann

Henry Zimmermann   Al Barrett   Kathy McCue

Cosmo Papa   Al Barrett

VI. Banquet Address

INTRODUCTION OF CHARLES H. TOWNES

BERNARD F. BURKE
Massachusetts Institute of Technology, Cambridge, MA

I think that the main purpose of tonight is the enjoyment of the evening and the opportunity for people to see old friends. Now, in honoring Al Barrett tonight we have a special old friend who is going to do most of the talking and that is the man who was Al's thesis supervisor, Charlie Townes, an extraordinary physicist who had an influence that I think touches many of the people in the audience tonight. I know that my own experience was as a graduate student of Woody Strandberg here at M.I.T., where through a combination of circumstances, it was obvious that I had to go to Charlie's lab at Columbia. I hauled my magnets and waveguides up to the 11th floor of Pupin Laboratory and got to know Charlie as I worked with his graduate student, Bob Hardy. Now it was a concentrated experience, but two things stand out very clearly that are completely relevant to our celebrations this week. The first is that I was shown around the laboratory and there were no secrets. Nobody tried to hide things. I was shown an unlikely apparatus that was in one of the back rooms of Charlie's lab. It was a device to separate the states in ammonia so that you could make something called a *maser*. I think we all know the consequences of that particular project. The interesting thing is that I was a graduate student from a competing laboratory and there were no secrets. I was shown everything and given explanations about every project. It was a marvelous and broadening experience. The other thing that sticks in my mind was an experiment that I had known as the *hot guide* experiment. The idea was that instead of looking at molecules in their ordinary gaseous state at room temperature, one could make a hot waveguide and look at alkali halides that had to be vaporized. One of my colleagues here at M.I.T. had been trying to do that and I understood what it was like. You build a piece of microwave apparatus that's wrapped with asbestos and heating wires and when you try to do the experiment, you get the vapors of alkali something-or-other corroding things at one end, and at the other end, where it is a little colder, an alkali something-else condenses out and upsets the electromagnetic properties of the apparatus. Knowing the difficulties of that experiment, I admired greatly the graduate student at Columbia who was laboring with it, and of course it was Al Barrett. While we are celebrating the various aspects of his career, I think it is entirely appropriate that the one who steered us in such unlikely directions, Charlie Townes, should be available to tell us about what things were and what things will be.

ALAN BARRETT: A CAREER IN ASTROPHYSICS

CHARLES H. TOWNES

University of California, Berkeley, CA U.S.A.

It's really a great pleasure to be here among so many friends and to help celebrate Alan Barrett, his career, and how much he has meant to all of us. When I was asked to speak I thought perhaps I was chosen because I had known Alan and Ginny longer than anybody else here. However, it turns out that Garry Hough is present and he knew him in grade school or maybe before, living two doors from him when they were boys. Alan and Ginny's children, Rick and Bonnie, are here and I find also that 14 out of Alan's 19 graduate students are present. They too probably know him well and from perspectives somewhat different from mine, so I don't regard myself as any expert. Nevertheless, I shall do what I can to try to help explain the phenomenon of Alan Barrett. Now, in trying to illustrate this phenomenon, I called a number of people to see if anyone had any really outlandish stories that I could tell to keep people alert after a very engaging and interesting day and a very good dinner. I called Ed Lilley, for example, who with Alan did the very first search for OH; I thought he must have some good stories. Ed said "Well, you know in those days we were awfully straight guys." With that remark about *those* days, I assume that he was primarily thinking of himself. I did find out from someone that Alan had a favorite story about the comparison between Harvard and M.I.T. students, but I decided that it wasn't fit for mixed company; by mixed company I mean both M.I.T. and Harvard people.

You may know that medical people have tried to classify individuals into two different types of personalities, A-type and B-type. The A-type people are the ones who charge ahead, are very ambitious, and work hard. They eventually give themselves heart attacks and probably give the people around them various kinds of nervous disorders. Then there is the B-type. They are more relaxed, take things as they come, and are not so ambitious. I thought we all had to be one of those or the other, but it turns out the T-type was recently invented, and I think that really better fits Alan's case. The T-type person is one who is looking for a challenge to enlarge his boundaries, who enjoys doing things that are difficult but important. It's the challenge that interests him, regardless of whether or not it makes any difference to other people. He figures things out, solves problems and tackles such challenges. That is what really interests the T-type, and that sounds like Alan. Although I couldn't find any really juicy stories about Alan, it's not because he hasn't been totally engaged and on the frontiers of things. Partly, it's because he is a New Englander, coming from Springfield, Massachusetts, and practices that sort of swaggering understatement of the New Englander. The perfect New Englander makes a point of not doing things to excess, just doing them right. In fact, in spite of his unassuming stance, Alan has been remarkably on the frontier of a number of things and has pushed hard enough to poke holes regularly through those frontiers.

I don't know whether everyone is properly aware of a very important astronomical event that took place in the early part of this century. I have tried to get some of the evidence of this event, and while I haven't been able to get the evidence to present its very earliest stage to you tonight, Figure 1 will show some of the evidence for this event in a fairly early aspect. There you see a very self-composed, cheery Alan Barrett. Further developments of this phenomenon are shown in Figure 2, at summer camp. There, the fourth person from the right is someone who at this point is not necessarily looking especially significant. As time went on there were further developments; Figure 3 shows a Navy phenomenon, where I believe Alan perhaps first ran into serious electronics.

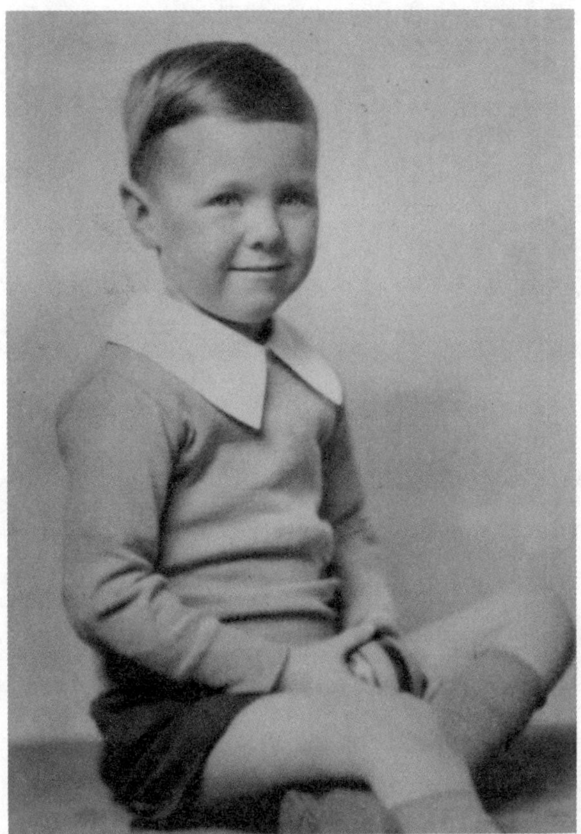

Figure 1. Alan Barrett, about 1930

As soon as he managed to escape from the Navy he went to college, and Figure 4 shows a very serious young man, still with a smile on his face.

There was another unique event that made this development particularly significant to astronomy and the history of this particular object especially significant. Figure 5 shows that event, the principal players being Ginny and Alan.

Figure 2. At summer camp, 1935 (AHB, 4th from right)

Ginny was a very vivacious, beautiful young lady whom Alan sagely singled out and luckily was able to marry. With all that preparation, Alan was then really ready for graduate school.

At graduate school Alan started research with me and, as Bernie Burke said, we were trying to do rather difficult experiments with the first hot waveguide system for microwave spectroscopy. At that time, the discovery of a new molecular microwave spectrum was a big event. If you found one line of one molecule, that was success. I remember the startled comments of one visitor who came into our building when one of my students who just preceded Alan burst out into the hallway shouting "Hey, I found it!" with all the enthusiasm of someone striking gold. But when Alan got to work, he wasn't just finding one molecule. For the first time, he and a fellow student opened up a whole new series of molecules. Their first paper was on essentially all the compounds of gallium, indium, and thallium with fluorine, chlorine, bromine, and iodine. Between those combinations and

Figure 3. 1944. Barrett enlisted in the Navy on his 17th birthday, just after graduating from high school, during World War II.

their isotopes there were 22 species. Large numbers of lines and much physics just fell out of Alan's work. I have some correspondence from Alan at about that time which I dug up out of my files. For the latter part of his work, I was away on sabbatical in France, and Alan wrote me in a way characteristic of his care: "I'm not sure I can write up my thesis. I find a discrepancy in the positions of a pair of lines." But also characteristically, he solved the problem, and the next letter said "...after all, now I think I can get my thesis. It's a good thing because of my financial condition. Now I can finish up." Part of my letter recommending Alan to Leo Goldberg, then at

Figure 4. Alan at Purdue University, 1948

Michigan, gives my observation of Alan as a graduate student. "He was one of our better students and is an intelligent and ambitious young physicist with a very sound approach to his own affairs and a friendly and cooperative attitude towards others. Barrett began work at the laboratory at Columbia with a more experienced student, but his ability and qualities of leadership showed up rather quickly and made him the spark plug of this team. Barrett showed intelligence, drive and ingenuity, which solved several rather difficult problems both of an experimental and of a theoretical nature. His ability and self-reliance have been particularly important in the latter stages of the work since I was at the time on sabbatical leave. From his performance, I

Figure 5. Alan and Virginia McCulloch were married on September 3, 1949, in Altamont, New York.

have no doubt he could lead a small group of research workers in an effective way on almost any type of problem which he considered practical to attack." I wouldn't take any word of that statement back. The last comment is, I think, a picture of Alan's very excellent judgment. He has always had the characteristic of choosing carefully, being both adventurous and at the same time able to carry through what he started.

What Alan did start immediately after graduate school, however, was by no means simple or easy. He wanted to be a radio astronomer, and in particular to look for OH. This was a time when microwave spectroscopy for physicists was beginning to play out and some of my students were going into various other fields. Alan decided he wanted to be a radio astronomer and the two of us talked over OH with some care. He planned to look for it in the galaxy. I remember very well how I explained to him, "You don't have much background in this field, and it's going to be hard to get a job." His reply was, "Well, in any case whenever there *is* a job in radio astronomy, Alan Barrett is going to be standing in line for it." He stuck with that, and he stuck with OH. Now OH was neither an obvious nor a simple kind of problem. For a young Ph.D. these days, one would well question whether such a project was an advisable thing to do; it was a problem that might have no solution at all. But for Alan, it was something that he wanted and he was determined to do it. So he went to the Naval Research Laboratory where there was one of the best antennas and radio astronomy groups of that time, and worked in John Hagen's section with Ed Lilley.

Here is a letter from Alan dated 1956. It says "...for reasons which escape me, I found that the OH experiment was merely awaiting my arrival. Most of the equipment was built and ready to go. I just had to design the microwave mixer and build a frequency doubler for the local oscillator." You see, the people at NRL were very friendly, and they were persuaded that Alan's project was worth the try. The letter continues: "The fifty foot antenna was assigned to Ed Lilley and me for a month beginning March 23 for the OH experiment. Actually we were ready prior to that and our impatience was too much for us so we made several crude attempts with a five foot horn stuck out a window. This helped us iron out the bugs in the receiver but it didn't get us any OH. From March 23 to April 23 we kept the antenna crew working 18 to 24 hours a day, seven days a week, but we never found OH. At the moment we're writing an NRL report on the experiment." My next letter from Alan, about a month later, involved a long calculation he and Ed Lilley had made on the probable population and temperature of OH. He noted that the excitation temperature should be very low, and that he'd been up to Harvard to talk to Ed Purcell and "he is in accord." He sent me his calculation, which of course was all quite in order under conditions envisioned by astronomers at the time. Alan concluded, "It seems to be that the time we spent looking for OH in the galactic center could have been more profitably spent on Cas. In view of this, our hope for better sensitivity (about 5 K or better), and your faith in the 1667 MHz calculation, we would like to repeat the OH experiment in September and October without waiting for the 84-foot antenna. There may be a good deal of opposition to our repeating the OH experiment, but we shall see." At that time, he had an approximate frequency for OH from our laboratory work at Columbia, not on this ground state line, but on other lines from which the frequency was predicted. Figure 6 shows Ed Lilley and Alan as well as the frequencies they had for the lines: 1665 and 1667 MHz.

Figure 6. A. Edward Lilley and Alan Barrett at NRL, 1956.

In the same letter, Alan let me in on a secret, namely that at NRL Venus had just been detected and found to be quite warm. That was Connie Mayer's work. Alan said not to tell anybody, it was not yet published but was obviously exciting to him. That surprisingly warm temperature piqued Alan's curiosity enough that sometime later he worked out, for the first time, the explanation of what made Venus hot. It required an understanding of the atmosphere to explain this high temperature. There were other things in planetary work that caught his interest. He was looking into possibly detecting OH in the earth's atmosphere, and he recognized that one could do important microwave spectroscopy on the atmosphere. That too bore fruit some years later in a program on microwave lines in the high atmosphere.

Alan took a job at Michigan in 1957. He was still eagerly pursuing OH and kept pressing me about it. The University of Michigan was in the process of building an 85-foot dish. I have letters from him suggesting that since we were working on masers and developing a maser for radio astronomy, we might bring the maser to Michigan. Would we consider using the maser there? Well, we had already settled on plans to take the maser to NRL. So Alan immediately turned to another person, Kikuchi, who was nearby and busy with an early maser based on ruby, whose virtues he had first recognized. But Alan kept writing me about the frequency of OH. In a letter to him I referred to a new student whom I was putting on the job of measuring this particular OH line. Immediately I started getting pressure from Alan: "What are you getting? What's the frequency?" We were of course delayed. Here is one of his letters. He says: "My curiosity has been put to quite a test because the results of the OH experiment have not been forthcoming." And it had only been two or three weeks since I last talked with him. A second letter says "I assume that the usual sort of difficulties prevented a measurement," and so on. But eventually the line was measured by Ehrenstein and Stevenson in our lab at Columbia. Again, let me remind you that the idea of looking for OH was not an enormously popular one in the astronomical community at that time, and Alan didn't receive much encouragement. However, he had looked at the situation very carefully, calculated everything, and kept pressing on it even though the community at large was not so hopeful. Alan's partner in the hot guide experiment by then had turned into a biologist. One of the two graduate students who had made the measurement on OH in the laboratory had also turned into a biologist and the other one into a financial type dealing with securities. Certainly not everyone was moving in Alan's direction.

By this time, the Electrical Engineering Department at M.I.T. had become interested in developing a program in radio astronomy; fortunately, they were able to hire Alan in a faculty position and pretty soon things really began to happen. Alan began organizing again, getting teamed up with the right people and the right kind of antenna, and he kept pushing. Figure 7 shows the discovery of OH in absorption towards Cas A.

Alan figured out that the way to look for OH, according to the best knowledge then available, was in absorption, and the figure shows absorption due to OH found in the direction of Cas A with the Millstone Hill antenna at Lincoln Lab. Figure 8 shows the group that did it: Sandy Weinreb, Alan, Lit Meeks, and Joe Henry.

On the blackboard there, at the top right, is the correct frequency of 1667.357 MHz. How much that really helped, I'm not sure, but at least they had the confidence of knowing just where to look while searching. M.I.T. made a news announcement that read in part as follows: "OH line observations were conducted on Sunday, Monday, and Tuesday. A series of careful tests were made which established, beyond a doubt, that the OH absorption line had been observed at the nominal frequency of 1667 and 1665 MHz.

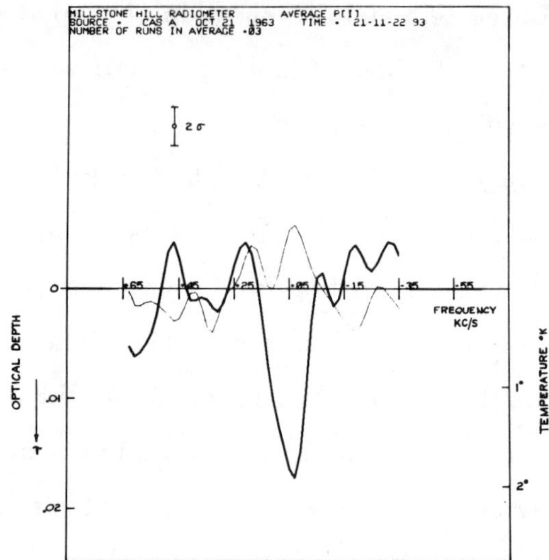

Figure 7. Absorption spectrum of OH in the 1667 MHz transition towards Cas A. The light line is an off-source spectrum. From the discovery paper. Reproduced by permission from *Nature*, Vol. 200, No. 4909, p. 830, copyright © 1963, Macmillan Journals Limited.

These observations, which began October 15, are the first in which the presence of the OH radical in interstellar space has been detected. A paper by Weinreb, Barrett, Meeks, and Henry, which gives the results, is to be submitted."

I happened to be in an administrative position at M.I.T. at that time, and of course M.I.T. wanted to make a public release about this discovery. I remember it very well because I was asked to represent the administration with this public release and to introduce Alan. I was very pleased to have the chance to do that and to try to explain to the press how important it was. I remember someone asked me, "Well, is this as important as the hydrogen line?" I said, "It is very important, but I don't think it is as important as the hydrogen line." I thought I was being carefully honest. Now 24 years later I can still say I was trying to be honest, but I may possibly have also been mistaken. Developments in molecular astronomy of the intervening decades

Figure 8. Sandy Weinreb, Alan Barrett, M. Littleton Meeks, and Joseph Henry, at Lincoln Laboratory, October 1963.

have made this an increasingly important initial discovery.

I want to read a letter from Sandy Weinreb in connection with the OH line discovery. Sandy says, "After not finding the deuterium line, I was convinced that the hydrogen line was the only radio astronomy spectral line. Considerable effort was involved to put together an OH line front-end, modify the deuterium correlator, and get the software right for the Millstone antenna. It was mainly Alan's confidence that the OH line would be found, and his persistence in doing the experiment that led us through all the preparatory work. I could not believe the results when the 1667 MHz absorption line was immediately visible in Cas A and was truly astounded when the 1665 MHz line appeared at precisely the correct frequency . . . I remember some ear-to-ear smiles on Alan's face. He was our leader and driving force in those early days and certainly the one who did the most to carry the work further."

Alan went on to various other things that I can only mention briefly. There was the Venus mission, Mariner II. Figure 9 shows a picture that showed up at that stage along with the publication of the work on Venus by Alan. He was also rather occupied with other kinds of things at the time; Figure 10 shows his children, Rick and Bonnie, who also gave him plenty to do.

Figure 9. Newspaper photo of Alan at the time of Mariner II Venus flyby, 1962.

But one of Alan's wonderful characteristics is his meticulous thoroughness. In one of my letters from him during that busy time I find that he points out that one of the formulae in my book was not correct, which also put me to work checking it. He was also continuing to examine OH, and noticed something that I think many people might have passed over, namely that the apparent intensity of OH changed a bit depending on the orientation of the source. After his usual thorough analysis, Alan was able to show, with his colleagues, that this effect was due to polarization. Polarization was by

Figure 10. Alan, Ginny, and children, Rick, 8, and Bonnie, 5, in Ann Arbor, Michigan, 1960.

no means expected; with just absorption and emission of a molecular line, how could the radiation possibly be polarized? It was the key and the first clear evidence of naturally occurring masers in space.

I must turn now to some other parts of Alan's career. On Alan's teaching I can't pretend to be an expert, but I have inquired about it. It's clear that Alan is a great favorite as a teacher. One of his undergraduate students wrote, "Barrett is my savior, I couldn't last without him." He was very down to earth and clear; he brought simple things into the classroom such as resistors to show people that a million ohms was not necessarily something large and frightening. He was very sympathetic with students and interested in their welfare. As for graduate students, again I'm no expert as I haven't had first hand experience working for Alan. I've worked *with* him, and know his powers. But I want to read some letters from students who had the privilege of being guided by Alan and working in his laboratory.

One student wrote, "I remember that he always gave us plenty of room to work, was always prepared to advise and discuss even the smallest details. He always found time for his students and never discouraged our ideas, even the dumb ones. We were encouraged to think for ourselves. In many ways, he led us by example, of working hard, thinking through a problem, and careful meticulous work. All of his students have a great fondness for him."

Another student said, "...but after a while I began to see Alan's approach to research. He would try to open a new area, and would spend considerable time and care estimating detectability, planning searches, and looking for possible pitfalls. He would know when some key information was lacking, and he wouldn't hesitate to identify experts and consult with them on his list of questions. He would turn the problem over and over, criticizing his own thinking well after I was satisfied that it made perfect sense. He would map out a plan of attack, would write proposals while delegating some work to others. He would reduce a problem to its essentials with keen practical logic. For example, he always liked to check equations by making sure the limiting cases produced understandable solutions." Now that is a picture of Alan as I know him too, and I think it is great that he has been able to do so much for so many students.

Another thing I found out was that Alan, as well as taking chances scientifically after a thorough analysis, also was willing to take chances on students. He has had a number of students of unusual background or style. One might note that Armstrong was an English major, a taxi cab driver, and a guitar player. Another student, Bill Lenoir, became an astronaut who gained some notoriety by bringing Jalapeño peppers on a space shuttle flight. Then there is Jackson and his exploit in scuba diving. He went to see the famous Australian yacht in the America's Cup race and got an underwater picture of its secret keel. That is the kind of student that Alan frequently was willing to take a chance on, and where both his judgment and guidance have worked.

Alan also has a very friendly, homey side that I know many people have seen up at Winnipesaukee as well as here in Cambridge. The Winnipesaukee aspect is illustrated a bit in Figure 11 and it is in such circumstances I hope Alan can now be with increasing probability.

Figure 11. Alan at his home on Lake Winnipesaukee, New Hampshire, 1970.

Now it's a bit questionable when scientists get together and congratulate themselves or one of their number. Perhaps we're prejudiced by our own mold and by the very likable people around us. So just to put ourselves outside of this possibly prejudiced scientific viewpoint, I want to mention something said by a person from a very different realm. What was said is this: "The greatest and noblest pleasure men can have in this world is to discover new truths, and the next is to shake off old prejudices." The person who said that was Frederick the Great, a military man, a great statesman, not a scientist. But to us scientists it rings true. We know that Alan has had those kinds of pleasures. There is another quotation that I would like to make from a person of still a very different background, someone with whom we feel a little more at home than with Frederick the Great. This quotation, too, applies to Alan. It's from Will Rogers, who said, "It is great to be great, but it's greater to be human."

I would like to propose a toast: Alan Barrett has discovered new truths for us, he has helped get rid of old prejudices, and he is human. Ginny has been a very important part of it all. She has encouraged and supported Alan through those years of hard work, as he struggled with results that were still aborning. Ginny has played a big part in the humanness factor. So with pleasure and respect, we drink a toast to Alan and Ginny for their discoveries and their humanness.

Rick Barrett Al Barrett

Nan Hough Garry Hough Bonnie Barrett

Bonnie Barrett

Al Barrett

Ginny Barrett

Jim Moran   Al Barrett

Phil Schwartz   Phil Myers   Al Barrett

Sara Beck   Paul Ho   Bob Martin

Barry Turner   Jim Jackson   Tom Armstrong   Tom Bania

VII. Alan H. Barrett:
Scientist, Colleague, and Teacher

MICROWAVE RADIOMETRY IN THE EARLY DETECTION OF BREAST CANCER

N. L. SADOWSKY
Faulkner Hospital, Jamaica Plain, MA

I feel privileged to have been asked to speak at this symposium honoring Professor Alan Barrett. You may wonder what connection a physician might have with Alan Barrett's work in astrophysics. I am a radiologist with a primary interest in the early detection of breast cancer. Professor Barrett's interest in microwave radiometry does have something in common with medical thermography, which has been used to aid the early detection of breast cancer.

Breast cancer is an important public health problem in the United States. It makes up one-fourth of all cancers in women, and in 1987 we estimate that 130,000 cases will be discovered in this country. Breast cancer is responsible for one-fifth of all cancer deaths in American women, with 41,000 expected in 1987. Lung cancer has just overtaken breast cancer as a killer of women in America, but there are striking differences between these two diseases. Whereas lung cancer is almost (90%) completely preventable by eliminating cigarette smoking, there is no single known cause of breast cancer so that primary prevention is not a viable strategy. On the other hand, early detection of breast cancer has been shown to save lives while detecting lung cancer early makes very little impact on survival in these patients. A woman born in the United States has a 10% lifetime chance of developing breast cancer, and if we were to use mammography and physical examination to screen 1,000 women over the age of forty selected at random we would find about 5 with detectable breast cancer.

If we can find breast cancer while it is in stage I (confined to the breast), there is an 80% chance of five-year survival, but if the disease is found after axillary lymph nodes are involved, the five-year survival rate is only 55%. The size of the tumor is directly related to the incidence of positive lymph nodes so that when a tumor is found accidentally by the patient herself it is usually about 3.5 cm in diameter and has a 60% chance of having positive lymph nodes; whereas, if the tumor is found by routine physical exam or by periodic breast self-examination, it is usually 1.5 cm in diameter and only 30% will have positive lymph nodes. Those tumors found by mammography alone (non-palpable) are often less than 1 cm in diameter and only 10% have positive lymph nodes. Mammography is capable of discovering tumors as small as 2 mm in diameter. Mammography is the most reliable method of detection of early breast cancer.

Mammography is an x-ray picture of the breast and, as such, is responsible for delivering a dose of ionizing radiation to the tissue through which it passes. X-rays have been shown to cause harmful effects when delivered in significant doses and these include the production of breast cancer. In 1976 a paper was published that claimed that x-ray mammography might cause more breast cancers than could be cured by early diagnosis. This paper caused a great deal of turmoil and had a negative impact on screening by mammography. At the time, a great deal of emphasis was placed on finding an alternative test that might substitute for mammography or at least to select patients who would have a high risk of having an abnormal mammogram. Thus, if a perfectly safe test could be developed that might separate out patients who did not need mammography from those who would benefit from mammography, we could avoid giving the radiation to a large part of the population. This was the hope for infrared thermography and microwave thermography.

Tumors can not grow larger than 3 mm in diameter without the development of a new blood supply. This neovasculature develops by virtue of a humoral substance elaborated by the tumor. It is these blood vessels that we believe are responsible for a local elevation in temperature associated with breast cancer. Increased metabolic activity related to rapidly growing cells may contribute something to this elevation in temperature, but it is mainly increased blood supply that is responsible for the temperature change. It is this phenomenon, the increase in temperature, that is the basis for medical thermography.

The physical bases of thermography are:

1. All objects at temperatures above 0 Kelvin emit and absorb electromagnetic radiation (e.g., infrared radiation and microwave radiation).
2. The intensity of the radiation is proportional to the temperature.
3. Infrared radiation is a surface phenomenon because human skin is opaque at these wavelengths. Microwave radiation can penetrate human tissue for distances of several centimeters so that microwave radiation is related to the temperature distribution at some distance below the surface.
4. The radiant energy can be detected quite easily and intensity differences can be displayed.
5. Thermal patterns in the body are generally symmetrical and are constant from one month to the next.
6. The pattern of heat distribution is determined by vascularity and underlying pathology such as inflammation and neoplasm.

Our research was conducted at the Sagoff Breast Evaluation Center at the Faulkner Hospital in Boston beginning in 1974 with a 3.3 GHz radiometer. The radiometer was a conventional superheterodyne receiver of the type used for radio astronomy applications, using a low-noise transistor amplifier switched at 10 Hz between a fixed-temperature termination and a dielectric-loaded waveguide antenna, designed to provide good impedance match when placed in contact with the skin. It was built under Barrett's supervision at the M.I.T. Research Laboratory of Electronics. All patients received a clinical examination, a mammogram, an infrared thermogram, and a microwave study. The microwave examination lasted approximately 20 minutes. Measurements were made on a nine-point grid representing nine specific locations in each breast. Each measurement was done for about ten seconds with approximately a 3 cm space between neighboring positions. The first measurement was made over the right nipple, with the left nipple areas measured next, so that the right and left breast were examined alternately in mirror image pairs. In 1976 a 1.3 GHz radiometer was added, and in 1978 a 6.0 GHz radiometer. A difference in temperature of 1 K was considered abnormal. The results of the microwave study were compared with the infrared thermogram, the mammogram and, in biopsied cases, the actual microscopic pathology. The comparative results among the three radiometers are given in Table 1.

It is obvious from the data in Table 1 that x-ray mammography is clearly the best method of detection. The results of microwave radiometry proved to be very comparable to those of infrared thermography. At 6 GHz the microwave thermography proved to be superior to the infrared thermography. The similarity in results of the two types of heat detection was not based on obtaining the same diagnostic information since many false negative microwave results were infrared true positive and vice versa. Indeed, when the data from infrared and microwave methods are combined, the true

Table 1

Results of Microwave Study

Radio-meter	Number examined	# CA[1]	Microwave[2] TP	Microwave[2] FP	Infrared TP	Infrared FP	Mammography TP	Mammography FP
1.3 GHz	1184	26	73	27	77	32	89	8
3.3 GHz	977	29	72	33	76	36	90	13
6.0 GHz	916	35	86	28	61	10	91	5

1. Number of Cancers
2. $\text{TP} \equiv \frac{\text{Cancers diagnosed as Cancer}}{\text{All Cancers}} \times 100 \equiv 100 - \text{FN}$

   $\text{FP} \equiv \frac{\text{Normals diagnosed as Cancer}}{\text{All Normals}} \times 100 \equiv 100 - \text{TN}$

   (TP = percentage true positive; FN = percentage false negative;
   FP = percentage false positive; TN = percentage true negative)

positive rate increases to that of mammography. However, the combined infrared and microwave data had a true negative rate of less than 70%. Using this combination of infrared and microwave, we could effectively screen for breast cancer and only do mammography in 30% of the population.

Since the time of our original research, developments in early detection of breast cancer have made the contribution of heat detection less alluring because:

a. The dosage from a two-view mammogram has been reduced significantly from well over 1 rad in 1974 to as low as 0.1 rad in 1987. At this level, there is virtually no danger from periodic screening mammography.
b. The resolution of mammography has improved remarkably so that even

smaller tumors can be detected. Ultimately, it is the inability of heat detection methods to find these very early cancers that make these methods unsuitable for screening.

c. Since we completed our study using heat detection methods, breast ultrasound has been developed into a reliable additional test. As an adjunct to mammography, heat detection does not add enough additional information to be significant, but breast ultrasound is a very reliable adjunct to mammography because it can distinguish solid from cystic (fluid-filled) lesions.

In conclusion, the work that we did under the guidance of Professor Alan Barrett seemed very promising at the time. However, the advances in mammography, which made it safer by lowering the dosage and more reliable by increasing resolution, have made it an ideal screening test. Mammographic screening of large populations has been shown to decrease the mortality from breast cancer by at least 35%, and there are indications from long-term studies it will decrease cancer mortality by 50%.

Again, let me thank the Conference Organizing Committee for inviting me to speak and give my best wishes to Professor Barrett and his wife for a relaxing and happy retirement. I would also like to thank Philip Myers who was our colleague in this research.

REFERENCES

1. A.H. Barrett and P.C. Myers, Medical Appl. of Microwave Imaging (IEEE Press, New York, 1986), p. 41.
2. B.R. Rosen, M.S. Thesis, M.I.T., 1980.

PASSIVE MICROWAVE TECHNIQUES FOR STUDYING PLANETARY ATMOSPHERES

DAVID H. STAELIN
Massachusetts Institute of Technology, Cambridge, MA, U.S.A.

Alan Barrett's early enthusiasm for microwave spectroscopy and its applications has played a major role in the development of passive microwave techniques for studying planetary atmospheres. This seminal role began with the publication in 1958 of his paper "Spectral Lines in Radio Astronomy," in a special radio astronomy issue of the *Proceedings of the IRE.*[1] This "initial petal in the pond," so to speak, produced waves with profound consequences, which I shall now describe.

The first impact of Alan's interest in microwave spectroscopy was in the area of planetary astronomy, with Venus as his first important target. The first opportunity came very early in his career shortly after the important discovery made at the Naval Research Laboratory that the brightness temperature of Venus appeared to be about 600 K at 3-cm wavelength. Not only was this temperature unexpectedly high, but subsequent measurements revealed temperatures near 300–400 K at wavelengths shorter than 1 cm; these lower temperatures were in better agreement with the infrared data. One of the first hypotheses to explain this puzzle was that the hot radio emission was originating from the ionosphere or from micro-discharges between charged atmospheric particles. Alan, however, proposed that microwave absorption at shorter wavelengths could also explain the results, if only the surface temperature were near 600 K. It was his important observation that if the atmosphere of Venus consisted of carbon dioxide at very high pressure, not only could nonresonant absorption in this high-pressure carbon dioxide produce the needed short-wavelength absorption, but also it could contribute to a substantial greenhouse effect consistent with the extremely hot surface. These important suggestions first appeared in a brief letter in 1960 to the *Journal of Geophysics Research*,[2] which was followed by more complete calculations in 1961 in the *Astrophysical Journal.*[3] Figure 1 contains diagrams from these early papers and illustrates how he fitted models to the observations available at that time. His conclusion that the surface pressure of Venus could be 20 atmospheres or more had some impact on plans for space exploration and dealt a damaging blow to the public's perception of Venus as a lush jungle.

In response to this controversy, NASA undertook the Mariner-2 mission to Venus, built in part around a two-channel passive microwave experiment designed by Alan, Ed Lilley, and others. This experiment, which was designed, built, and launched in less time than most missions today require for approval, was sharply focused on resolving the controversy concerning the potential habitability of Venus. The instrument was designed to measure whether Venus was limb-brightened or limb-darkened at two wavelengths in the temperature transition region between 300 and 600 K. In particular, if the surface of Venus were cold, then the hot 3-cm radiation would have come from the ionosphere or perhaps atmospheric discharges, which would appear brighter near the limbs of the planet simply because the path length that the radiation must travel through the emission region is greater there. The opposite hypothesis called for a hot surface and cold atmosphere, which would exhibit lower brightness temperatures near the limb, called limb darkening, for the same reason – near the limb the brightness temperature approaches that of the upper atmosphere because of the longer effective path

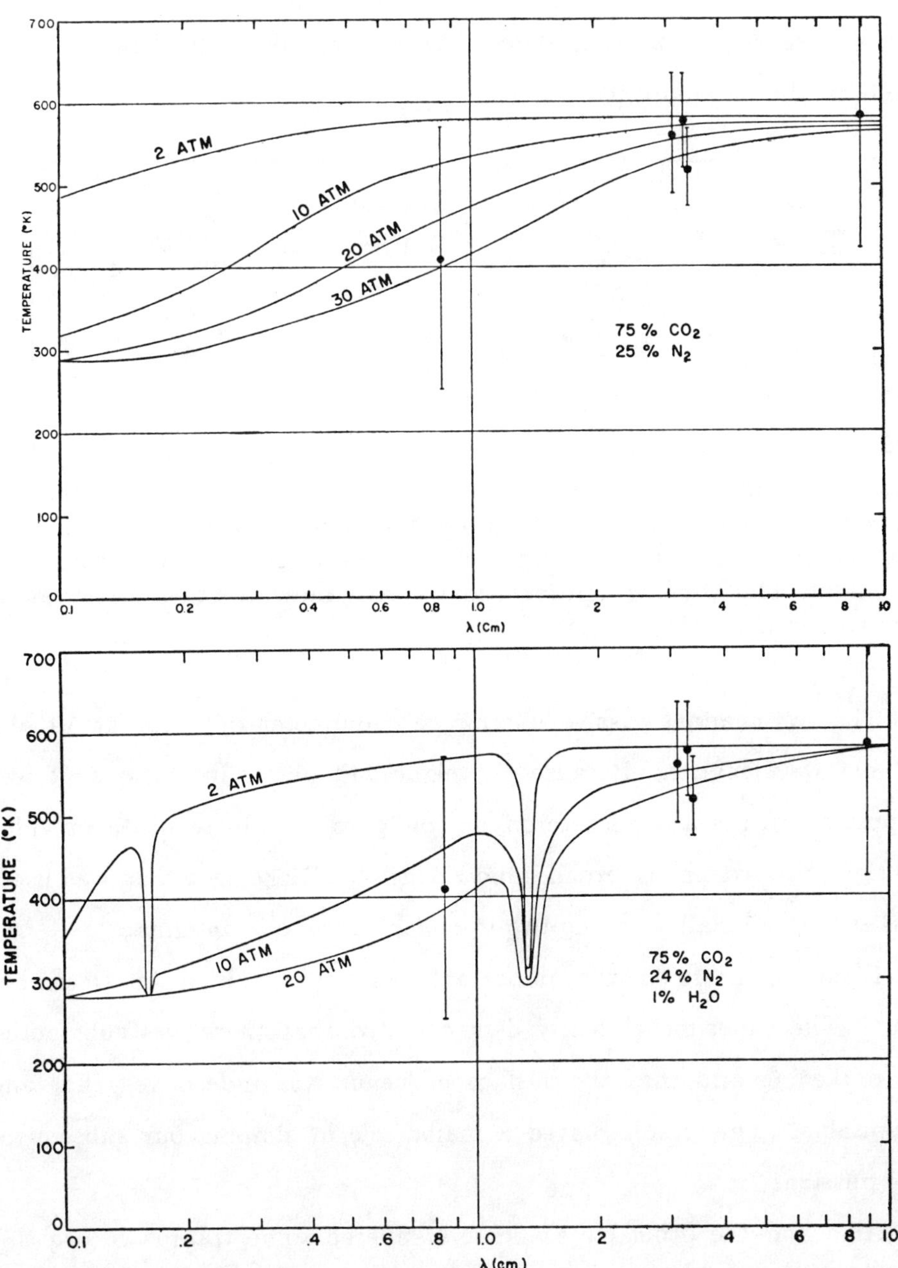

Figure 1. (top) Brightness temperature vs. wavelength calculated for a model atmosphere of 75 percent CO_2, 25 percent N_2, and surface pressures of 2, 10, 20, and 30 atm. (bottom) Brightness temperature vs. wavelength calculated for a model atmosphere of 75 percent CO_2, 24 percent N_2, 1 percent H_2O, and surface pressures of 2, 10, and 20 atm [from ref. 3].

lengths there. Figure 2, from *Sky and Telescope*,[4] illustrates how these two models can be distinguished.

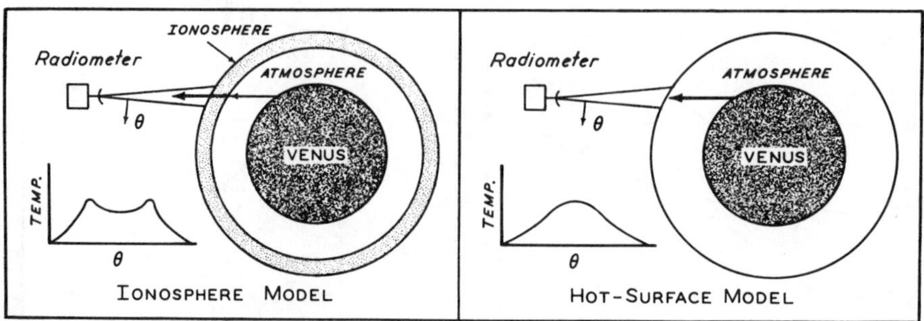

Figure 2. Two models of Venus. In each case, a small graph shows the results that would be expected from a high-resolution radiometer scan across the planetary disk. Limb brightening is shown at the left, and limb darkening at the right. The Mariner-2 observations clearly indicated the correctness of the model on the right. [From ref. 4.]

The two-channel passive microwave radiometer operated at 19 and 13.5 mm wavelengths. Because of weight and power limitations, it was not possible to use a superheterodyne configuration. Instead, direct video detection with extremely broad bandwidths and Dicke switching was used, together with a small scanning two-frequency parabolic antenna.

Figure 3 presents the important results, obtained in December 1962.[5,6] The experiment clearly demonstrated that there was substantial limb darkening and that the surface of Venus was indeed very hot and inhospitable. This result played a major role in shaping our subsequent space missions.

Alan and the other experimenters suffered some trauma before this result was obtained because of an unexpected malfunction in the instrument. Due to limitations in NASA's ability to point spacecraft accurately at that time, they had elected to use the raw radiometer data to control the antenna scanning mechanism. When the antenna sensed the high brightness

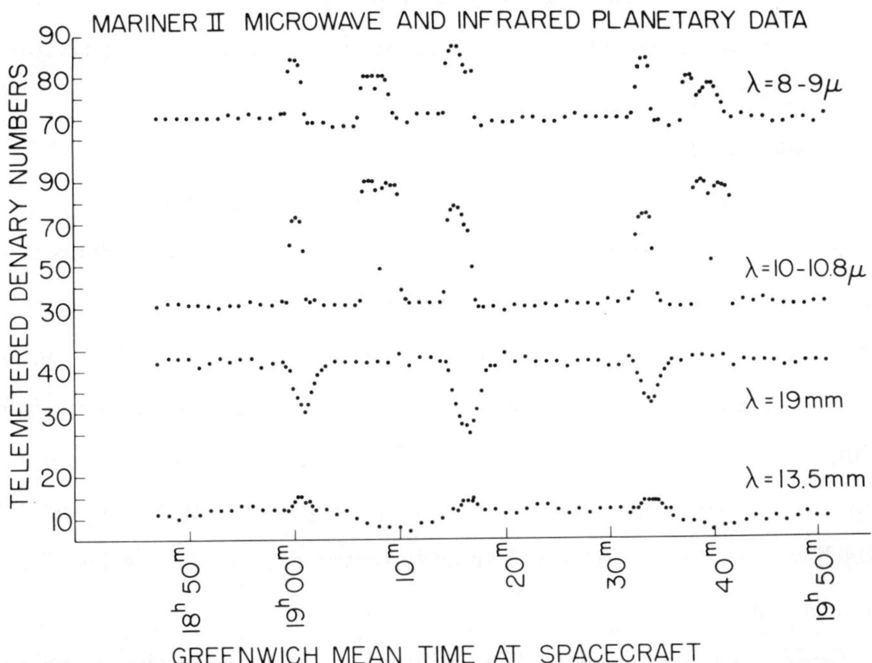

Figure 3. The three scans across Venus exhibit limb darkening at 19-mm wavelength and also near 10 μm.

temperature of the planet, it was supposed to rapidly scan in order to produce microwave brightness temperature images. The problem was that the radiometer electronics had drifted so as to invert the polarity of the output: cool had become hot and vice versa. However, fate was kind and due to a concatenation of engineering accidents the three good scans in Figure 3 were obtained, to everyone's relief.

The hot surface result was obtained principally from the 19-mm channel. Unfortunately, the same gain drift problem left the 13.5-mm channel, centered on a water vapor resonance, with a gain so low that no strong statements could be made about humidity, and the role of water vapor as a resonant absorber, in defining the microwave spectrum of Venus. The question of water vapor abundance was important, because the more resonant absorption contributed to the low brightness temperatures at short wavelengths, the lower was the surface pressure required for the additional nonresonant CO_2 absorption.

While Mariner-2 was being developed, Alan arrived at M.I.T. and started some of the students, staff, and myself on the development of a multichannel microwave radiometer that would measure the emission spectrum of Venus near the 13.5-mm water vapor resonance. This project, which was part of my thesis as Alan's first doctoral student, put an upper limit on water vapor abundance,[7] which we lowered further during the following inferior conjunction of Venus. These results made it clear that Venus was not only hot but it was also dry – the hopes for tropical growth suffered again. These limits have since been extended sharply downwards with a series of in-situ measurements.

Today we know that Al's original estimates of the atmospheric constituents of Venus were essentially correct. In-situ measurements of carbon dioxide suggest abundances close to 96% with only 4% nitrogen. The surface temperature has been found to be closer to 750 K, while the surface pressure is nearly 100 atmospheres. Although there might be 100 parts per million of water vapor and perhaps 180 per million of SO_2, another resonant absorber, the other trace constituents are even less abundant. The clouds appear to be highly concentrated sulfuric acid perhaps mixed with other constitutents and could, together with the SO_2, contribute some moderate portion of short wavelength microwave absorption.

Alan wrote an important paper speculating on other planetary microwave spectral lines.[8] However, the only detected mircrowave spectral line on Venus is carbon monoxide, which has also been studied on Mars. I understand that HCN has also been observed on Titan at 88 GHz. The reason so few microwave spectral lines have been studied in other planetary atmospheres, such as those of Jupiter and Venus, is that the lines are so pressure broadened that they are very difficult to observe. The atmospheric regions at lower pressures are often above the clouds where infrared techniques have been more successful because of the greater line strengths in this spectral region.

The planet Earth, however, is a much more promising target for passive microwave probing and was a major priority of Alan's when he arrived at M.I.T. Figure 4 presents the transmission spectrum of the terrestrial atmosphere up to 300 GHz and shows that both water vapor and oxygen resonances were quite strong and easy to observe. Weak resonances due to trace constituents are omitted here; their opacities are significant principally above 200 GHz, beyond the easy reach of radiometric technology available in the early 1960s.

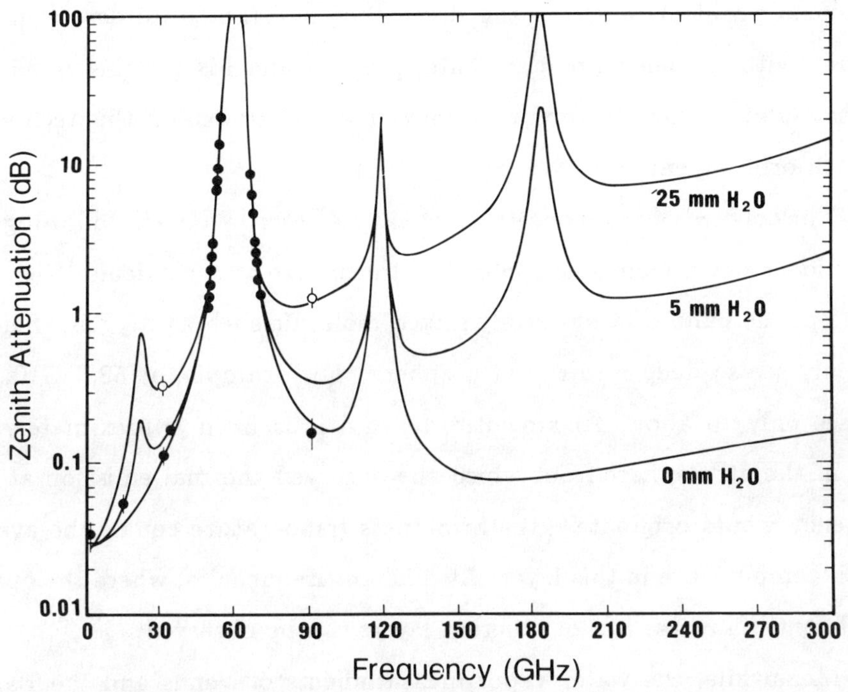

Figure 4. Zenith attenuation versus frequency for a model of the earth's atmosphere with three different values of precipitable water vapor.

The effort to study the microwave spectrum of the Earth began with two instrument development efforts and two theoretical studies. I began designing the multi-channel K-band radiometer needed for both Venus and terrestrial water vapor, while Alan Barrett, Bill Graham, Jack Barrett, Cosmo Papa, Jim Kuiper, and others began developing the 5-mm spectrometer for observing oxygen. Alan's motivation was not only to confirm directly the theoretical predictions of oxygen absorption, but also to study and exploit the suggestion made by Meeks and Lilley[9] that the atmospheric temperature profile could be sounded from space using the oxygen spectrum. Bill Lenoir's doctoral thesis involved development of this 5-mm radiometer and its flight from high altitude balloons. Small parabolic antennas looked upwards and downwards at 5-mm wavelength and generally confirmed the theoretical predictions.[10] These successes, marred only by occasional problems such as the descent of our tape recorder from high altitude without benefit of parachute, gave Al and his growing number of students and colleagues comfort in moving ahead to exploit this technology in earth-orbiting satellites.

The concept of these observations can be described in the following way. If one looks down from space into the atmosphere at a particular frequency in an opaque band of a uniformly mixed molecule such as oxygen, then one can only see so deep into the atmosphere. For example, at 58.5 GHz, one can see only to about 20 km altitude, and it is from approximately this level in the atmosphere from which the observed thermal emission at that frequency would originate – the brightness temperature equals the average kinetic temperature in this layer. At different frequencies, where the opacity is different, temperatures of different layers can be observed.

Meanwhile, the water vapor measurements of Venus and the Earth's atmosphere were under way at Lincoln Laboratory. By comparing the observed spectral shape near 1-cm wavelength with the shape predicted from humidity and temperature profiles measured using nearby radiosondes, it was discovered that not only could hourly variations in water vapor abundance be accurately tracked even when light clouds were present, but also some limited information about the water vapor distribution with altitude could be deduced from the known pressure broadening characteristics of the spectral line.[11]

Alan's first paper on terrestrial water vapor, written with Victor Chung, noted that the water vapor abundance observed in the stratosphere and lower mesosphere by high altitude probes could yield a moderately intense very narrow water vapor line at 13.5-mm wavelength.[12] Despite repeated efforts over many years, we were unable to observe this line in the terrestrial atmosphere. Alan's later students and others ultimately succeeded with the help of maser amplifiers and more advanced instrumentation.

The general interest in water vapor then shifted to its significance as a tool for terrestrial observations from earth-orbiting meteorological satellites. The way for this was paved not only by the water vapor observations that were part of my thesis but also by the much more extensive observations performed by Norman Gaut, a Ph.D. student of Barrett's from the Department of Meteorology. These early measurements confirmed not only that water vapor could be measured but also that the nonresonant absorption by clouds could be separately determined, even if only two frequencies were employed.[11]

Armed with these early data and growing troops of eager graduate students and staff, Alan turned again to NASA with missionary zeal and helped encourage them to make the development of passive microwave systems in earth orbit a research priority. He encouraged the efforts of Bill Nordberg to establish a passive microwave working group that included representatives from NASA and both the industrial and university communities. Out of this series of meetings, in which Alan played a leading technical role, two spacecraft experiments were eventually designed and successfully proposed, after a few false starts.

One spacecraft experiment involved the Nimbus-5 passive microwave spectrometer,[13] which was built by the M.I.T. group – myself, Al, Norm Gaut, Bill Lenoir, Phil Rosenkranz, and Joe Waters – plus Frank Barath, John Blinn, and Jim Johnston from JPL. This five-channel instrument (three channels on the wings of the 5-mm oxygen complex, one on the water vapor resonance, and one in the 8-mm window) was launched in 1972 and clearly demonstrated the ability of such systems to measure atmospheric temperature profiles, water vapor abundance, and precipitation, even in the presence of a wide variety of clouds.[13] The motivation for this research was the hope that advanced instrumentation could map such data over the entire globe every 12 hours (for one satellite) and could thus drive numerical weather prediction models. Most exciting to the meteorological community was the realization that data could be obtained over oceans where there were no radiosondes and a decreasing number of commissioned meteorological ships.

The second instrument that grew out of that same series of meetings that Bill Nordberg organized was a 19-GHz imager that was about the size of a card table and that mapped the earth with approximately 25-km resolution. The initial technical leadership for this project came from Pat Thaddeus, a microwave spectroscopist and student of Charlie Townes (Al's thesis supervisor), who was followed by Tom Wilheit, the Ph.D. student of Al's who continued Bill Lenoir's work.

During this period, Alan was turning his scientific attention more to radio astronomy, but his atmospheric program had been launched and was gathering a momentum of its own. Both the Air Force and NOAA launched a series of passive microwave sounders on operational meteorological satellites. Although these initial sounders were used principally as a supplement to the infrared sounders, the much greater insensitivity to clouds of the passive microwave systems, as well as their improved altitude resolution in the stratosphere and above, led to the decision by NOAA to make the next-generation passive microwave spectrometer the primary sounder on future operational meteorological satellites, starting in the early 1990s. These instruments will have 20 channels and exploit both the 60-GHz oxygen complex and the 22.2- and 183-GHz water vapor resonances. Although these will be operational instruments, their unique capabilities will also almost certainly reveal a variety of new scientific results. Another instrument is now being conceptually developed for use in geosynchronous orbit, operating near the 118-GHz oxygen complex and the 183-GHz water vapor resonance.

One of my students, Joe Waters, has extended Alan's interest in atmospheric spectroscopy to a whole range of trace constituents in the terrestrial stratosphere and mesosphere. Such trace constituents include H_2O, O_3, CO, CH_4, NO, N_2O, HNO_3, NO_2, ClO, and many more. Bill Wilson and Phil Schwartz, two radio astronomy Ph.D. students of Al's, have also studied carbon monoxide and other atmospheric molecules. In fact it is fair to say that the great majority of all work in this field has been done by Alan's first- or second-generation students and that the waves from Alan's "initial petal in the pond" have indeed traveled far and wide.

REFERENCES

1. A. H. Barrett, Proc. I.R.E., 46, 250 (1958).
2. A. H. Barrett, J. of Geophys. Res., 65, 1835 (1960).
3. A. H. Barrett, Astrophys. J., 133, 281 (1961).
4. A. H. Barrett and A. E. Lilley, Sky and Telescope, 25, 192 (1963).
5. F. T. Barath, A. H. Barrett, J. Copeland, D. E. Jones, and A. E. Lilley, Science, 139, 908 (1963).
6. F. T. Barath, A. H. Barrett, J. Copeland, D. E. Jones, and A. E. Lilley, Astron. J., 69, 49 (1964).
7. D. H. Staelin and A. H. Barrett, Astrophys. J., 144, 352 (1966).
8. A. H. Barrett, Mem. Soc. Roy. Sci. Liege, Cinq. Ser. 7, 197 (1962).
9. M. L. Meeks and A. E. Lilley, J. of Geophys. Res., 68, 1683 (1963).
10. A. H. Barrett, J. W. Kuiper, and W. B. Lenoir, J. of Geophys. Res., 71, 4723 (1966).
11. D. H. Staelin, J. of Geophys. Res., 71, 2875 (1966).
12. A. H. Barrett and V. K. Chung, J. of Geophys. Res., 67, 4259 (1962).
13. D. H. Staelin, A. H. Barrett, J. W. Waters, F. T. Barath, E. J. Johnston, P. W. Rosenkranz, N. E. Gaut, and W. B. Lenoir, Science, 182, 1339 (1973).

ALAN BARRETT'S CONTRIBUTIONS TO INTERSTELLAR SPECTROSCOPY

PATRICK THADDEUS
Harvard-Smithsonian Center for Astrophysics, Cambridge, MA

There are some people whose almost superstitious dread of retrospective occasions such as this was well summed up by the famous illustrator Charles Addams, when in one of his *New Yorker* cartoons he depicted a dreadful television show a generation ago with a name something like "This is Your Life." The announcer is saying to the victim, "*Now*, Mr. Barrett, for the girl you haven't seen since you were in the Navy thirty years ago...." The man has a worried expression on his face, and standing in the shadows is this girl with a grim expression, pulling an enormous pistol out of her purse. Although I can't vouch for Alan Barrett's career in the Navy, with respect to his subsequent activities as a radio astronomer, he is one of those who have least to fear from such scrutiny.

One way to begin a sketch of his many contributions to the study of the interstellar medium is to remind you of the kind of qualifying exam problem given to graduate students who aspired to a Ph.D. in astronomy ca. 1955-60, which also offers a fairly good illustration of the conventional wisdom confronting Al Barrett when he began research in astrophysics. The question had two parts and went something like this: What are two ways in which radio astronomy is always likely to remain inferior to optical astronomy? Answer: (1) Angular resolution: an antenna as big as a battleship is required to equal the resolution of the human eye, and (2) Spectral lines:

the 21 cm line is unique, so there will be no radio equivalent to the rich Fraunhofer spectrum of the sun and stars on which most of astrophysics rests. We are indebted to two MIT professors for major contributions in planting the petards that hoisted those flimsy generalities: Bernie Burke, for the development of radio interferometry which now exceeds in angular resolution the best optical telescopes by at least 3 orders of magnitude, and Alan H. Barrett, for unleashing the flood of radio molecular lines that has transformed the study of interstellar space and star formation.

Alan Barrett was well prepared to look for molecules in space, having done a thesis at Columbia University in the middle 1950s on microwave spectroscopy under the rigorous tutelage of Charlie Townes. I couldn't find the exact title of your thesis, Al, in the copy of your resume put into my hands by organizers of this meeting, so I telephoned the Pupin Physics Library, and the librarian on duty read it to me: "The Microwave Spectrum of InCl, InBr, InI, and GaCl." As a graduate student in the same group shortly afterward, I can testify that the education we received in microwave electronics and molecular spectroscopy, and in molecular physics generally, was sound preparation for radio astronomy and that the idea that what we were doing in the laboratory had possible astronomical applications was very much in the air, stimulated by Townes himself.

AHB and I overlapped at Columbia by about one year, but I knew him little then: he was writing his dissertation and I was mainly taking courses, but I distinctly recollect him coming into the laboratory occasionally with a worried, preoccupied expression—something I soon recognized as the standard demeanor of a Columbia physics student in the terminal agonies of his typically protracted thesis (a contemporary of ours actually spanned three decades in the course of his research—in the mathematical sense of the term). [AHB from audience: We senior graduate students didn't talk much to the juniors....] About the time I joined the group, or shortly after, another student arrived named Gerry Ehrenstein, to whom Townes assigned the job of measuring the ground-state lambda doublets of the hydroxyl radical OH. This surprised some of the intermediate students in Townes' large group, since Dousmanis and Sanders had done a rather thorough study of the radio spectrum of OH at Columbia only a short time before, until they realized that the main purpose of the new study was astronomical: to measure directly the frequencies of the most astronomically relevant transitions, so they could be sought in space with a radio telescope without a laborious hunt in frequency.

After the discovery of the 21 cm line of atomic hydrogen by Ewen and Purcell in 1952, Townes, Shklovsky, Barrett[1] (and possibly others) in the late 1950s considered the possibility of detecting microwave lines of simple molecules in the interstellar gas, and they pointed out that the lowest-lying lambda-doublet transitions of OH near 1600 MHz were among the most favorable candidates. On completing his Ph.D. at Columbia and taking up a postdoctoral position at the Naval Research Laboratory in Washington, D.C., AHB, in collaboration with Ed Lilley, as we heard from Townes last night, attempted to find those lines with the NRL 50 ft radio telescope, but failed, at least in part owing to the uncertainty in frequency (of order 5 MHz—not a small region to scan with the instrumentation then available). As the notes on the blackboard in the photo of Alan and Ed that we have seen clearly indicate, however, they did cover the correct frequencies in their search, so this initial failure seems essentially to have been a failure of sensitivity.

What was needed were exact laboratory frequencies, and these were provided by Ehrenstein, Townes, and Stevenson[2] in 1959. AHB did not return immediately to the OH problem when the new laboratory data were published, but when he did in 1963—having in the meantime arrived at MIT—he had the benefit of the new autocorrelation spectrometer which

Sandy Weinreb had developed there while a thesis student under the supervision of Jerome Wiesner. I believe the successful detection of OH at 1665 and 1667 MHz in absorption toward the well-known supernova remnant Cas A[3] was the first successful application of the new autocorrelator, which has of course become one of the crucially important devices of radio astronomy. [AHB: Sandy tried two other searches, but on both occasions failed.]

Although the discovery of OH had the great merit of breaking the monopoly of the 21 cm line as the only spectral transition in the radio astronomical band and so turned the thoughts of radio astronomers to molecules, it had its discouraging side as well. The amount of OH detected[4,5] was not large with respect to the three other simple molecules long known from optical observations to exist in the interstellar gas, CH, CH^+, and CN, whose abundances relative to atomic hydrogen were thought to be at most only about 10^{-6}. As a consequence, OH did not seem to contradict the widespread view that, owing to the corrosive action of starlight and other factors, only simple transient diatomics were likely to be found in the interstellar gas and that the prospect of finding larger stable molecules was poor. There was no indication in AHB's early OH observations that a significant fraction of the interstellar medium is molecular—as we now know it to be—or that the dense pockets of gas we now call molecular clouds even existed.

The first hint of really new things came a few years later, with the discovery by others of OH emission. Alan and his colleagues at MIT—including Bernie Burke, Jim Moran, and Alan Rogers—immediately took up the chase, and in a remarkably short period their crucial experiments demonstrated the existence of the first cosmic masers—point sources of polarized,[6,7] time-varying OH emission so small in angular size[8–10] that baselines of continental dimensions are required to resolve them. Today, the maser interpretation of the OH point sources is taken for granted, since no other interpretation comes close to explaining their astonishing properties, but it was a struggle whose outcome largely depended on the work just cited, which constitutes *in toto* one of the most dramatic and influential investigations in modern astrophysics. In 1971 AHB was co-recipient of the Rumford Medal for his contribution to this splendid work.

Another discovery that Alan made about this time in collaboration with Alan Rogers was the first step in what has come to be a fascinating direction of research with wide-reaching implications: the study of isotopic ratios in the interstellar gas. Let me remind you that in 1966 little was known about the abundances of isotopes beyond the solar system, and, in particular, about those in the interstellar gas almost nothing was known. As a consequence, there was no way of knowing whether the enormous table of isotopic abundances arrived at by studying the earth and meteorites represented a fair sample of the Galaxy or was instead the result of local fluctuations, with little general significance. Leaving aside deuterium (a special case because it readily fractionates) subsequent workers have found what Barrett and Rogers[11] found: interstellar isotopic ratios not too different from the terrestrial. This result, which to some investigators may have appeared disappointing, is of enormous general importance, because it *validates* the entire table of isotopic abundances, indicating that the record of nuclear evolution sampled by the earth and solar system is a fair sample of our Galaxy at large.

David Staelin has mentioned AHB's interest in studying water in the Venus atmosphere by means of its 22 GHz radiofrequency transition. Twenty years ago, almost no one imagined that this transition—the famous line that wrecked K-band radar during World War II—could be detected in astronomical sources, since it lies very far above the rotational ground state of the water molecule (about 650 K in units of temperature) and therefore will not be excited under "normal" interstellar conditions. As the discoverer of astronomical OH, and with the possibility of detecting this very H_2O line in planetary atmospheres much on his mind, AHB must have kicked himself when his canny thesis supervisor Charlie Townes, working with Jack Welch and others at the University of California, first discovered the spectacular point sources of 22 GHz maser emission that exist in dense "protostellar" condensations near certain galactic H II regions. This was the second cosmic maser, and it turned out to be a beauty. Alan and his students and colleagues subsequently made important early extensions of this work of Townes and Welch, particularly toward stars,[12–16] which I had best leave to Snyder to describe.

Years later, AHB and Phil Myers had the idea of looking for maser emission from vibrationally excited water.[17] Nothing was found, but the basic idea is sound and remains of considerable interest. Water is so stable and undoubtedly so abundant in many cool stars and other astronomical sources, and its rotational spectrum is so rich, that the idea is worth pursuing. Surprises may be in store when we are able to explore the infrared spectrum free from the obscuration of telluric water vapor.

AHB and his students wrote six papers in the period of 1971-77 on the methyl alcohol or methanol molecule,[18–23] and have demonstrated that, at least in certain sources, some of the lines of this organic molecule too are the result of a cosmic maser.

Collectively, these papers illustrate the high standards that have generally characterized research under AHB's direction. Time and again maser action has been invoked by less fastidious workers to explain things away: why line frequencies in claimed identifications of new molecules weren't quite right, why intensities didn't make sense, why results couldn't be reproduced, and so on. AHB would have none of that facile reasoning! He has made severe use of Occam's razor and refused to accept exotic interpretations of his observations until simpler ones could be excluded. This meant that maser action could be claimed only after polarization, time dependence, small source size, etc., had been rigorously demonstrated.

Finally, let me cite briefly a complicated series of investigations AHB and his students have undertaken in the last ten years or so, the use of known molecules to study a variety of objects: dark nebulae, Bok globules, giant molecular clouds, Herbig-Haro objects, H II regions, etc.[24–41] Because its well-known inversion transitions lie in a band where the antenna and receivers at Haystack are particularly effective, ammonia has been one of the most widely used molecules in these studies, but other molecules and facilities have been used as well. A dominant theme of this work has been the attempt to understand how modest stars like the sun form, and how star formation proceeds in molecular clouds under conditions simple enough to analyze. This research is an open-ended program, not a well-rounded, finished investigation easily summarized. Of the seeds planted by AHB, however, it is likely to prove one of the most fruitful and influential, owing to the central role of star formation in astrophysics and, for that matter, planetary sciences.

Various eminent physicists have been credited with the crack that someone's idea was "not even wrong," and Bohr once complained of one of Pauli's ideas that it wasn't crazy enough. Honest mistakes in science are not the worst offenses and can even be said to have their place—but it is still better not to make them. In rereading AHB's papers in radio astronomy, I have been struck at how few errors of reasoning or execution they contain, and how well they have stood the test of time. They are clear, to the point, and characterized by soundness, common sense, and proportion—the work of a craftsman and a professional. Their influence can be judged by the circumstance that not a few members of this audience now spend their days and nights exploring the ramifications of one or more of the discoveries of AHB. It has been a great pleasure, Alan, to have gone over this ground again, to have witnessed you at work, so to speak, over the last quarter of a century, and to have been invited to this well-merited celebration.

REFERENCES

1. A.H. Barrett, Proc. I.R.E., 46, 250 (1958).
2. G. Ehrenstein, C.H. Townes, and M.J. Stevenson, Phy. Rev. (Lett.), 3, 40 (1959).
3. S. Weinreb, A.H. Barrett, M.L. Meeks, and J.C. Henry, Nature, 200, 829 (1963).
4. A.H. Barrett, M.L. Meeks, and S. Weinreb, Nature, 202, 475 (1964).
5. A.H. Barrett, and A.E.E. Rogers, Nature, 204, 62 (1964).
6. S. Weinreb, M.L. Meeks, J.C. Carter, A.H. Barrett, and A.E.E. Rogers, Nature 208, 440 (1965).
7. A.H. Barrett, and A.E.E. Rogers, Nature, 210, 188 (1966).
8. J.M. Moran, A.H. Barrett, A.E.E. Rogers, B.F. Burke, B. Zuckerman, H. Penfield, and M.L. Meeks, Ap. J. (Lett.), 148, L69 (1967).
9. J.M. Moran, P.P. Crowther, B.F. Burke, A.H. Barrett, A.E.E. Rogers, J.A. Ball, J.C. Carter, and C.C. Bare, Science, 157, 676 (1967).
10. J.M. Moran, B.F. Burke, A.H. Barrett, J.A. Ball, J.C. Carter, and D.D. Cudaback, Ap. J. (Lett.), 152, L97 (1968).
11. A.E.E. Rogers, and A.H. Barrett, Ap. J., 151, 163 (1968).
12. M.L. Meeks, J.C. Carter, A.H. Barrett, P.R. Schwartz, J.W. Waters, and W.E. Brown III, Science, 165, 180 (1969).
13. D. Buhl, L.E. Snyder, P.R. Schwartz, and A.H. Barrett, Ap. J. (Lett.), 158, L97 (1969).
14. P.R. Schwartz, and A.H. Barrett, Ap. J. (Lett.), 159, L123 (1970).
15. D.F. Dickinson, K.P. Bechis, and A.H. Barrett, Ap. J., 180, 831 (1973).
16. P.R. Schwartz, P.M. Harvey, and A.H. Barrett, Ap. J., 187, 491 (1974).
17. P.C. Myers, and A.H. Barrett, Ap. J., 263, 716 (1982).
18. A.H. Barrett, P.R. Schwartz, and J.W. Waters, Ap. J. (Lett.), 168, L101 (1971).
19. A.H. Barrett, R.N. Martin, P.C. Myers and P.R. Schwartz, Ap. J. (Lett.), 178, L23 (1972).
20. A.H. Barrett, P.T.P. Ho, and R.N. Martin, Ap. J. (Lett.), 198, L119 (1975).
21. A.H. Barrett, J.M. Bologna, A.C. Cheung, M.F. Chui, P.T.P. Ho, K.J.

Johnston, R.N. Martin, D. Matsakis, J.M. Moran, and P.R. Schwartz, Ap. (Lett)., 18, 13 (1976).
22. R.B. Buxton, A.H. Barrett, P.T.P. Ho, and M.H. Schneps, Astr. J., 82, 985 (1977).
23. R.N. Martin, and A.H. Barrett, Ap. J. (Lett.), 202 L83 (1975).
24. A.H. Barrett, P.T.P. Ho, and P.C. Myers, Ap. J. (Lett.), 211, L39 (1977).
25. P.T.P. Ho, R.N. Martin, P.C. Myers, and A.H. Barrett, Ap. J. (Lett.), 215, L29 (1977).
26. R.N. Martin, and A.H. Barrett, Ap. J. (Supp.), 36, 1 (1978).
27. M.H. Schneps, R.N. Martin, P.T.P. Ho, and A.H. Barrett, Ap. J., 221, 124 (1978).
28. P.T.P. Ho, R.N. Martin, and A.H. Barrett, Ap. J. (Lett.), 221, L117 (1978).
29. P.T.P. Ho, and A.H. Barrett, M.N.R.A.S., 184, 93P (1978).
30. P.T.P. Ho, and A.H. Barrett, Ap. J. (Lett.), 224, L23 (1978).
31. M.H. Schneps, P.T.P. Ho, A.H. Barrett, R.B. Buxton, and P.C. Myers, Ap. J., 225, 808 (1978).
32. P.T.P. Ho, A.H. Barrett, P.C. Myers, D.N. Matsakis, A.C. Cheung, M.F. Chui, C.H. Townes, and K.S. Yngvesson, Ap. J., 234, 912 (1979).
33. P.T.P. Ho, and A.H. Barrett, Ap. J., 237, 38 (1980).
34. M.H. Schneps, P.T.P. Ho, and A.H. Barrett, Ap. J., 240, 84 (1980).
35. M.H. Schneps, A.D. Haschick, E.L. Wright, and A.H. Barrett, Ap. J., 243, 184 (1981).
36. P.T.P. Ho, R.N. Martin, and A.H. Barrett, Ap. J., 246, 761 (1981).
37. J.M. Jackson, J.T. Armstrong, and A.H. Barrett, Ap. J., 280, 608 (1984).
38. J. Stutzki, J.M. Jackson, M. Olberg, A.H. Barrett, and G. Winnewisser, Astr. Ap., 139, 258 (1984).
39. J.T. Armstrong, and A.H. Barrett, Ap. J. (Supp.), 57, 535 (1985).
40. J.T. Armstrong, P.T.P. Ho, and A.H. Barrett, Ap. J., 288, 159 (1985).
41. P.T.P. Ho, J.M. Jackson, A.H. Barrett, and J.T. Armstrong, Ap. J., 288, 575 (1985).

THE SCIENTIFIC CONTRIBUTIONS OF ALAN H. BARRETT TO THE STUDY OF CIRCUMSTELLAR MOLECULES

LEWIS E. SNYDER
Astronomy Department, University of Illinois, Urbana, Illinois

Abstract The scientific contributions of Alan H. Barrett to the study of circumstellar molecules are discussed in detail. Key papers by Barrett and his students and collaborators are shown to have contributed to many of the subsequent discoveries and measurements which have shaped this field of study.

INTRODUCTION

In the large majority of cases, the circumstellar molecules studied by radio astronomers belong to the shells surrounding red giants or supergiants with spectral types of late M or N. Most are pulsating variables of the Mira, semiregular, or irregular class. It is generally believed that these objects are on the asymptotic giant branch (AGB) of the Hertzsprung-Russell diagram, which denotes an evolutionary phase characterized by helium and hydrogen shell burning. A few stars are much hotter than red giants, with spectral types ranging from B to F. Their evolutionary states are unknown, but the similarity of the spectra from their circumstellar shells with the spectra of red giants suggests that they may have undergone a recent, rapid evolution off the asymptotic giant branch. In modern astronomy, attention was drawn to evolved objects by the infrared sky surveys of the late 1960's and early 1970's of, for example, Neugebauer and Leighton[1], and Price and Walker.[2] Many well-known long period variables and a large number of similar objects never before seen at optical wavelengths appeared as strong infrared sources in these surveys. The infrared flux was interpreted as the re-emission of radiation by dust grains heated by stellar flux. Spectrophotometry by Woolf and Ney[3] and by Merrill and Stein[4] subsequently revealed infrared absorption bands characteristic of silicate grains. The first discoveries of radio molecular line radiation from these objects were the detection of OH maser emission by Wilson and Barrett[5] in 1968 and H_2O maser emission in 1969 by Knowles, Mayer, Cheung, Rank, and Townes[6], and in 1970 by Schwartz and Barrett.[7] A few years later, SiO maser emission, previously observed only in Orion[8], was discovered coming from these objects by Snyder and Buhl working with Kaifu[9,10] on stellar surveys. Each of these species, OH, H_2O, and SiO, emanates from a different region of the circumstellar shell and, as a result, each has provided unique information about shell properties. Of the circumstellar sources known to support one or more of this famous oxygen compound trilogy, more than 350 have OH, more than 150 have detectable H_2O, and at least 150 have detectable SiO. Most of these circumstellar objects are believed to be oxygen-rich sources ($[C]/[O] \leq 1$) with only a few exceptions such as the S-type star χ Cygni where $[C]/[O] \approx 1$. In addition to the observed maser lines, several molecules have been observed in thermal or nonmasering emission. Nonmasering SiO in the vibrational ground state was found in 1975 by Buhl, Snyder, Lovas, and Johnson.[11] Emission from CO was observed in oxygen-rich objects in 1977 by Zuckerman, Palmer, Morris, Turner, Gilra, Bowers, and Gilmore.[12] It has been found that the strongest nonmaser molecular emission often arises from the so-called carbon-rich sources or stars where $[C]/[O] \geq 1$. Carbon-rich sources are characterized by strong emission from CO and, in some cases, many other molecules. Although there are fewer carbon-rich sources than oxygen-rich sources, one carbon-rich star, IRC+10216, has become very famous and has been mentioned many times during this meeting. First discovered in CO emission by Solomon, Jefferts, Penzias, and Wilson[13], currently it is known to possess several dozen molecules in its circumstellar shell and generally it is believed that many of these molecules may exist with detectable abundances in numerous other circumstellar sources.

To fully appreciate how this very successful area of research got started, it is necessary to examine some key papers written by Alan Barrett, his students, and his collaborators.

SOME KEY PAPERS

The literature contains at least six key papers by Alan Barrett, his students, and his collaborators which were very instrumental in shaping the research which is ongoing today on circumstellar molecules. These are listed below with the year of publication and an asterisk by the name of each Barrett graduate student:

1. "Discovery of Hydroxyl Radio Emission from Infrared Stars", by W. J. Wilson* and A. H. Barrett (1968).[5]
2. "Observations of Water-Vapor Emission Associated with Infrared Stars", by P. R. Schwartz* and A. H. Barrett (1970).[7]
3. "OH Radio Emission Associated with Infrared Stars", by W. J. Wilson*, A. H. Barrett, and J. M. Moran* (1970).[14]
4. "Characteristics of OH Emission from Infrared Stars", by W. J. Wilson* and A. H. Barrett (1972).[15]
5. "New H_2O Sources Associated with Infrared Stars", by D. F. Dickinson, K. P. Bechis, and A. H. Barrett (1973).[16]
6. "Time Variation of the H_2O Maser and Infrared Continuum in Late-Type Stars", by P. R. Schwartz*, P. M. Harvey, and A. H. Barrett (1974).[17]

In paper 1, Wilson and Barrett reported the discovery of OH maser stars but the observations were tedious. They searched 20 stars and found only 4 that were OH emitters: NML Cyg, CIT 3, CIT 7, and NML Tau. It turned out that there were many interesting results that came from this discovery. For example, they found that

1612 MHz emission was predominant with no circular or linear polarization, but 1665 MHz was weaker, and predominantly left circularly polarized. The polarization properties of circumstellar masers are still an active area of study in both theoretical and observational astronomy. Perhaps the most significant aspect of this paper from the standpoint of an observer was the detection of the first of many of the famous twin-peaked 1612 MHz OH maser spectra, i.e. maser emission concentrated in two distinct velocity ranges (sometimes called "horned" or "rabbit-eared" spectra), which were to become the observer's generic signpost for identifying circumstellar OH maser emission. Their 1612 MHz spectrum of NML Cyg taken with left circular polarization is shown in Figure 1. The 1665 MHz spectrum of NML Cyg showing both the left and right

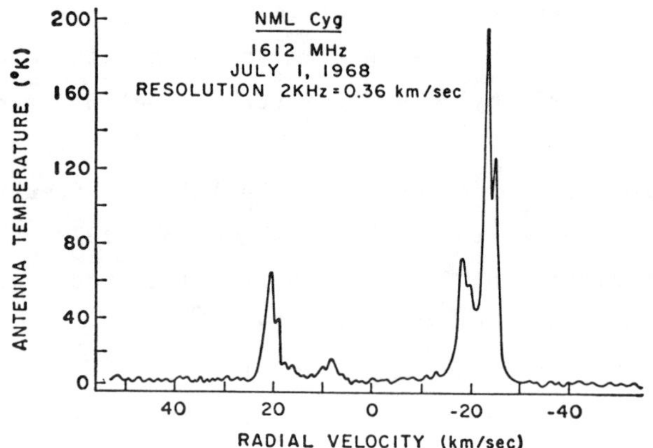

FIGURE 1 The 1612 MHz spectrum of NML Cyg taken with left circular polarization. Note that the emission is concentrated in two velocity ranges. This distinctive shape of the emission line envelope became a guide for quickly identifying circumstellar OH maser emission. Both Figures 1 and 2 were reproduced from reference 5 with permission of the authors and the American Association for the Advancement of Science.

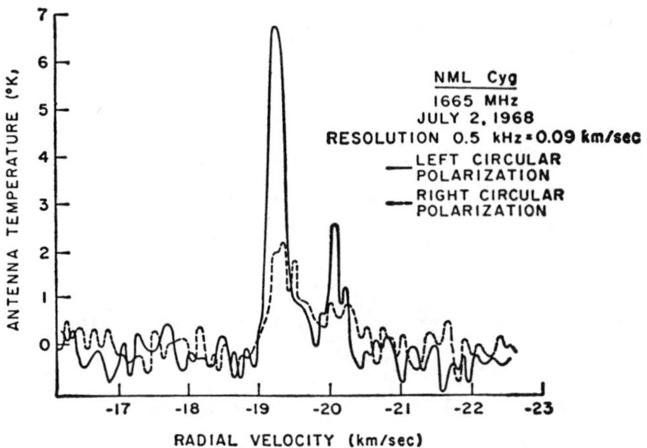

FIGURE 2 The 1665 MHz spectrum of NML Cyg showing both the left (solid) and right (dotted) circular polarization. Copyright 1968 by the American Association for the Advancement of Science

circular polarization is shown in Figure 2. In paper 1, Wilson and Barrett pointed out that the distinctive shape of the OH emission spectrum suggested that the circumstellar shell was expanding, contracting, or rotating. At the time, the only detailed theoretical treatment which predicted many of the observed circumstellar OH maser properties reported in paper 1 was the infrared pumping theory developed by Litvak[18] to explain interstellar OH masers (now, after many years, several different theoretical approaches, and many different observations, it is believed that expansion is the correct mode which explains most of the circumstellar shell observations). Wilson and Barrett also pointed out that the strength of its OH emission made NML Cyg an excellent candidate for VLBI. In retrospect, perhaps one of the most remarkable legacies of paper 1 is that it jolted many stellar astronomers into beginning to recognize the vast potential of radio spectroscopy for stellar studies (for example, see the discussion by Wallerstein[19]).

Paper 2, published in 1970 by Schwartz and Barrett, was instrumental in helping to establish H_2O maser stars as a class of astronomical objects. Schwartz and Barrett searched 134 IR stars for H_2O and found 3 new sources: NML Cyg, U Her, and W Hya. The H_2O maser from VY CMa had been reported previously by Knowles et al.[9] in 1969 and the R Aql H_2O maser was reported in 1970 by Turner, Buhl, Churchwell, Mezger, and Snyder[20], so paper 2 brought the total number of known H_2O maser stars to 5.

In paper 3, Wilson, Barrett, and Moran reported that 60 infrared stars had been searched for 1612 MHz OH maser emission and 7 had been detected. Thus paper 3 brought the total number of known OH/IR stars to 9 (excluding Orion). This paper also gave the results from the first measurements ever made with a very-long-baseline interferometer (VLBI) between Green Bank and Haystack on the 1612 MHz OH star NML Cyg. As a result of the data reported in paper 3, Wilson, Barrett, and Moran were able to identify OH/IR sources with red-giant or supergiant stars and correctly predict that the OH maser emission would become an important tool for calculating mass loss from red giants.

Papers 4, 5, and 6 were concerned with establishing the characteristics of OH and H_2O maser emission from infrared stars. In 1972, Wilson and Barrett used paper 4 to report that they searched 456 IR stars and 18 other objects and found OH emission from 25 IR stars. These results brought the OH/IR star grand total for 1972 to 29 OH/IR sources plus 8 other OH sources with spectra that "looked" like OH/IR stars. Dickinson, Bechis, and Barrett used Haystack to search 138 infrared stars from the Caltech 2-μ survey and 50 other objects (including all known OH/IR stars) and found 8 new H_2O sources. When they reported their results in paper 5, this brought the total number of identified H_2O/IR stars for 1973 to 19! With paper 6 in 1974, Schwartz, Harvey, and Barrett showed that the time variations in the H_2O maser emission from late type stars were periodic.

WHAT WAS KNOWN IN THE MID-1970'S

The work that had begun in 1968 with the Wilson and Barrett detection report of hydroxyl radio emission from infrared stars had made remarkable progress by the mid-1970's when the clarification of the structure and physical processes in these objects began to emerge. An almost complete list of what was known by the mid-1970's includes the following:

1. OH/H_2O/IR stars had been found to be typically of late M type, often (but not always) M6 through M8.

2. H_2O emission from circumstellar shells was surmised to be a maser but VY CMa was the only infrared star reported to have been measured by means of a series of experiments using a very-long-baseline interferometer (VLBI) in H_2O. The VLBI results of Burke, Papa, Papadopoulos, Schwartz, Knowles, Sullivan, Meeks, and Moran[21] showed that VY CMa had a brightness temperature of 10^{11} - 10^{12} K in the 6_{16}-5_{23} rotational transition of H_2O at 1.35 cm wavelength, which is far in excess of any physical temperature associated with the source.

3. Almost all H_2O stars were also OH stars, and in velocity space the H_2O emission velocities were generally bounded by the OH emission velocities.

4. Many OH/H_2O/IR sources were also known variable stars, such as Miras or semiregulars, at infrared or visual wavelengths, which suggested pumping by an infrared transition.

5. The expanding shell model had given a simple theoretical explanation for the distinctive twin-peaked shape of the OH emission line envelope. In this model the two OH emission complexes originate from opposite sides of an expanding shell so that their velocity separation is twice the shell expansion velocity. The strongest masers should be situated near the line of sight through the center of the shell, along which the gain paths may be the longest. The basic characteristics of the expanding shell model were presented in papers by Goldreich and

Scoville[22], by Elitzur, Goldreich, and Scoville[23], and by Kwok.[24]

Given the above list, it is clear that many of the accepted ideas about the structure and nature of circumstellar shells were in place, but not necessarily well-confirmed, in the mid-1970's. As will be discussed in the next section, the particularly persuasive observational results which support the expanding shell model and use the OH twin-peak emission symmetry to determine the stellar radial velocity were beginning to be developed and accepted.

SUBSEQUENT DISCOVERIES AND MEASUREMENTS

Subsequent discoveries and measurements have refined the ideas of the mid-1970's into the current thinking about the basic characteristics of evolved objects. Current research topics of interest such as the molecular composition, expansion velocity, diameter, magnetic field, and structure of circumstellar envelopes and the distance, systemic velocity and the mass-loss rate of the central star will be discussed below as they relate to the earlier works of Alan Barrett, his students, and his collaborators.

Other Circumstellar Masers

The SiO maser, the third member of the famous oxygen compound circumstellar maser trilogy of OH, H_2O, and SiO, partially owes its identification to the earlier circumstellar OH maser research of Wilson and Barrett. In 1973 December, Snyder and Buhl were observing Orion with the NRAO 36-ft telescope in order to study the region around 89.189 GHz for satellites of the X-ogen line (X-ogen is now known to be HCO^+) when a surprisingly strong group of unidentified emission lines was detected in the lower sideband of the receiver at 86.2 GHz.[8] As shown in the 250 kHz resolution spectrum illustrated in Figure 3, the emission was concentrated in two velocity ranges or clusters. When examined with 100 kHz spectral resolution, seven distinct lines could be seen: three were

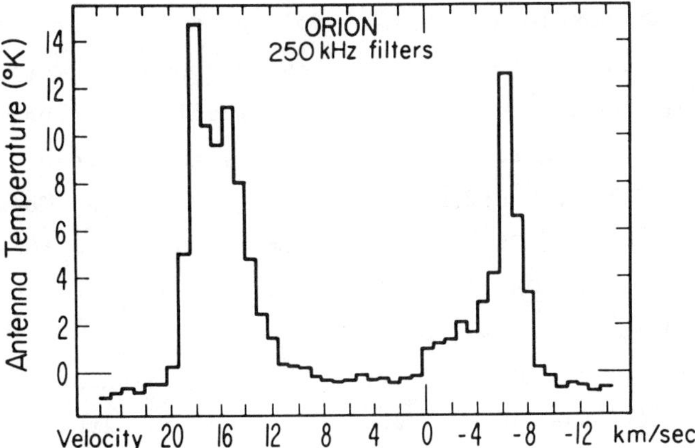

FIGURE 3 The unresolved, unidentified emission spectrum from the central region of the Orion Nebula molecular cloud which was the first detection of the SiO maser. This spectrum was obtained with 250 kHz filter spacing during 1973 December 19 on the NRAO 36-ft telescope. The radial velocity (with respect to the local standard of rest) is calculated for a rest frequency of 86,243.4 MHz. Note the remarkable resemblance to the Wilson and Barrett 1612 MHz OH circumstellar maser spectrum shown in Figure 1.

in the high velocity cluster and four were in the low velocity cluster*. To make a long story short, two possibilities emerged for the identification. The first possibility was that seven new molecular lines, each with its own astronomical rest frequency, had been detected. Although this possibility encompassed everything

---

*It should be noted that there was absolutely no uncertainty about which sideband contained the emission lines because the local oscillator frequency was carefully shifted several times. As far as I have been able to determine, most radio observers who know enough to use this clever frequency identification technique either learned it from Alan Barrett or learned it from somebody else who had learned it from Alan Barrett.

from multiple hyperfine transitions of one or two molecular species to seven different species, no likely molecular candidates with transitions matching the seven calculated astronomical rest frequencies could be found. The second possibility was that the two emission clusters in Figure 3 were in reality generated by a single molecular transition with different Doppler velocities. Then the two emission clusters in Figure 3 would be analogous to the distinctive twin-peaked OH circumstellar maser emission clusters originally found by Wilson and Barrett[5] (for example,see Figure 1). As discussed by Snyder and Buhl[8], this line of reasoning led to an astronomically determined rest frequency which was within 0.2 MHz of the 86,243.28 MHz rest frequency of the J=2-1 transition of the first vibrationally excited state of SiO. Thus the earlier groundwork on circumstellar OH masers contributed heavily to the later correct identification of the SiO maser.

Until recently, OH, H_2O, and SiO were the only circumstellar masers known and these strong masers were almost exclusively found in oxygen-rich stars. However, with the development of more sensitive telescopes and equipment, weak maser emission from carbon-rich circumstellar envelopes has been found. For example, weak SiS maser emission has been reported by Grasshoff, Tiemann, and Henkel[25], Rieu, Bujarrabal, Olofsson, Johansson, and Turner[26], and Sahai, Wootten, and Clegg[27]. Weak CO maser emission possibly has been detected by Zuckerman and Dyck[28]. Only this year, the first detection of a relatively strong (≈75 Jy) HCN maser at 89,087.90 MHz (the rest frequency of the vibrationally excited (0,2°,0) J=1-0 transition of HCN) was reported by Guilloteau, Omont and Lucas[29]. They found this HCN maser in the carbon-rich envelope of CIT 6 (RW LMi) with intensity comparable to but weaker than SiO masers. Thus it appears that the detection of circumstellar masers, a research project started 20 years ago by Alan Barrett and his students, appears to be entering a productive new phase with the detection of masers from the envelopes of carbon stars.

Stellar Radial Velocities

The distinctive twin-peaked shape of the OH emission line envelope which was originally found by Wilson and Barrett turned out to be a feature of numerous evolved stars, as shown in Figure 4. The OH

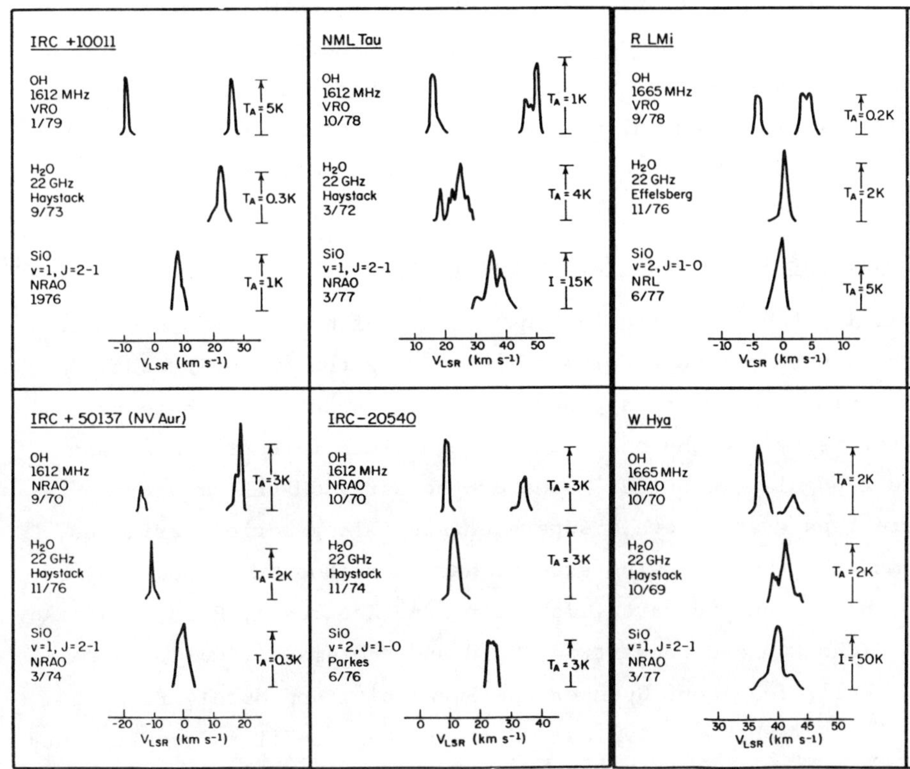

FIGURE 4 Representative OH, H_2O, and SiO maser emission spectra for six Miras and OH/IR stars. Note the distinctive twin-peaked shape of the OH emission line envelope of each star.

envelope became an important astrophysical tool when it was shown by Reid[30], Reid and Dickinson[31], and Dickinson, Reid, Morris, and Redman[32] that the stellar radial velocity of a typical long-period variable star lies between the optical absorption and emission lines near the midpoint of the OH radial velocity pattern. This work changed the traditional view, which had been held for over 35 years, that the stellar radial velocity was associated with the velocity of the optical absorption lines. It also presented the first strong observational argument for the expanding shell model of circumstellar envelopes rather than the shock model because it was more likely that OH maser emission originates in the front and back halves of an expanding shell of circumstellar material rather than from shock waves which give rise to blueshifted OH and from more distant quiescent gas which supports OH emission near the absorption line and stellar radial velocity.

Circumstellar Shell Sizes of OH Maser Stars

Determining the circumstellar shell sizes of maser stars could be viewed as part of the problem of confirming the expanding shell model. This problem was approached via two separate techniques: interferometry and the phase lag method. Interferometry has been the most popular technique. Some of the early OH VLBI measurements of late-type stars, both M supergiants and long-period variables, were made by the following groups: Masheder, Booth, and Davies[33]; Moran, Ball, Yen, Schwartz, Johnston, and Knowles[34]; Reid, Muhleman, Moran, Johnston and Schwartz[35]; Reid and Muhleman[36]; Reid, Moran, Leach, Ball, Johnston, Spencer and Swenson[37]; and Mutel, Fix, Benson, and Webber.[38] Typical OH maser shell radii for supergiants were found to be $\approx 10^{17}$ cm for 1612 MHz emission and apparently somewhat less for the main line emission. Typical Mira variables stars were found to have 1612 MHz radii of $\geq 3 \times 10^{15}$ cm. Early H_2O VLBI measurements were made by Rosen, Moran, Reid, Walker, Burke, Johnston, and Spencer[39] and by Spencer, Johnston, Moran, Reid, and Walker.[40] They found that typical late-type stars and the supergiant VY CMa have an H_2O masering region with radius $\approx 10^{15}$ cm. The compactness of the H_2O maser distribution relative to the 1612 MHz OH maser distribution in the envelope of VX Sgr is shown

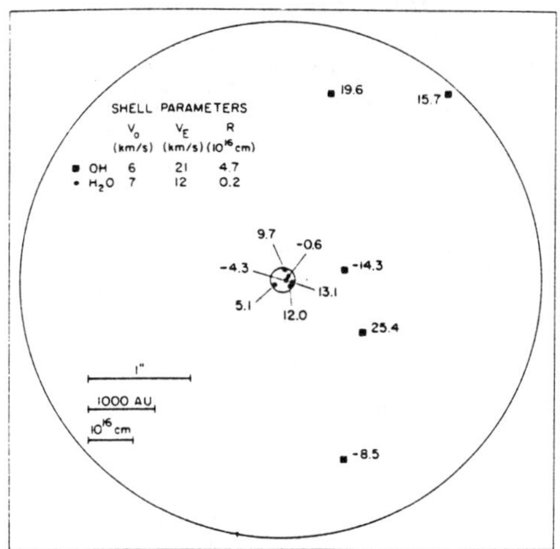

FIGURE 5 The 1612 MHz OH masers and H_2O masers located in the circumstellar envelope of VX Sgr. Note the compactness of the distribution of H_2O masers relative to the distribution of 1612 MHz OH masers. This figure was reproduced from reference 41 with permission of the authors.

in Figure 5, taken from Moran, Lichten, Reid, Huguenin and Predmore[41]. To complete the picture, Moran, Ball, Predmore, Lane, Huguenin, Reid, and Hansen[42] performed the first successful SiO VLBI measurements of evolved stars and found the radius of the masering region to be $\approx 10^{14}$ cm for the Mira variable R Cas and $\approx 10^{15}$ cm for VY CMa. Probably the definitive H_2O and SiO interferometry of circumstellar shells will be done after the VLBA network is completed, but the partial picture that is available now shows that the H_2O and SiO masers are excited much closer to the central star than the OH masers, just as would be expected from an examination of the excitation energy required for each masering transition. This would also help explain the relative compactness of the H_2O and SiO velocity distributions relative to the OH velocity distribution for each star in Figure 4.

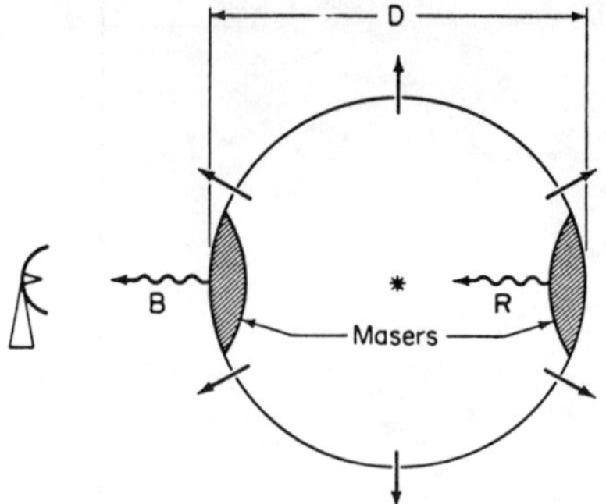

(a) Expanding Circumstellar Shell

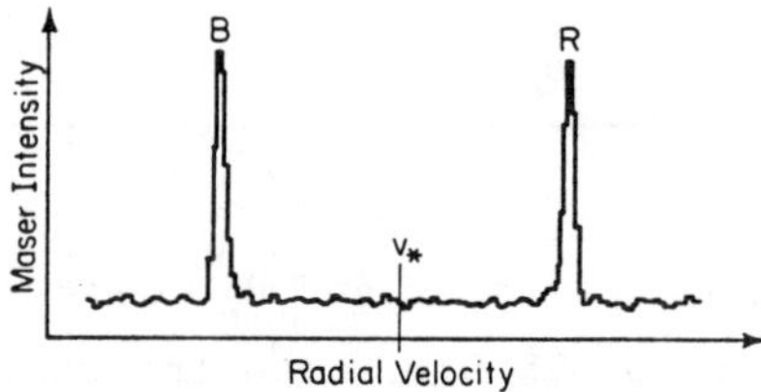

(b) Blue and Redshifted Maser Emission

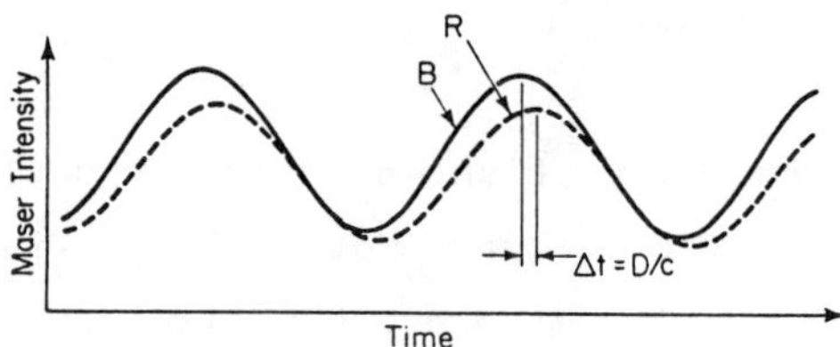

(c) Phase Relation between Blue and Redshifted Emission

FIGURE 6 Some of the details of the phase lag method for determining the linear diameter of circumstellar shells. This figure was reproduced from reference 43 with permission of the author.

As illustrated in Figure 6, the phase lag method is based on the expanding shell model which in turn was inspired by the distinctive twin-peaked shape of the OH emission line envelope. Since the two emission complexes in the spherically expanding shell model arise from opposite sides of the circumstellar shell (Figure 6a), the strongest maser emission is expected along the line-of-sight through the star where the velocity gradients are smallest and gain paths the longest. As shown in Figure 6b, the stellar radial velocity is midway between the two OH emission complexes and the expansion velocity of the circumstellar material is given by half of the complex separation. If the OH maser emission intensity in the shell varies smoothly and periodically because of variations in the pumping flux from the central star, then at the position of the star the intensity of the OH masers located in the front and back of the shell would be exactly in phase. However, as illustrated in Figure 6c an observer on Earth would see the OH maser emission from the back of the shell (the red-shifted emission) lag behind the emission from the front of the shell (the blue-shifted emission) by phase lag Δt, which is the light travel time across the shell. Hence the linear shell diameter would be given by $D = \Delta t \cdot c$, where c is the speed of light. Through careful monitoring, it should be possible to determine the linear shell diameters of circumstellar OH maser shells. Inspired by the various expanding shell models mentioned earlier (references 22-24) and by the successes of the early VLBI measurements (for example, references 33-38), Elitzur urged various radio observers to try the phase lag method. Several attempts by different groups were made before the first statistically significant measurement was obtained with a high density of data points over a stellar cycle. Jewell, Webber, and Snyder[44] measured the line-of-sight diameter of the 1612 MHz circumstellar shell of IRC+10011** and obtained $(6.6 \pm 1.4) \times 10^{16}$ cm. At the present time, Herman and Habing[45] have used the 25 m Dwingeloo telescope to monitor the phase lags of 90 maser stars.

**IRC+10011 is of course CIT 3 which was originally found to be an OH maser by Wilson and Barrett.

Distances To Maser Stars

For objects such as OH 26.5+0.6 (also AFGL 2205 and IRAS 18348-0526) which agree well with the ideal spherically expanding shell model, shell diameters determined by the phase lag method may be combined with two-dimensional interferometer maps of the emitting regions to provide a three-dimensional view of the circumstellar shell and to estimate the distance to the source. Herman and Habing[45] have combined their phase lag measurements with both VLA and MERLIN maps to determine distances to 16 stars so far. One of the goals of their program is to measure enough stellar distances to determine the scale length of the galaxy, either directly from OH maser sources in the galactic center or indirectly from OH maser sources at the tangential point.

Mass Loss from Maser Stars

The mass loss rates in the asymptotic red giant phase are important for both the stellar evolution models discussed by Iben and Renzini[46] and for the planetary nebulae formation models discussed by Kwok[47]. Usually measurements of mass loss from evolved stars utilize CO emission spectra from carbon-rich sources (for example, see Knapp and Morris[48]). However, in recent years OH maser spectra from OH/IR objects have also produced extremely interesting mass loss results, which suggest that their recycling contribution to the interstellar medium has been underestimated. Netzer and Knapp[49] have combined H_2O photodissociation and radiative transfer models with observations of OH maser shell radii and wind outflow velocities to derive mass loss rates for 42 stars. They find that the total mass return to the galaxy from luminous OH/IR stars is ≈ 0.2 solar masses per year. To put the mass loss from evolved stars in perspective, Bowers[50] estimates that the contribution from merely the long-period, oxygen-rich, Mira-type variable stars is at least an astounding 35% of all material recycled to the interstellar medium!

Mapping the Gravitational Potential Near the Galactic Nucleus

Winnberg, Baud, Matthews, Habing, and Olnon[51] have initiated a VLA search for OH/IR stars close to the galactic center. Their goal is to detect a large enough number of stars to allow the construction of high resolution (≤10 pc) maps of the gravitational potential close to the galactic nucleus. Their method would be free of the systematic uncertainties which occurred with earlier methods.

Recent Maser Polarization Studies

The final topic in this discussion, maser polarization studies, was selected because it is very contemporary even though the first OH maser star polarization measurements were initiated almost 20 years ago in papers 1,3 and 4 by Alan Barrett and his students. With the exception of only U Ori and IRC+10420, until recently the 1612 MHz OH circumstellar maser emission was commonly thought to be unpolarized. This was a bit puzzling because it has been known ever since the earliest maser star surveys (see the discussion by Wilson and Barrett in paper 4, for example) that the supergiant OH/IR stars exhibit significant main-line circular polarization such as that shown in Figure 2 for the 1665 MHz OH emission from NML Cyg. New 1612 MHz measurements of the supergiants VY CMa, VX Sgr, IRC+10420, and NML Cyg by Cohen, Downs, Emerson, Grimm, Gulkis, Stevens, and Tarter[52] that were made with a frequency resolution of 300 Hz (higher than ever previously available) have revealed many narrow components in the 1612 MHz spectra which have significant degrees of circular polarization. Based on the earlier work of Davies[53] and on the work of Deguchi and Watson[54], Cohen *et al.*[52] are able to assign magnetic field strengths of ≈ 1 mG to the 1612 MHz maser regions. Clearly this new evidence for relatively strong magnetic fields in the outer envelopes of supergiants will revitalize studies of magnetic fields in circumstellar envelopes as polarization observations with both high frequency and spatial resolution are extended to other types of OH/IR stars.

CONCLUSIONS

It is evident that the scientific contributions of Alan H. Barrett to the study of circumstellar molecules have influenced many of the subsequent discoveries and measurements which have shaped the field. I have briefly discussed our debt to Barrett, his students, and his collaborators in the detection and identification of circumstellar masers, studies of stellar radial velocities, measurement of circumstellar shell sizes, measurement of stellar distances, determination of stellar mass loss, mapping of the gravitational potential, and outer envelope magnetic field measurements. Even here at this conference several poster papers clearly show the Barrett legacy such as "Water Masers Around Two Short Period Miras: R CETI and RZ SCORPI", by Little-Marenin, Benson, and Dickinson, or "Linear Polarization of HCN Maser Emission in CIT 6", by Goldsmith, Lis, Guilloteau, Lucas, and Omont. But perhaps something that is not so evident as the published papers is another legacy that Alan Barrett has tried to pass on to his students and his collaborators: That is, the ability to use correct observational techniques while pursuing the most important observational problems. For example, I have yet to hear of a Barrett student pointing at a galactic maser with the sidelobe of a radio telescope or misidentifying an autocorrelator glitch as a spectral line and furthermore, I don't expect that I ever will.

ACKNOWLEDGMENTS

I wish to thank Dr. P. R. Jewell for helpful discussions.

REFERENCES

1. G. Neugebauer and R. B. Leighton, Two-Micron Sky Survey (NASA Publ. SP-3047, 1969).
2. S. D. Price and R. G. Walker, AFGL Four Color Infrared Sky Survey (U. S. Air Force Publ. AFGL-TR-76-0208, 1976).
3. N. J. Woolf and E. P. Ney, Astrophys. J. Lett., 155, L181 (1969).

4. K. M. Merrill and W. A. Stein, Publ. Astron. Soc. Pac., 88, 847 (1976).
5. W. J. Wilson and A. H. Barrett, Science, 161, 778 (1968).
6. S. H. Knowles, C. H. Mayer, A. C. Cheung, D. M. Rank, and C. H. Townes, Science, 163, 1055 (1969).
7. P. R. Schwartz and A. H. Barrett, Astrophys. J. Lett., 159, L123 (1970).
8. L. E. Snyder and D. Buhl, Astrophys. J. Lett., 189, L31 (1974).
9. N. Kaifu, D. Buhl, and L. E. Snyder, Astrophys. J., 195, 359 (1975).
10. L. E. Snyder and D. Buhl, Astrophys. J., 197, 329 (1975).
11. D. Buhl, L. E. Snyder, F. J. Lovas, and D. R. Johnson, Astrophys. J. Lett., 201, L29 (1975).
12. B. Zuckerman, P. Palmer, M. Morris, B. E. Turner, D. P. Gilra, P. F. Bowers, and W. Gilmore, Astrophys. J. Lett., 211, L97 (1977).
13. P. M. Solomon, K. B. Jefferts, A. A. Penzias, and R. W. Wilson, Astrophys. J. Lett., 163, L53 (1971).
14. W. J. Wilson, A. H. Barrett, and J. M. Moran, Astrophys. J., 160, 545 (1970).
15. W. J. Wilson and A. H. Barrett, Astron. Astrophys., 17, 385 (1972).
16. D. F. Dickinson, K. P. Bechis, and A. H. Barrett, Astrophys. J., 180, 831 (1973).
17. P. R. Schwartz, P. M. Harvey, and A. H. Barrett, Astrophys. J., 187, 491 (1974).
18. M. M. Litvak, Astrophys. J., 156, 471 (1969).
19. G. Wallerstein, in Molecules in the Galactic Environment, edited by M. A. Gordon and L. E. Snyder (John Wiley & Sons, New York, 1973), p. 125.
20. B. E. Turner, D. Buhl, E. B. Churchwell, P. G. Mezger, and L. E. Snyder, Astron. Astrophys., 4, 165 (1970).
21. B. F.Burke, D. C. Papa, G. D. Papadopoulos, P. R. Schwartz, S. H. Knowles, W. T. Sullivan, M. L. Meeks, and J. M. Moran, Astrophys. J. Lett., 160, L63 (1970).
22. P. Goldreich and N. Scoville, Astrophys. J., 205, 144 (1976).
23. M. Elitzur, P. Goldreich, and N. Scoville, Astrophys. J., 205, 384 (1976).
24. S. Kwok, J. R. Astron. Soc. Canada, 70, 49 (1976).
25. M. Grasshoff, E. Tiemann, and C. Henkel, Astron. Astrophys., 101, 238 (1981).
26. N.-Q-Rieu, V. Bujarrabal, H. Olofsson, L. E. B. Johansson, and B. E. Turner, Astrophys. J., 286, 276 (1984).
27. R. Sahai, A. Wootten, and R. E. S. Clegg, Astrophys. J., 284, 144 (1984).
28. B. Zuckerman and H. M. Dyck, Astrophys. J., 311, 345 (1986).
29. S. Guilloteau, A. Omont, and R. Lucas, Astron. Astrophys., 176, L24 (1987).
30. M. J. Reid, Astrophys. J., 207, 784 (1976).

31. M. J. Reid and D. F. Dickinson, Astrophys. J., 209, 505 (1976).
32. D. F. Dickinson, M. J. Reid, M. Morris, and R. Redman, Astrophys. J. Lett., 220, L113 (1978).
33. M. R. W. Masheder, R. S. Booth, and R. D. Davies, Mon. Not. R. Astron. Soc., 166, 561 (1974).
34. J. M. Moran, J. A. Ball, J. L. Yen, P. R. Schwartz, K. J. Johnston, and S. H. Knowles, Astrophys. J., 211, 160 (1977).
35. M. J. Reid, D.O. Muhleman, J. M. Moran, K. J. Johnston, and P. R. Schwartz, Astrophys. J., 214, 60 (1977).
36. M. J. Reid and D.O. Muhleman, Astrophys. J., 220, 229 (1978).
37. M. J. Reid, J. M. Moran, R. W. Leach, J. A. Ball, K. J. Johnston, J. H. Spencer, and G. W. Swenson, Astrophys. J. Lett., 227, L89 (1979).
38. R. L. Mutel, J. D. Fix, J. M. Benson, and J. C. Webber, Astrophys. J., 228, 771 (1979).
39. B. R. Rosen, J. M. Moran, M. J. Reid, R. C. Walker, B. F. Burke, K. J. Johnston, and J. H. Spencer, Astrophys. J., 222, 132 (1978).
40. J. H. Spencer, K. J. Johnston, J. M. Moran, M. J. Reid, and R. C. Walker, Astrophys. J., 230, 449 (1979).
41. J. M. Moran, S. Lichten, M. J. Reid, R. Huguenin and R. Predmore, in preparation.
42. J. M. Moran, J. A. Ball, C. R. Predmore, A. P. Lane, G. R. Huguenin, M. J. Reid, and S. S. Hansen, Astrophys. J. Lett., 231, L67 (1979).
43. P. R. Jewell, Ph.D. thesis, University of Illinois, (1982).
44. P. R. Jewell, J. C. Webber, and L. E. Snyder, Astrophys. J. Lett., 242, L29 (1980).
45. J. Herman and H. J. Habing, in Late Stages of Stellar Evolution, edited by S. Kwok and S. R. Pottasch (Reidel, Dordrecht, 1987), p. 55.
46. I. Iben and A. Renzini, Annu. Rev. Astron. Astrophys., 21, 271, (1983).
47. S. Kwok, in Planetary Nebulae, edited by W. D. Flower (Reidel, Dordrecht, 1983), p. 293.
48. G. R. Knapp and M. Morris, Astrophys. J., 292, 640 (1985).
49. N. Netzer and G. R. Knapp, preprint (1987).
50. P. F. Bowers, in Mass Loss from Red Giants, edited by M. Morris and B. Zuckerman (Reidel, Dordrecht, 1985), p. 189.
51. A. Winnberg, B. Baud, H. E. Matthews, H. J. Habing, and F. M. Olnon, Astrophys. J. Lett., 291, L45 (1985).
52. R. J. Cohen, G. Downs, R. Emerson, M. Grimm, S. Gulkis, G. Stevens, and J. Tarter, Mon. Not. R. Astron. Soc., 225, 491 (1987).
53. R. D. Davies, in Galactic Radio Astronomy, edited by F. J. Kerr and S. C. Simonson (Reidel, Dordrecht, 1974), p. 275.
54. S. Deguchi and W. D. Watson, Astrophys. J. Lett., 300, L15 (1986).

RECOLLECTIONS OF ALAN BARRETT'S STAY AT NRL

CORNELL H. MAYER
Naval Research Laboratory, Washington, D.C.

I was given a difficult assignment for this meeting. It was to recall incidents and anecdotes from the brief time Al spent as a postdoc 31 years ago. This does not seem so hard for a dynamic individual like Al, who surrounds himself with incidents, but 31 years is a long time. It would be much easier to recount his many significant scientific accomplishments, or his remarkable ability to inspire exceptional graduate students to greater things. In any case, I am very glad to be able to be here for this very nice tribute to Al.

Al and I worked in different sections of the Radio Astronomy Branch at NRL. He worked mainly on spectral lines, in part with Ed Lilley, while I worked on short wavelength continuum radiation. It may not be generally appreciated that observing time on the 50-foot reflector, which at the time was the only large reflector in the world capable of operation at short radio wavelengths, was tough to come by, and competition for observing time was intense. Many of you have seen the 50-foot reflector at least from airplanes landing at National Airport.

A problem arose in 1956 when Al and Ed were keen to pursue the detection of OH. At the same time we were confident that with our 3-cm instrumentation on the 50-foot telescope we could determine the surface temperature of Venus. We also wanted to measure the polarization of the radio radiation of the Crab Nebula in order to confirm that the radio emission was due to the synchrotron radiation process and to provide another parameter for study. Fortunately, a plan was arranged by Ed McClain, who was then Head of the Radio Astronomy Branch, so that we were allowed two weeks of observing time near the inferior conjunction of Venus, after which the OH search would proceed. Meanwhile, Al and Ed were hard at work with detailed planning of the OH experiment. Unfortunately, as I recall it, the experiment failed because the frequency predictions weren't accurate enough and possibly because of instrumentation problems.

My main contacts with Al and Ginny were social, and they were very enjoyable. For example, I recall the weekly bridge club gatherings with the Lilleys and the Coates. Our most enjoyable trip to the West Coast was by train, where Ginny, Al, my wife, Carey, and I played bridge the whole trip. I don't remember if we slept at all.

I also recall trips to view the nighttime wonders of the great city of Baltimore, accompanied by our wives, of course.

My main recollection is that Al's stay at NRL was much too short. His intelligence, insight, and good humor were deeply appreciated at NRL. We were very sorry when he left.

I personally consider Al the number one radio astronomer in the country, considering his many contributions to radio astronomy and other related areas of radio science.

I envy Ginny and Al's plans for the future, but they deserve them.

ALAN BARRETT AND RADIO ASTRONOMY AT THE UNIVERSITY OF MICHIGAN

W.E. HOWARD III

U.S. Naval Space Command, Dahlgreen, VA

Good afternoon, friends and colleagues. It's a great pleasure to be with you this afternoon to tell you some tales of "Uncle Al" and of our Michigan days together.

When I look back over almost thirty years, my memory tends to dim, but in certain instances—those we'll take up here—I can still remember the high points of our relationship and the high points of what Alan Barrett accomplished for Michigan Radio Astronomy in those days. And those accomplishments were numerous. When I was asked by Phil Myers, Jim Moran and Paul Ho to make some remarks about Al's contribution during his Michigan days, I sat down and wrote an *Ode to Alan Barrett* that I shall read to you in a few minutes. I've taken to writing these odes over the past few years and I find that, whenever I deliver one, the audience needs a "road map" to understand its contents better. So my remarks now will be directed two ways—first to fill in the background of those Michigan days, and also to fill in your data bank so that the ode will be more meaningful. Each serves the purpose of the other!

BUILDING THE 85-FOOT TELESCOPE

The first recollection I want to mention is that Al's principal task when

I arrived at Michigan was to supervise the construction of the new Michigan 85-footer. Al had joined the University a year or so before I did in the very late 1950's. That dish, for its time, was a very unusual undertaking. It was a twin to the 85-1 antenna at Green Bank, which was called the Howard E. Tatel Telescope and which is now a part of the Navy-funded NRAO interferometer there. The Michigan twin wound up having a better frequency performance due to a more precise surface—a tribute to Al's oversight and management. The antenna was a gamble in those days. It was to operate at a wavelength of 3 cm, although the receivers available at that time could only detect a handful of radio sources at that high a frequency. I think it was a tribute to Arnold Shostak—Seth Shostak's father—and to the funding agency, the Office of Naval Research, that the scientific potential of short centimeter radio astronomy was seen so early. A calculated risk was made by ONR to open up a new frequency frontier. And it was Al's job to bring that telescope on line. There were lots of problems as the construction was going on, but most of them were behind Al by the time I came on the scene.

THE SETTING

We shared an office in the old Observatory building up near the U of M Hospital and the women's dorms. Conditions were crowded; Al smoked like a chimney but those were in the days when no one minded. Michigan was exciting then, for the Department had Leo Goldberg as chairman and there were Aller, Dodson-Prince, Elste, Haddock, Kundu, Liller, Malville, McLaughlin, McMath, Miller, Mohler, Muller, Walsh, Wentzel, and several other luminaries—all of whom were dwarfed by the reputation among the undergraduates of Hazel Losh—Doc Losh—who was still teaching undergraduate astronomy from a 1936 textbook and spending half the lecture time taking roll!

UFO STORY

I now must tell you of an incident that happened as Al Barrett was in the final stages of testing the Blaw-Knox antenna late one weekend night. The setting is at Peach Mountain, located in Dexter Michigan, several miles outside of Ann Arbor. It was about one o'clock on a Sunday morning and Al was wrapping up the night's testing. The antenna is located at the top of a rise. It can easily be seen from the surrounding countryside and it was lit up from the underside by a rather bright set of lights that were on while Al was doing his work. Located nearby is the radio tower of WUOM, the University's FM station, and it has a blinking red light on top of it to warn aircraft. At the foot of the hill there is a paved road that runs by the antenna and a traveller on that road during the daytime could look up and see the antenna on the hill with the WUOM antenna located behind the antenna, along the line of sight. That early morning, just as Al was completing the night's work, a slightly inebriated couple coming home from a party were travelling by car along that road. They looked up to see a flying birdcage hovering at the top of the hill and it had a blinking red light on it! After several seconds of hovering, the giant birdcage zoomed off into the night! They were so frightened that they telephoned the Ann Arbor News which published the story of a UFO in the next morning's edition. Al read the account that next day and let out a whoop, for it was just at the time that the birdcage zoomed off into the night that Al turned off the antenna illumination light to go home! As I remember the story, Al immediately called the editor and explained what happened. The editor was amused too, but asked Al for a favor—and that was to call the couple himself and explain what happened, since they would never believe the story second hand. That story was told in one of Donald Menzel's books debunking UFOs.

CATASTROPHE AVOIDED: "NOW IT CAN BE TOLD"

One of the tasks that Al and I performed together was calibrating the pointing of the antenna after it had been built, but before it had been formally accepted. To do this properly meant a stint at the telescope that would go on for 24 hours. The measurements consisted of observing as many strong radio sources at as many declinations as possible. The best results were obtained when those sources were also observed over as wide an hour angle range as possible. But the limit switches on the hour angle motion had not yet been activated since we were still in a testing mode. We had been at the telescope doing that calibration for most of those 24 hours and it was in the very early morning. We were listening to the American Airlines radio show—excellent, soft, dreamy music for observing in those days—when Al sat bolt upright! He leaned over and quickly turned off the hour angle scan switch, looked at me pale and speechless, and pointed to the hour angle indicator which was *very* far west—virtually pinned on the dial! We turned on the intercom and Al rushed out and climbed up to the hour angle axis to see how bad the situation really was. His report was that about two of the five gear teeth that normally made contact were beyond that normal point of contact. Seconds later the antenna would probably have had a short history indeed! Slowly we scanned the antenna eastward and saved the day. The trauma was so intense that, to the best of my knowledge, no one ever knew how close we came to catastrophe. This is the first time this story has become public!

Now with the antenna up and operating, Al and I decided to try to detect the radio radiation from the planet Mercury. It hadn't been done at that time, but the combination of an 85-footer and a short observing wavelength meant that the Michigan antenna was among the best in the world. We remain indebted to Connie Mayer and his colleagues at NRL for teaching us their "tricks of the trade", because they had a lot of experience with observations of the brighter planets at that time, particularly Venus and Jupiter. It was a tricky project because radio telescopes are like airplanes—they behave best at night when conditions are uniform and very stable. But observing Mercury, located so near the sun, meant that observing had to be done near sunrise or sunset, just when atmospheric and temperature conditions are the most unstable. We were successful, but the best runs we had were when Peach Mountain was enveloped in an early morning, very uniform fog or inversion layer that was thick enough to protect the antenna from thermal transients but not thick enough to attenuate the radio signal coming through it. The detection was reliable, but the signal strength was not what we predicted. It turned out that the planet was rotating while we believed the optical astronomers whose observations at that time seemed to indicate that the Mercury kept one face to the sun just as the moon keeps one face to the earth. A few years later Gordon Pettengill showed by radar that Mercury actually rotates. Our own observations made more sense with that new model.

RADIO UFO

Even in these days of investigating unexplained phenomena, I doubt that Al or I have been brave enough to tell the tale of the time we found a 3 cm radio UFO with the Michigan antenna. It was after the dish was calibrated and we had made the Mercury detection. Lawrence Aller's interest in planetary nebulae was at its height then and he and Don Osterbrock had each provided us with some predicted 3 cm wavelength flux densities. Al and I had developed a search technique on the antenna for faint radio sources called a "spiral search" where you first point to the coordinates where the object has the highest probability of being located. Then you move the antenna in slow scan along lines of constant right ascension and then of constant declination, widening the search in a spiral pattern along scanning lines separated by half a beamwidth, integrating the signal as you go, and looking for a hump in the record. Once found, you determine the center of the hump along the scanning coordinate from the analogue record, set up the telescope at the position of the maximum, and scan in the orthogonal direction. The original position of the hump, combined with the position of the hump in the orthogonal scanning direction then give the "first cut" at the dial coordinates of the source that must then be verified by repeating the process again. So one night Al and I were at the telescope trying to observe this very elusive class of low signal-to-noise objects—planetary nebulae. While in the midst of this, we found a hump whose signal characteristics were not particularly unusual. But the problem came when we tried to verify the position of the hump. It had moved! Every time we set on the previous right ascension or the previous declination and redetermined it, it was different! Moreover, its angular motion seemed to be accelerating and, as I remember, its signal strength was increasing too! As I recall, we followed it for about four or five re-settings, but then the acceleration was too much for us and we lost it! It was the first radio UFO that I can remember, aside from rediscoveries of the moon that every young radio astronomer has to experience at some point in his career!

RICK BARRETT STORY

Although this next story is not in the Ode, its telling more adequately reflects the honest, straightforward, no-nonsense Al Barrett who has come to mean so much to all of us here today. In Ann Arbor, Al was senior to me in years, in experience, in engineering sense, and in worldly goods. He had a beautiful red brick house in Ann Arbor Hills, a pretty posh neighborhood in those days, while Miriam and I were renting a small apartment near Bill Liller about a mile and a half from the stadium. One day Al dropped in with his son, Ricky—who was then about five years old. Ricky took one look at our apartment, turned to Al and said, "Our house is better than this house, Dad!" Al didn't blink an eye, but he turned to Miriam and me and said, "Well, we've taught him honesty, but we haven't started on tact!" That's Alan Barrett!

THE ATMOSPHERE OF VENUS

All the time Al was working on the observing part of his career at Michigan, he was secretly yearning to do a theoretical study of the atmosphere of Venus. After the antenna was up and going and after we had all moved into more spacious, renovated offices at Michigan, Al began that study. He used the observations from NRL and other institutions, and he determined that the incredible heat at the surface of that planet had to be accompanied by an enormous atmospheric pressure. His atmospheric composition was basically confirmed by the Venus landings that took place after his study was finished.

THE PATIENT TEACHER

Al and I would often get into some interesting professional conversations and much of what I learned in the early days of my career in radio astronomy, I learned from him. I can remember one time when I asked him for an explanation of some technical point at which he was the obvious expert and I was all ears as he was explaining how it all worked. But the explanation just wasn't getting through to me and Al knew it. In some frustration, he looked me right in the eye and said, "Which word don't you understand?" —That's Alan Barrett.

GINNY

But our story wouldn't be complete without Ginny, because Ginny has been Al's wife and companion in adventure through all these years—raising kids, not all of whom were their own, teaching youngsters, and roughing it in New Hampshire and in the Florida Keys. She and Al haven't changed one bit since I met them almost thirty years ago. Her outlook on life is so refreshing, it's a model that all of us could well live by, and her sparkle of life has reached out to us all. Al's a very lucky guy to have such a great gal as Ginny. If I had to choose one word that exemplifies Al and Ginny's human contribution to us all, I would choose the word "laughter". And if I could choose modifiers to that word I would choose the adjectives "infectious" and "warm"—that is, "warm, infectious laughter"—That's Al and Ginny.

Well that's enough reminiscing of the Michigan days for now. Let's turn to the ode for which I hope you are now adequately prepared!

ODE TO ALAN H. BARRETT

So Alan's retiring from Cambridge, they say,
And he's finding it hard keeping all 'us at bay.
It's that big warm smile without hidden agenda
That brings us all here to partake in this bender.*

A score and eight years have now gone by
Since first I met Uncle Al, eye-to-eye.
We shared a tiny office in Ann Arbor town
And started on a friendship that'll never wind down.

One night when Al was testing the Blaw-Knox scope
A couple saw a "birdcage" up a-hovering on the slope.
They watched it all with horror as it quickly zoomed away.
They thought they'd seen a UFO to end their boozy day!

They called Ann Arbor News which ran the tale at dawning
Where Uncle Al then saw it as he read the News that morning.
"That's when I put the light out, and it really was quite late!",
Said Al to the local newsman as he put the matter straight.

Peach Mountain beckoned both of us, and to this very day
We'll ne'er forget the time, with mortified dismay-
We almost ran the dish beyond its hour angle gears
But we caught it in the nick of time and saved our joint careers!

We pulled the planet Mercury out from all the noise
While a stable foggy inversion layer added to our joys.
And another crazy memory of those times so long ago
Was when we both discovered the first radio UFO!

While working in Ann Arbor town, our Al then studied Venus
Long before unmanned spacecraft found its conditions heinous.
He told us of the presence of great pressure and much heat
And the very weird components of its gasses, so complete.

Al left for the Charles and I for Green Bank.
But for memories of Michigan, we've Al and Ginny to thank.
It's friendships like these that really do last
And that feed us recollections and warm thoughts of the past.

Bill Howard 1987

* Note that "bender" must be pronounced in New Englandese (which shouldn't be too hard for Uncle Al to do!)

ALAN BARRETT AND THE RESEARCH LABORATORY OF ELECTRONICS

JONATHAN ALLEN
Massachusetts Institute of Technology, Cambridge, MA

I am very pleased to be here on such a special occasion. At these events, it is appropriate to pause and think back to where we have been and what has happened over time. This symposium also stimulates me to think about the interaction between people like Al and the institutions that are here at MIT.

As many of you know, the Research Laboratory of Electronics (RLE), where Al has been all these years, was formed in 1946, about 40 years ago. The characteristics of the Laboratory are very strongly those characteristics of Al himself, which have contributed to his fruitful career. Let me mention some of these facets.

> We've had, over the years, an insistence on fundamental understanding of the phenomena we study, and Al's use of radio probe techniques has been an essential tool for gaining that understanding.
>
> We've had an openness to multiple viewpoints and disciplines. Al, you fit that to a "t." As many of you know, Al started out at MIT as a professor of electrical engineering, and then became a professor of physics. Those are the two dominant disciplines within RLE, and, of course, the whole reason for RLE is to promote interdisciplinary research.
>
> We've always provided a tight coupling between theory and experiment. The use of radio techniques by Al for astrophysical research has shown a constant interplay between theory and new data to extend that understanding, in the very best scientific tradition.

We've usually gotten our motivation through a focus on applications. The expanding exploration of space, so much a part of our times, has been well served by Al's research into the nature of that environment.

In RLE, we've seen an ongoing interaction between the development of probe techniques and the discovery of new phenomena. In this regard, Al's work provides a perfect example. On the one hand, new microwave techniques have evolved, including new methods for the generation of microwaves and their reception. On the other hand, we constantly use these abilities to probe and discover new phenomena, which in turn lead to new probe techniques. In Al's work, we've seen this experimental loop traversed successfully many times.

Finally, we have always maintained in RLE a significant part of our effort in the direct service of society. Here Al has maintained this important tradition with his work in microwave thermography, building on his technical and scientific skills to help his fellow man.

Clearly all of these attributes come together cohesively and effectively with wonderful results in Al's work. I am delighted to have within RLE and within MIT such a strong effort in radio astronomy, and I thank you very much, Al, for all of your contributions to it.

ALAN BARRETT AND THE M.I.T. PHYSICS DEPARTMENT

JEROME I. FRIEDMAN
Massachusetts Institute of Technology, Cambridge, MA

It is a pleasure to have the opportunity to say a few words on behalf of the Physics Department at this symposium to honor Alan Barrett.

Without a doubt, Al exemplifies the very best characteristics of Scientist, Colleague, and Teacher. This symposium is in itself a reflection of the world-wide impact that his work has had in radio astronomy and astrophysics in general. It is also a demonstration of the high esteem in which he is held by his colleagues. This esteem is also indicated by the numerous science advisory committees on which Al has served during his career.

As co-discoverer of the radio emission from interstellar OH, the first molecular spectral line of radio astronomy, Al was instrumental in initiating one of the most productive fields of contemporary astronomical research. This led to the discovery of giant molecular clouds and the elucidation of the complex processes that occur in these clouds. His work was critical in establishing the existence of cosmic masers and helped establish the currently accepted models of their origin. In recognition of his role in the development of VLBI for the study of masers, Al was awarded the Count Rumford Medal by the American Academy of Arts and Sciences.

In addition of his investigations of the interstellar medium, a good part of his career was directed to the study of planetary atmospheres. As co-experimentor in the microwave radiometer experiment on the Mariner-2 Venus spacecraft, he helped clarify the properties and structure of the atmosphere of Venus. There is more that can be said about his achievements in a wide range of projects, including his work in medical physics. The speakers on this program will, I am sure, tell you about these. But suffice it to say that we in the Physics Department take great pride in what Al has accomplished.

Al has been one of the stalwarts of our educational program; he is among our best teachers. Our undergraduates are tough in their evaluations of teaching performance, but Al has consistently received very high ratings. He coauthored an excellent text entitled *Electromagnetic Vibrations, Waves, and Radiation* published by the MIT Press. Over the years, this book has been used often as a text for our sophomore physics course on waves and oscillations. In the 26 years that Al has been at MIT, he supervised 19 Ph.D. graduate students, many of whom have gone on to become leaders in their field. One senses that Al has had a special closeness to his students. He was understanding of their needs and was simultaneously their mentor, colleague, and friend. That some of his students played a major role in organizing this symposium is a reflection of this special relationship.

Al also has been very helpful in departmental matters. For example, a few years ago, when we were starting our distinguished lecture series, I asked Al to chair a committee to set up the first series, a week long set of lectures and social events. Without a moment's hesitation, he consented and personally took responsibility for the whole thing, from choosing the speakers to making the arrangements. In his typical fashion, he did a marvelous job and got the series off to a very successful start.

We will miss Al as a colleague. His friendly manner, his sense of humor, and his willingness to help have been greatly appreciated. We are indeed fortunate to have had Al as a colleague, and we hope that he and Ginny will have a retirement that exceeds all of their expectations.

HOW WE HIRED ALAN BARRETT

H. J. ZIMMERMANN

Massachusetts Institute of Technology, Cambridge, MA

Last year the M.I.T. Research Laboratory of Electronics (RLE) celebrated its fortieth anniversary. When Alan Barrett came to M.I.T. in 1961, RLE was a mere 15 years old. There was no radio astronomy program in RLE, but there were strong feelings among several staff members that there should be.

Since 1946, when RLE started as an outgrowth of the M.I.T. Radiation Laboratory, there were strong components of research in various aspects of microwave physics, microwave systems and signal processing, all pertinent to radio astronomy. In the early fifties, the work in RLE played a significant role in the evolution of the M.I.T. Lincoln Laboratory. Lincoln's work on the air defense system led quite naturally to large microwave antennas and these of course were eminently suited to radar and radio astronomy. Lincoln had been bouncing radar signals off the moon and was doing some work in radio astronomy as well.

There had been discussions with Lincoln staff members about the desirability of coupling M.I.T. graduate students and faculty to some of their large antennas, and a number of RLE staff members were interested. In fact, Professor M. W. P. Strandberg of the Physics Department had written a memo urging that research in radio astronomy be undertaken at RLE. What was really needed was a catalyst, so the stage was set for someone like Alan Barrett.

One day (I think it must have been early in 1961), Ed Lilley telephoned from Harvard. He said there was a millimeter wave radio astronomer at the University of Michigan who was interested in coming to Cambridge. Ed wanted him here but there were no openings available at the Smithsonian Astrophysical Observatory or Harvard. Since RLE had an interest in millimeter wave research, this seemed like it might be a good match. I met with Ed Lilley and Alan Barrett, and after listening to Ed's glowing praises, this seemed to be the catalyst RLE had been seeking.

Alan wanted a faculty appointment, and for maximum effectiveness with graduate students, that was what we wanted as well. However, a faculty appointment could only be made by an academic department. Alan wanted to be in the Physics Department but they were tight on budget or had other ideas – so the answer was negative. Fortunately, the Electrical Engineering Department was interested in promoting collaboration with Lincoln Laboratory and also managed to come up with some funds. So in July 1961, Alan Barrett became an Associate Professor of Electrical Engineering.

Alan started a graduate course on Radio Astronomy that proved to be quite popular, and he had numerous graduate students interested in Ph.D. theses. Since RLE is an interdepartmental laboratory, he had access to students in Physics as well as Electrical Engineering. His research program took off like a shot, with OH radical observations in the interstellar medium, planetary atmosphere studies, Venus probes and all the rest. By July 1965, Alan's track record looked so good that he was promoted to Professor of Electrical Engineering.

At about that same time (1965), the Physics Department brought Bernard F. Burke to M.I.T. as Professor of Physics, and his presence broadened the scope of the radio astronomy program in RLE. Then in 1967,

Alan Barrett shifted to the Physics Department since he had always felt that to be a more appropriate academic home. However, during the time he spent in Electrical Engineering, Alan made important contributions to the department, both via his classroom teaching and via thesis research. Through the years, he has supervised about 20 doctoral theses in Electrical Engineering and Physics.

Among his varied research activities, I have always felt the application of millimeter wave radiometry to medical diagnostics was particularly interesting. It provided an excellent example of the fact that basic research often leads to unexpected applications. That, of course, is one of the goals of a basic research laboratory such as RLE.

Alan has had a distinguished career at M.I.T. He has published more than 80 papers on a variety of subjects and has launched numerous graduate students on successful careers. We are grateful to Ed Lilley for steering Alan in our direction, and I am pleased to have had a hand in bringing Alan to M.I.T.

For more than 25 years, Alan has been following the stars. Now, he and Ginny plan to spend more time following the Sun.

BEGINNINGS OF RESEARCH IN RADIO ASTRONOMY AT M.I.T.

Jerome B. Wiesner
Massachusetts Institute of Technology
Cambridge, MA

Alan Barrett's first MIT years occurred while I was away--one could say that our careers at MIT sort of butted together; as he came, I left. I don't think that there was a causal relationship here. Actually, I was very sorry to leave at that time because I had been very much involved in the early years of radio astronomy, both here at MIT and at the National Radio Astronomy Observatory in Green Bank. I became involved with radio astronomy through Ed Purcell and Lloyd Berkner, both close friends of mine. Purcell and I worked with Lloyd Berkner, whom many of you will recall in the creation of the Green Bank Observatory, including helping to locate it.

Last night, as I thought about today's symposium, I wondered why I never became a radio astronomer in spite of the many opportunities and temptations to do so. In fact, I must admit that I have been one from time to time. My conclusion was that I must have been born to be an engineer and not a scientist. I obviously enjoyed inventing and making things--low noise amplifiers, correlators, big antennas, and radomes--rather than using them. I was actually interested in communications and noise theory, particularly in scatter communication, and radio provided me with an exciting challange. The system problems of radio astronomy receivers, radio telescopes, and microwave communication systems were the same. Scatter is no longer a communication medium but then it was a good excuse for building enormous antennas. As I already said, it also provided a good excuse for exploring low noise receivers. I worked with cooled diodes, in liquid helium as well as in other coolants. That was before the availability of radiometers and just after Ed Purcell and his partner, Harold Ewen, had done their pioneering hydrogen line experiment. At that time, Sandy Weinreb, a graduate student, someone you know very well, had been looking for a thesis and came to Professor Yuk-Wing Lee and me with a proposal for a radio astronomy receiver using auto correlation techniques. Professor Lee and I had been working on correlation devices for detecting signals--mind you, we didn't have radio astronomy in mind--when Weinreb said, "you know this would make a wonderful receiver for radio astronomy." He explained that radio astronomers needed flexible spectrometers that covered a range of bandwidths with different resolutions and that also achieved the theoretical sensitivity after long integration times. The traditional approach involved banks of analog filters. Sandy suggested building a one-bit correlator. It was an exciting idea and we agreed to the proposal enthusiastically.

I asked the NSF for $60,000, to build an experimental receiver. Those were the days when the NSF would say yes quickly to a good idea. Believe it or not, within a couple of weeks Sandy had $60,000, a lot of money for a graduate student, particularly in those days. This must have been 1958 or 1959 so you would have to multiply those dollars by a factor of 10 (or so) to find its equivalent for today. He built the receiver and we were not greatly surprised that it worked. We then had to decide what to do with it. We had the big antennas at Lincoln Lab for our tropospheric scatter research, and we considered putting the receiver on one of them. We finally decided to put the receiver on the original 25-m Tatel telescope at NRAO and to look for the 327 MHz line of deuterium. We didn't find the line and to this day no one else has produced a convincing detection.

At that point in time, as they say, I was diverted from radio astronomy by an invitation from Jack Kennedy to be his Science Advisor. Al turned up at just about that time, so I didn't get to know him until I returned three and one-half years later. Sandy went to work at Lincoln Lab to build instrumentation for the new Haystack telescope at that point. He also expanded and upgraded his correlator for use on the Millstone antenna. Al got involved and they discovered the 18 cm transition of OH from galactic gas in the Fall of 1963.

You have just heard people give Al credit for tremendous persistence and observation of all those shells around late type stars that he looked at--hundreds and hundreds were mentioned earlier this morning--well let me tell you a secret. When I came back from Washington after about three years, Sandy said to me, "you ought to go out and take a look at our new telescope and see how we do it now" because I had only seen drawings of Haystack when I had left.

Herbie Weiss was also involved, and I used to get telephone calls from him that told me about the progress of the dish and the radome in which he was particularly interested, but I had never seen the whole thing together and running. As it turned out, one of Al's graduate students was on duty there the day I went, and he had the job of showing the new system to me. To my great surprise, the central feature was not the large steerable antenna or the radome but a UNIVAC computer. I got the impression there was an antenna somewhere but I didn't see it.

Al's student sat me down in front of the computer, punched a button, and the computer asked "where do you want to look", so I quickly made up some azimuth and altitude numbers and we punched them in and it asked, "what integration time do you want?" We punched in some more numbers and then it asked "in what form do you want this data presented?", and after it listed all the possible formats, we decided to record it on paper tape. Dozens of other such questions were asked, and finally we punched another key and then everything started to whirl. As we watched, the antenna came into position and settled in on the proper direction. After this very interesting demonstration I said, "What a hell of a way to do research" and the kid looked at me and asked, "what other way is there?"

That reminded me of a particularly painful incident I had in Washington. Some of you may recall that the Atomic Energy Commission conducted the Hardtack experiments in which it exploded two very high altitude nuclear bombs in the Pacific. The spectacular effects of the explosion over Eniwetok were visible in Hawaii. The newspapers reported that we broke Van Allen's Radiation Belt and astronomers all over the world were raising cain. We were anxious to know how badly we had broken Van Allen's Belt, so I persuaded the Air Force to send up two sounding rockets to penetrate the belt and measure the electron densities as they transited the belt. A couple of days after the rockets were launched, I called up the Air Force to find out what the experiment showed and they said, "We don't know because we haven't run the codes yet." And I said, "Well, did the meters go up or down?" and they said, "What meters?" So you can see that I was conditioned to be upset when the computer became a substitute for man as the observer.

But as we talked about the beauties of the new Haystack telescope, for example, I learned that one could pick 200 or more observation points and the telescope would systematically work its way through them. My image of what happened with those shells Al observed is that some time during the course of an afternoon Al gave a graduate student a list of all the objects he wanted to examine and the amount of time he wanted to spend observing each one of them, and how he wanted the data recorded, maybe even how he wanted it analyzed, because the computer looked as though it could do that too, as well as print out spectra and other fancy things. The student then went home and came back a couple of days later and found his data thoroughly analyzed showing those peaks. I don't know whether that really happened, but if it didn't Al and his student weren't making full use of the technology that was available to them.

It was certainly a lucky day for MIT when Al turned up and joined the department and the radio astronomy activities. He, Bernie Burke, and other collegues have created one of the most interesting, exciting, and productive programs in the field. They have involved me from time to time. I don't know if that is the proper word, but whenever there was a need for money I think they recruited me, and that has been relatively often.

At a later time we created NEROC to make the local facilities available to all of the scientists in the Northeast. We also attempted to build a big antenna which I recall as a 440 foot diameter parabola inside a 700 foot diameter radome. We had almost convinced the NSF that they should fund it when the budgets tightened up and multi-dish interferometers became the rage, and so did millimeter wavelength instruments. After considerable effort to scale the big antenna down and make it more accurate, it died. There were, of course, valid arguments about whether or not our plan was the best way to use money or whether a smaller antenna, say 120 feet in diameter and good to 3-mm wavelength, wasn't better.

Through all of this turmoil we were fortunate to have the clearheaded guidance of Al. So, it is with very great regret and concern for the future of our highly respected research and teaching activities, that I come to this gathering. I won't say congratulations because I don't believe in retirement. I am sure that Al doesn't either and will go off and do something very exciting elsewhere. For all of the MIT community and particularly for the Research Laboratory of Electronics and the Electrical Engineering and Computer Science department, all I can say is "thank you" for being you and for many years of leadership and friendship.

EARLY DAYS OF OH OBSERVATION

M. L. MEEKS
Meeks Associates Inc., Stonehedge, Lincoln, Massachusetts, USA

Abstract Recollections of the circumstances surrounding the first detection of the microwave OH lines and their early observations at Millstone Hill and Haystack Observatory.

When I see all of my old friends here at the Barrett Symposium, I am filled with nostalgia for the old days of radio astronomy. In the winter of 1961 I came to Millstone Hill for an interview with Gordon Pettengill about a job at M.I.T. Lincoln Laboratory. Driving up the hill after a heavy snow, we found the road blocked: the 60-ft diameter Westford antenna had broken off its mount when the operators attempted to dump out snow, and the antenna lay face down in the roadway. This accident fortunately had no long–term effect: Walter Morrow, the Group Leader for the Westford Project at that time, is now the Director of Lincoln Laboratory.

When I came to work at the Laboratory in the summer of 1961, Haystack was under construction; the 150–ft diameter radome was a shell without an antenna inside. This was to be the first radome–enclosed antenna intended for astronomical use, and there were problems to be solved: (1) how to calibrate the pointing of an antenna inside an opaque structure, and (2) how to measure the transmission loss of signals passing through the radome. It was my job to solve these problems, and I proposed to use astronomical radio sources for the calibrations. The possibility of using the Haystack antenna as a radio telescope had in fact attracted me to M.I.T. from Georgia Tech. Although lunar and planetary radar astronomy were strongly supported at Lincoln Laboratory at that time, radio astronomy lay outside the scope of Laboratory interests, and I hoped that scope might be expanded.

We were able to calibrate the pointing in a straightforward way, but to measure the radome loss required absolute measurements of the flux of several radio sources. For this purpose we used a horn antenna with very low sidelobes, a so-called Hogg horn, which the telephone company used for microwave relays. In the process of calibrating the Hogg horn, we received a telephone call from Arno Penzias at the Bell Telephone Laboratory in Holmdel, N.J. Arno asked whether in the calibration process we might have measured the background temperature of the sky. Sander Weinreb, who made our calibration measurements, replied that strangely enough the data showed a few degrees of antenna temperature looking at the cold sky, where of course we expected to measure zero degrees. Without recognizing the significance, we had independently observed the Three–Degree Background a few months before Penzias and Wilson published their results in the Nobel–Prize–winning paper.

Haystack was conceived as part of a project to extend radio communication well beyond the horizon through the use of a cloud of resonant dipoles in earth orbit. Because the scatterers were short lengths of fine copper wire, the project had unfortunately been named Project Needles. When this project was proposed, radio astronomers expressed concern about an artificial medium in space, and the press judged that this project might pollute interplanetary space with needles. The orbit for this resonant belt was in fact carefully calculated so that the lifetime of the dipoles in space would be limited. I recall a Colloquium at the Harvard College Observatory in which radio astronomers presented results of calculations suggesting that the dipole belt would be detectable with radio telescopes and could interfere with observations. At the conclusion of this Colloquium, Irwin Shapiro, who had calculated the orbital motion of the scatterers at Lincoln Laboratory, spoke up from the audience, questioning the detectability of the belt and pointing out that radiation pressure from the Sun would bring down the dipoles in a short time. Fortunately the astronomical impact of Project Needles, renamed Westford Project, was temporary: Irwin Shapiro now directs the Harvard College Observatory.

Although many radio astronomers viewed the activities at Lincoln Laboratory with alarm, the Millstone Radar led the way in integrating digital computers with large–scale radar systems. While the National Radio Astronomy Observatory was building a 140-ft radio telescope on a polar mount, the 84–ft Millstone antenna was steered under computer control on a much simpler, less expensive altitude–azimuth mount. At the time of early OH observations, the altitude-azimuth mounts and computer steering of the Millstone and Haystack antennas were unique among radio telescopes (That turned out to be important for OH observations as you will see.) The radar data at Millstone Hill were processed and pointing was controlled by the first all-solid-state computer, the CG-24, which had been designed and built at Lincoln Laboratory. At this time in the early 1960's no other radio telescopes were equipped with digital computers. My goal was to find ways in which this technology could be applied in radio astronomy. Since I lacked a reputation as an astronomer, and since I was the only radio astronomer at Lincoln Laboratory, I turned to the M.I.T. campus for support. Alan Barrett had arrived in the Electrical Engineering Department at the same time that I arrived at Lincoln Laboratory, and Charles Townes was an important figure on the campus. I encouraged Alan to participate in planning for radio astronomy at Haystack, although that was hardly necessary, and I hoped that Professor Townes' interest would give support in making Haystack a facility for radio as well as radar astronomy. Although in those days there was close contact between solid-state physicists at the Laboratory and on campus, there were no corresponding ties in radiophysics and astronomy. Alan was the first radio astronomer to join the M.I.T. faculty, and he forged ties with the campus by sending his graduate students out to the Millstone and Haystack sites.

While Haystack was under construction, Sandy Weinreb was completing this doctoral research under Jerome Wiesner, who had gone to Washington to serve as President Kennedy's Science Advisor. Sandy's thesis involved the design and construction of what amounted to a digital spectrometer. His key contribution was an early example of parallel processing: an autocorrelator which accumulated the autocorrelation function of a signal in real time. Given the autocorrelation function, a Fourier transform yielded the spectrum of the input signal, but a separate computer was necessary to perform this calculation. In the course of his thesis research Sandy successfully developed the correlator and used it at the National Radio Astronomy Observatory in Green Bank, West Virginia in an unsuccessful search for the deuterium line. This search required that the autocorrelation data be transferred to punched cards for input to Green Bank's IBM 1610 computer.

Alan was of course eager to see the autocorrelator used in a search for OH; the long integration times required for a spectral-line search could be obtained with this spectrometer. In February, 1963 when Sandy completed his graduate work, the Haystack system was still two years from completion, and someone was needed to engineer the radiometric system for the required calibration. It was clear that the Laboratory should offer Sandy this job, and as a part of the deal he could install his autocorrelation system on the Millstone antenna where it could be integrated with CG–24. Nevertheless Sandy turned down the Laboratory's offer, and accepted a job with industry. We were stunned. But fortunately after a few 16–hour days writing proposals for contracts, Sandy resigned from industry and came to the Laboratory. It should be noted that in those days the Millstone Radar was heavily engaged in satellite tracking, and we were fortunate that Gordon Pettengill, the Group Leader, encouraged and supported the OH search.

To integrate a new piece of hardware into a complex system with a heavy operating schedule became the responsibility of John Henry, who knew every detail of the hardware and software of the CG–24 computer. We learned how much John knew about his computer during OH observations when the water boiled out of an electric teapot in the control room. The smell of the overheated teapot was so strong we decided to play a little joke on John. We opened the back of one of the computer racks and put the teapot inside. Soon the smell of hot–teapot insulation filled the control room. Then one of us ran down the hall and told John that the CG-24 smelled hot. Rushing into the control room, John took a deep breath, relaxed, and said, "That doesn't smell like the CG-24!"

Although the staff and facilities of Lincoln Laboratory played a key role in the discovery of the OH absorption in the interstellar medium, Alan was the one who put it all together. For him the search for OH was not a new idea. He knew the astronomy and the physics: where to look and how to interpret the data. In October, 1963 the OH transitions at 1665 and 1667 MHz were detected in absorption against Cassiopeia A.

The level at which the spectral–line system at Millstone Hill performed in 1963 can be judged by examining the Poloroid photograph of the CRT display, which is reproduced in Fig. 1. Immediately after the data were taken, the CG–24 generated displays like this one showing the spectrum and relevant information about the observation. This figure shows the spectrum observed in a single 2000–sec integration. Successive observations could be averaged together on-line to increase the signal-to-noise ratio.

In the summer of 1965, shortly after the bright sources of OH emission had been discovered, the Haystack system went into operation as a radio observatory. The 120–ft Haystack antenna was the largest then available for OH observations, and this turned out to be important when we observed the galactic source W3. With the higher angular resolution of Haystack, we quickly discovered that the OH emission was offset by about 14 minutes of arc from the continuum source W3. During the first observations at a frequency of 1665–MHz, as we watched the computer

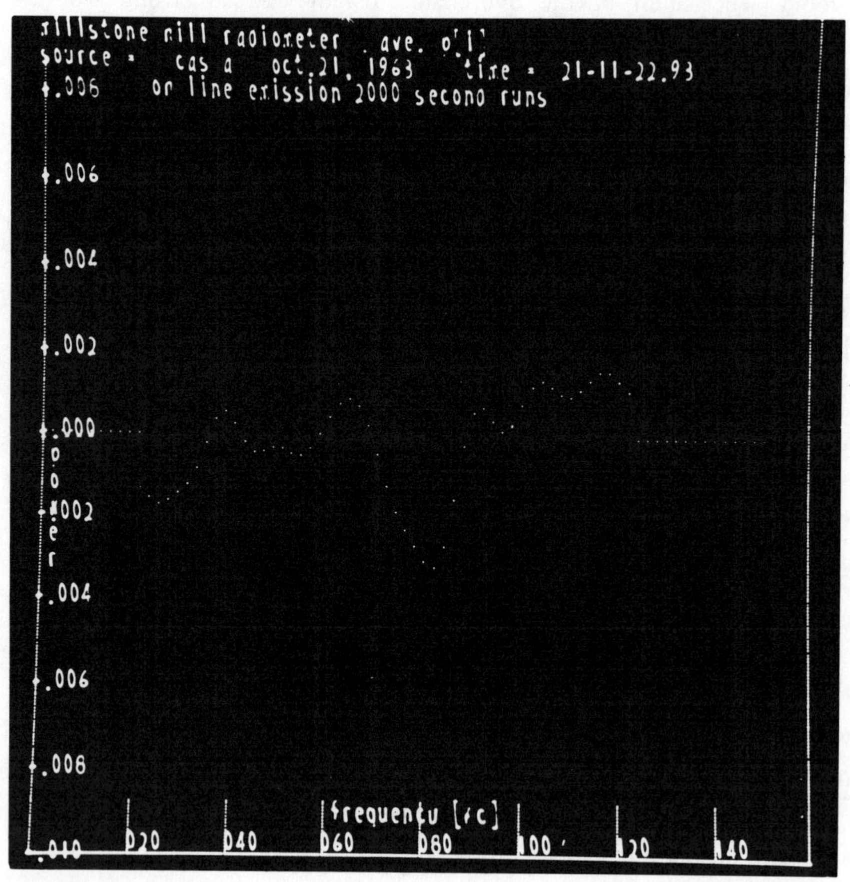

FIGURE 1 A photograph of the CRT display generated by the CG–24 computer showing the spectrum of the 1667–MHz transition of OH in absorption against Cassiopeia A. This observation was made with the 84–ft paraboloid at the Millstone Hill Observatory of M.I.T. Lincoln Laboratory on October 21, 1963; the integration time was 2000 sec.

displays of successive spectra measured with linear polarization, *one of the features appeared to change markedly from one 2000–sec integration to the next.* After an hour or so the measured intensity had decreased by more than half! How could it be that no other observatory had reported this? Why, we asked ourselves, should we be seeing this at Haystack? We remembered that the position angle of our linearly polarized antenna feed was rotating rapidly as we tracked this circumpolar source with our alt–azimuth mount. Could it be that this emission feature was *linearly polarized*? We quickly confirmed this conjecture by climbing up and rotating the antenna feed. This was the most exciting moment in my career as a radio astronomer.

After the discovery of linear polarization, Alan Barrett and Alan Rogers then found circular polarization at Green Bank before we could assemble a circular feed to look for it with Haystack. It should have been clearer to us that this intense, narrow–band, polarized emission indicated unambiguously that we had detected maser action in the interstellar medium. However we continued to think of the Zeeman effect as a possible cause of the polarization. If we had known more about masers, we might have predicted that the angular size of the OH source when measured would be very small.

A series of measurements began in 1966 to determine the angular size of the OH sources; the Haystack antenna, became one element for interferometric measurements with increasingly long baselines, leading of course to VLBI measurements, which finally resolved the maser sources. Alan and his students led the way, and twenty years later Haystack continues to be a center for very-long-baseline observations and for the development of digital hardware and software for VLBI. Looking back it seems to me that there were in fact two important events in October, 1963: radio astronomers discovered the OH lines and radio astronomy discovered the digital computer.

SIDETRACKED

Recollections of an introduction to research in astronomy under Alan H. Barrett

RONALD J. ALLEN

Astronomy Department, University of Illinois, Urbana, IL

1. INTRODUCTION

Alan, it is a great pleasure to be here, and a great honor to have this opportunity to recall some of the things that we did together during those few years in the mid 1960's when I was one of your Ph.D. students at MIT. I have nevertheless wondered why the organizers asked me to do this, since our paths diverged after my thesis was finished and have crossed only infrequently in the intervening years. Your main interest is spectroscopy of interstellar molecules, whereas my thesis work was on the continuum emission of radio sources. However, your experience with me apparently cured you of developing further in the direction of continuum radio astronomy, since I think I was not only your first but also your last student of that subject.

In thinking of how best to go about presenting these recollections, some old skeletons fell out of my mental closet. They don't seem nearly as threatening now as they did then, and I think it's about time we buried them. But since all this happened over 20 years ago and the subject was not in the main stream of your work, I'm sure you have forgotten a lot about it. In a moment I will try to refresh your memory by showing a few old slides from those times.

2. ASTRONOMY APLENTY

Although you had done some continuum radio astronomy at Michigan before you came to MIT, your interest in those sorts of observations seemed to be driven mostly by the practical need to calibrate the Haystack Antenna in the centimeter wavelength range. As Lit Meeks mentioned earlier today, the people who were to do the work were, of course, the graduate students. I remember it being described to me how continuum observations of discrete sources with known flux densities could help to calibrate the antenna. But the first problem that came up was that no one knew the absolute flux densities of any sources at such high frequencies. So an antenna of known gain had been obtained, an antenna in the shape of a cornucopia. This cornucopia horn was installed outside but near the Haystack radome (it was removed some time ago), and in the winter at the end of 1965 I started observing a few of the strongest sources in the radio sky in order to measure their absolute flux densities at 3.6 and 1.9 centimeters. These bright sources, Cygnus A, Cassiopeia A, and Taurus A, would later be mapped with the Haystack telescope. The absolute flux densities of these sources, as measured with the cornucopia horn, would thereby permit a radio-astronomical determination of the Haystack antenna gain.

Figure 1 shows the cornucopia horn that was my introduction to astronomy. In fact, I had come to MIT in order to study experimental plasma physics, and how I actually got into astronomy is an anecdote I'm going to keep for the end of this talk. I had a great deal to learn about astronomy, and I'm pleased to be able to publicly thank two of my good friends from those days, Alan Rogers and John Ball, for sharing their knowledge and experience with me. I didn't even know about right ascension and declination, and observing with the cornucopia horn was a wonderful opportunity to learn about astronomy in a very practical way. The horn was a transit

FIGURE 1. The cornucopia horn antenna connected to the radiometer instrument box. The mounting permitted motion of the antenna beam in the meridian plane for transit observations.

instrument; the beam could be steered only in elevation along the meridian, and the observation proceeded by simply letting the source drift through the response pattern. The time for carrying out the experiment could not be freely chosen; it was dictated by the heavens! It often meant getting up in the middle of the night, as astronomers were supposed to do, setting the telescope elevation, and waiting patiently for the source to come along. Besides an introduction to certain aspects of classical astronomy, this experience was also my introduction to the use of advanced electronics and computers, and in particular to the marvellous tunnel diode receivers that Sandy Weinreb had built.

Figure 2 shows a typical drift scan, starting with a noise diode calibration signal, followed by some baseline, and finally the source drifting through. This was Taurus A in mid-January 1966; the stable winter weather with its low atmospheric water vapor content was essential. The noise diode signal was in turn calibrated against thermal loads, and along with the known gain

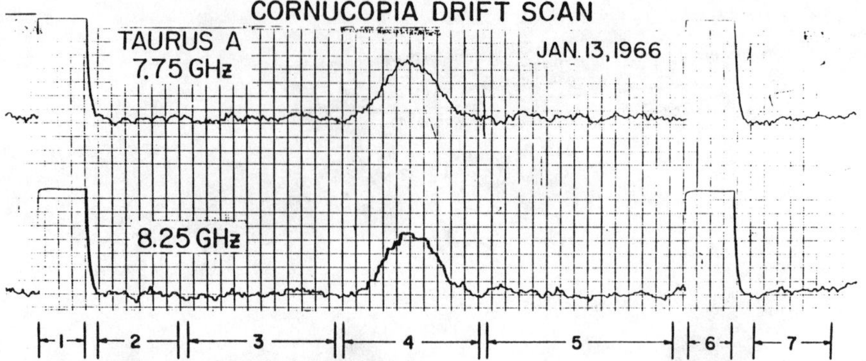

FIGURE 2. Drift scan of Taurus A with the cornucopia horn at 3.6 cm in mid-January 1966.

of the horn we could finally determine the flux densities on an absolute scale.

3. VARIOUS SECULAR VARIATIONS

So, along with Alan Rogers and John Ball, using Sandy Weinreb's magnificent electronics, and with benevolent supervision by Lit Meeks and Paul Sebring, we began the calibration of the telescope. During this process, one of the interesting things that turned up was the observation of secular variations in the radio flux density from several quasars and the radio galaxy 3C84. Now these secular variations had already been discovered by Bill Dent at Michigan in 1965, but it was clear that Haystack was even more ideally suited to doing this kind of work. Compared to Dent's equipment, the effective area of the Haystack antenna was larger, and the sensitivity and stability of Weinreb's tunnel diode receivers at 3.6 and 1.9 centimeters were considerably better. We had made a few preliminary measurements of these sources during a test period in the spring and summer of 1965, but a severe instability in the pointing of the antenna had prevented us from making regular observations. Some of the old-timers from Haystack may remember this problem; every so often the telescope would develop St. Vitus's dance, and the whole structure would jitter for no apparent reason. Someone was able to tone down the servo system enough to allow us to begin a program of periodic observations in about March 1966. When we compared the observations of 3C273 with the preliminary data taken nearly a year earlier, we discovered immediately that the radio flux density had increased substantially. The increase was greatest at our shortest observing wavelength of 1.9 cm. From suggestions made in the literature, we had expected that the amplitude of the time variations was going to be larger at the shorter wavelengths, and indeed it was. Figure 3 shows that the flux density of 3C273 was going up rapidly in the spring of 1966, and I recall amusing myself by explaining to my colleagues Alan Rogers and John Ball just how long we would have to wait before, at this rate of increase, this source would eventually fry the receiver. We may even have had a problem with living on this earth. However, the increase flattened out in mid-1966; fortunately for the world the danger passed, and we did not have to warn all humanity about the impending disaster of being cooked in a celestial microwave oven.

We did all our observing of these sources using a computer program that Patty Crowther had written to Lit Meeks' specifications. It was called the Discrete Source Scan (DSS) program, and at that time it was a major advance in computer control of the telescope and in on-line data analysis. Figure 4 shows the operation of this program. It scanned the telescope first in declination and then in right ascension across the radio source, followed by a noise diode calibration. The whole thing was analyzed automatically in the on-line computer; out of it we got the peak antenna temperatures and the angular widths of the source, in this case the telescope beam width. During all this time we were using Virgo A as a reference source. This is a relatively nearby galaxy in the Virgo cluster, and we did not expect it to vary. Well, lo and behold, Virgo A did remain at about 2.3 K antenna temperature at 1.9 cm during the early part of 1966, but then it began to

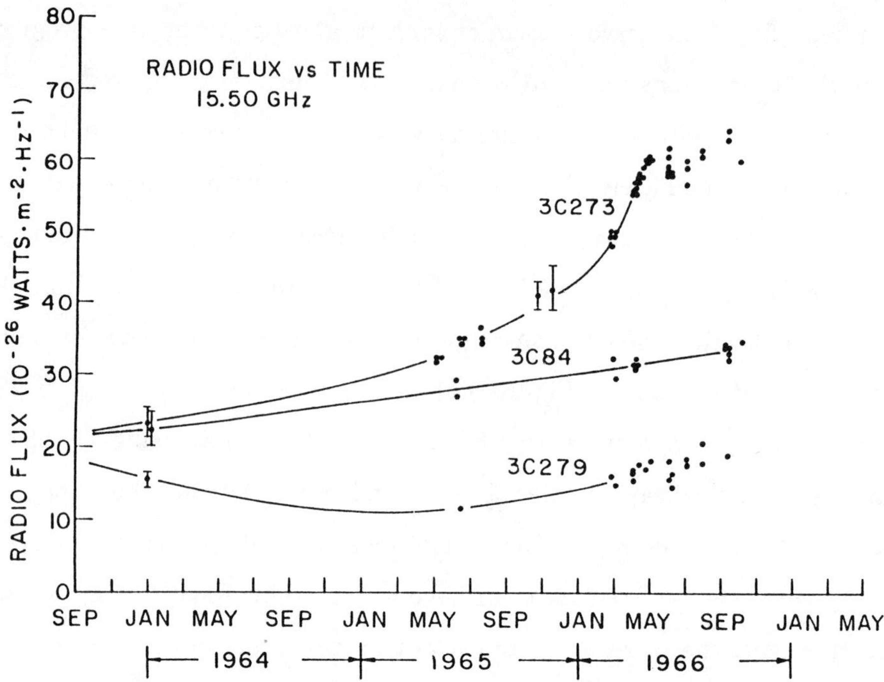

FIGURE 3. Secular variations in the radio flux density of several extragalactic sources at 1.9 cm as measured with the Haystack antenna in 1965 and 1966. The points with error bars were taken from other authors.

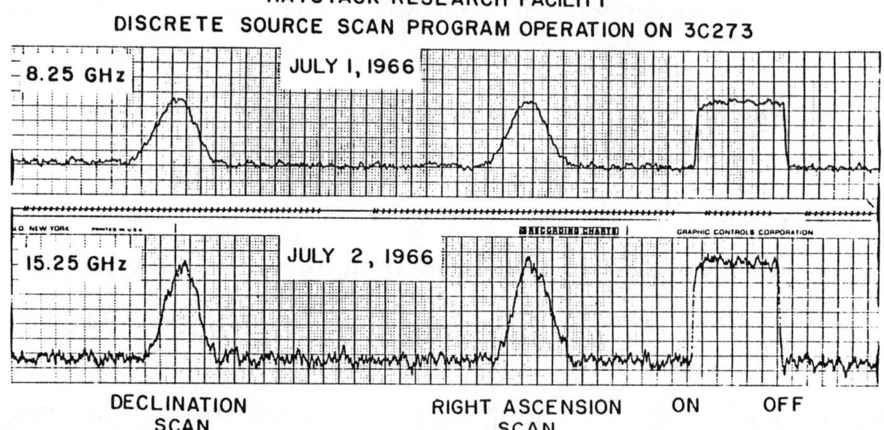

FIGURE 4. The operation of the Haystack Discrete Source Scan program on 3C273. Two separate scans are shown, the first (upper) at 3.6 cm on July 1, 1966 and the second (lower) at 1.9 cm on July 2, 1966.

increase slightly during June, reaching about 2.6 K in early July. Although small, this increase was well above the uncertainties in the measurement. On the other hand, we knew that the radio emission from another galaxy 3C84 was also varying, so there seemed to be no reason to get upset. We simply included Virgo A as one of the variable sources of our sample. So I wrote an abstract of a paper for the summer 1966 meeting of the American Astronomical Society, which was to be held in Ithaca from July 24 to 28. In that paper, we reported on the time variations of the quasi-stellar radio sources 3C273 and 3C279, and we also pointed out that Virgo A was showing a small but measurable variation. However, I think that Alan Barrett must have heard on the telephone from Bill Dent or perhaps from Ken Kellermann, that they were unable to confirm this change in the flux density of Virgo A. It seemed to be a Haystack phenomenon, and it was not observable anywhere else. The plot thickened when I noticed that the Haystack beamwidth, as recorded in the summer of 1966 by Patty Crowther's DSS program, actually increased over the winter values. So we had a very curious situation, where the peak temperatures on Virgo A had gone up and the beamwidth had become fatter too. There seemed to be no obvious way of understanding this. Fortunately, the people at Haystack got very worried about this problem and began to suspect that there was something wrong with the antenna. An investigation was started by two members of the engineering staff; I think it was Cal Frerichs and Dave Stuart who undertook a study of the response of the Haystack antenna to thermal shock. The radiometric data that we had been collecting on the continuum sources turned out to provide an interesting verification of their suggestion about what was really happening.

Figure 5 shows a familiar diagram of the backup structure of the Haystack telescope. There is a very thick ring, number 3 in the figure, that holds up much of the structure. It appeared from the mechanical studies that

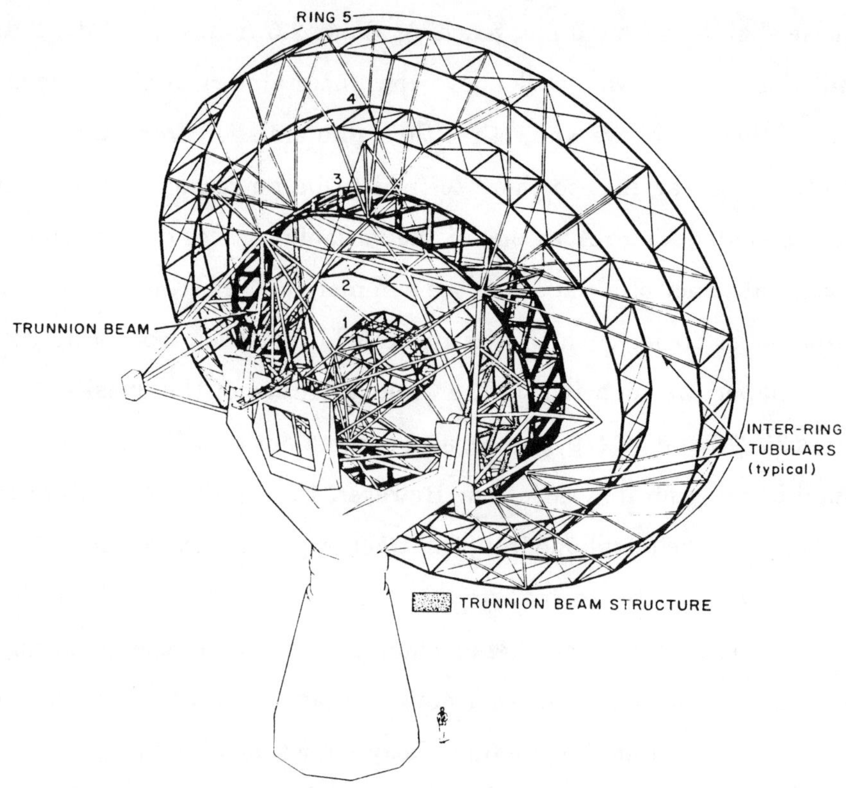

FIGURE 5. Sketch of the Haystack backup structure. Ring 3 is a relatively massive construction, called the splice plate.

when the temperature increased, the inner and outer parts of the surface did not both expand at the same rate owing to the long response time of this very massive ring. Figure 6 shows this schematically in cross section. The steady state of the antenna appeared to be a more-or-less annular aperture, with the center part of the surface not contributing fully. As the temperature increased, the inner parts apparently deformed in such a way as to make a better overall parabolic shape. Both the beamwidth and the peak signal strength increased, because as we go from an annular to a filled aperture the aperture efficiency will increase (you have more effective collecting area), and the beamwidth will also get larger (because annular apertures have narrower beamwidths than filled apertures).

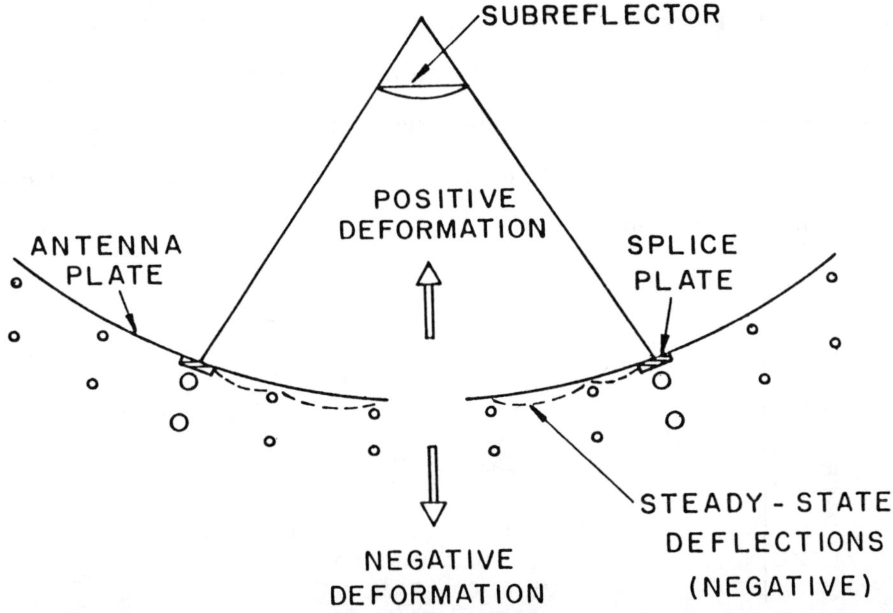

FIGURE 6. Sketch showing the proposed changes in the figure of the Haystack parabolic surface with temperature.

This all sounds very plausible, but why were we seeing this effect as a systematic increase on Virgo A? Well, the reason was because we were trying to be very careful with these observations. Alan Barrett had taught me to be very suspicious of a new instrument, and we were expecting all kinds of systematic effects especially as a function of antenna elevation. So we were trying to observe preferentially around transit; the telescope gain was greatest and varied the least at that time. However, another thing I learned about astronomy is that there is a difference between civil time and sidereal time. When we started observing Virgo A it was a nighttime object in the winter of 1966. But as the year progressed it became an early evening object, and in June and July the transit observations were occurring in the late afternoon. Now as the sun came up on that big Haystack dome in the midsummer, the temperature increased and eventually ran away from the temperature control equipment. Figure 7 shows what happened; the peak antenna temperature on Virgo A turned out to be a function of the local time of day! Figure 8 shows that the beamwidth also went up about the same time. These increases lagged behind the outside temperature variations by many hours, but the implications were clear; the variations of Virgo A were not intrinsic to the source.

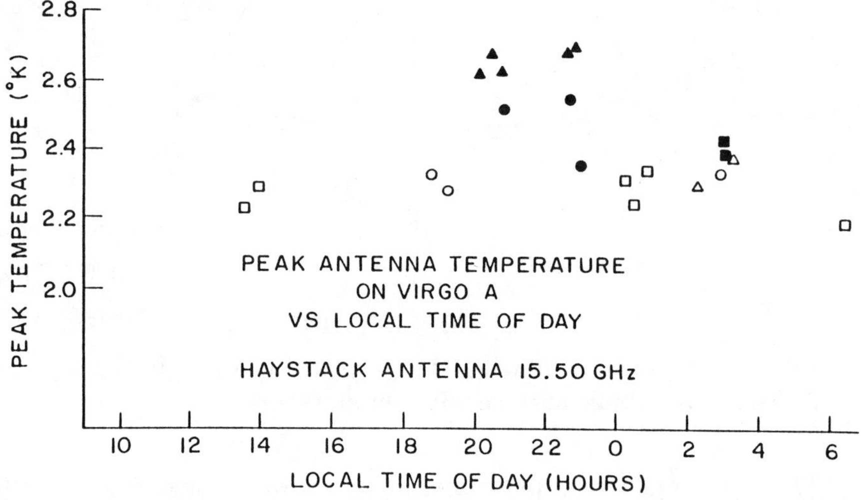

FIGURE 7. Measurements of the peak antenna temperature recorded by the DSS program on Virgo A as a function of the local time of day.

Now all this was a very nice demonstration of how radio astronomy techniques could contribute to the understanding of the thermal behavior of the antenna, but what about the Ithaca paper? We had already sent in the abstract, and it said that Virgo A was varying. What could we do? Time was very short, and there seemed to be only two possibilities; either I would get up at the Ithaca meeting and publicly acknowledge this blunder, or we could try to coerce the organizers into removing the abstract entirely from the BAAS proceedings and leave a blank space there. The latter option would also have been embarrassing, because I would eventually have to explain

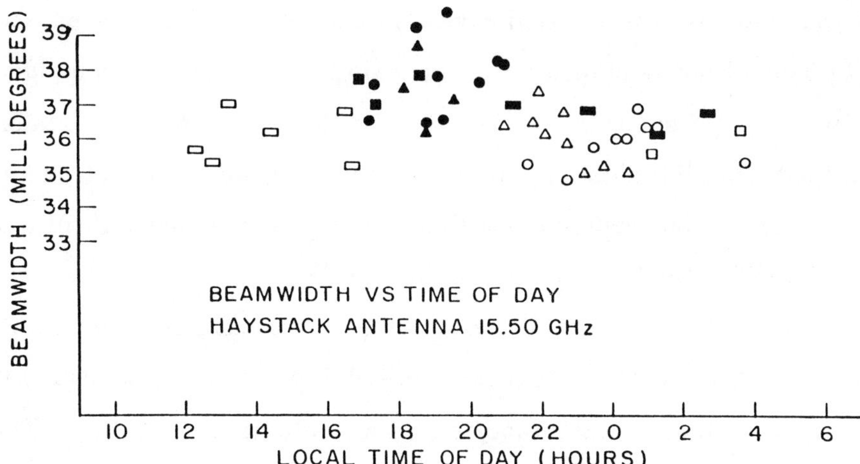

FIGURE 8. Changes in the Haystack beamwidth as measured by the DSS program as a function of the local time of day.

that blank space. Well, Alan Barrett saved my skin with a few well-placed phone calls from his office early in July only a few weeks before the meeting; we were allowed to change the abstract. Here is what actually came out in print: "A small increase in the antenna temperature of Virgo A during June and July 1966 has been traced to unusual distortions of the antenna from temperature gradients in the structure." *Magna est veritas et praevalet.*

4. SIDETRACKED

Well, Alan, the years have gone by and we have gone our separate scientific ways. My work has been much more on the kinematics and dynamics of galaxies and spiral structure, yours in the study of molecules and the physics of the interstellar medium. I have often wondered what it was about you that attracted me to do radio astronomy; after all, I had originally come to MIT in order to study plasma physics. I think it must have been your capacity to instill confidence that impressed me, and I needed that. Here I was, a small-town kid from the boondocks of Saskatchewan, appearing in a big eastern city and on the campus of this prestigious school. Did I ever tell you that up until my senior year in college I had never even heard of MIT? The initials reminded me of some kind of soap, and I had no idea that this was a very famous institution of higher learning. So when I realized all this, I felt quite intimidated. But when I talked to you about astronomy and about science, you made me feel that it wasn't really that difficult, that it was a lot of fun, and that even I could do it.

And now it's time to tell my last story, about how I got into radio astronomy, because it really was Alan's doing. When I came to MIT in September of 1962, I was supported on a graduate assistantship. The GA duties involved working 16 hours a week for one of the physics or electrical engineering professors on a research project. I remember going to the physics department offices in Building 6 and getting a list of faculty members with whom I was supposed to discuss possible projects that they might have for me. Now remember that I came to MIT to do experimental plasma physics, so it is not surprising that the first name on the list was Sanborn Brown. I walked from Building 6 to those barracks buildings, known collectively as Building 20, on the other side of Building 26 to look for Dr. Brown. His secretary informed me that he was away at a symposium. So I looked at the second name on the list. That was Dr. Allis, a theoretical plasma physicist. The secretary said that, unfortunately for me, Dr. Allis was away at the same conference as Dr. Brown. Well, it was rather late in the afternoon by that time, and I thought I would try at least to catch the third person on that list. I no longer remember the name of the third person on the list of four, but I do remember that this person's office was way back where I had just come from. In any case, I had to traverse Building 26 in the reverse direction. I noticed that the office number of the fourth person on the list was in that building, so I thought: let's be efficient, let's drop in and talk to this guy and get this out of the way so I can go on to the next one. Well, yes, Alan was in his office, and he began to tell me about a project in which he was involved to study the emission from the atmosphere of Venus using a small radio telescope on the Mariner 2 satellite. He was looking for a student to do some computer calculations on the sort of scan profile he might expect to measure under various assumptions about the atmosphere of Venus. Now this had nothing to do with plasma physics, as far as I could see. But as we sat there and talked, Alan's enthusiasm became infectious. I began to think that even I could do this kind of work, and it really sounded like fun. Besides, I thought, it was only a job for money, 16 hours a week, and after that I could do what I wanted. So I agreed to work for him, at least for that year. Well, by the end of the first semester of interacting with Alan Barrett I was excited about doing a thesis in radio astronomy, and I never went back to plasma physics.

Alan, it was very much your enthusiasm and your ability to instill confidence that brought me into radio astronomy. Over the years, I have told many people that if Alan Barrett had been an oceanographer, then I'd be doing oceanography right now!

Comments by Alan Barrett

This seems to be the time to tell stories about me, but maybe Ron won't mind if I tell a story about him. He says he needed confidence-building when he first came to MIT, and maybe I can give you a reason for that. He came from Canada, Saskatchewan, wasn't it, Ron? Now the schools in the northeast have intramural hockey, and other students will drag out the student phone book and run down the list looking for all the students from Canada and call them up to try to get them on their team, because they are vastly superior hockey players. And Ron had to admit that he didn't know how to skate.

GETTING STARTED

JAMES M. MORAN

Harvard–Smithsonian Center for Astrophysics, Cambridge, MA

I can remember very clearly the day I met Alan Barrett. It was in March 1963. I was a senior at a small college in Indiana (Notre Dame), majoring in electrical engineering, and I was trying to figure out where to go to graduate school. My father had gotten me interested in radio astronomy in high school, and I had been a summer student at NRAO in 1962. I was quite intent on becoming a radio astronomer, and I had applied to four places. I considered Ohio State and wrote to John Kraus. I remember taking the bus to Columbus, Ohio, where John Kraus showed me around his radio telescope. A few weeks later, he wrote and offered me a research assistantship, but he stipulated that I had only until Christmas to decide whether to accept his offer. Perhaps against my better judgment I turned his offer down because I wasn't going to hear from any place else so soon. I was also interested in the University of Michigan and wrote to Fred Haddock. He invited me to visit and look the place over, so I went up there. He began probing my technical background by asking how a klystron worked. I didn't know anything about klystrons, but he told me he would take me. I felt uneasy about the grilling I had gotten. I applied to Stanford and wrote to Ron Bracewell. He never responded. I got a fellowship there, and while I was dreaming about the California sunshine, I wrote to Bracewell again and still got no answer. I also applied to M.I.T. and wrote to Alan Barrett. If my files were as well organized as those of Charlie Townes, I could probably read you the letter that he wrote back to me, but I don't know where it is. Anyway, I remember that it was a very warm response, and he invited me to visit. So I came to M.I.T. in March of 1963, and I can remember him sitting in his fourth floor office in Building 26. He leaned back in his chair, fiddled with his pipe, and started talking about the Mariner 2 experiment, which we have heard described by Dave Staelin this morning. The Mariner 2 probe had just flown in December 1962, and although the experiment was sort of a technical disaster, they managed to obtain the decisive data that showed that the surface of Venus was hot. I thought this was tremendously exciting stuff, and so I decided that I wanted to come to M.I.T. and do radio astronomy with Alan Barrett. And I did.

I showed up in the fall of 1963 and enrolled in course 6.625, which was Alan Barrett's electrical engineering course in radio astronomy. This was really a wonderful course; two decades of students have learned the basic principles of radio astronomy from it. He handed out a set of mimeographed notes that was about three inches thick. I still refer, to this day, to the 1961 copyrighted version. The notes were very clear and informative. There was a great section on radiative transfer and radiation processes, with a really excellent in-depth discussion of synchrotron emission and thermal bremsstrahlung. Of course, the spectral line material was quite comprehensive, with tables of frequencies of fine structural lines of hydrogen and other atoms; many of these lines still haven't been detected in outer space. Alan worked on those class notes a lot and completed a major revision in 1970, but they were never published. I don't know why. The notes contained a comprehensive summary of most of the observations, and perhaps the data were pouring in so fast that he thought he couldn't keep up with the rapid advance of the field. Alan, of course, wrote a successful book on electricity and magnetism a few years later. As he mentioned last night, he thought

M.I.T. students were like blotters and I surely fit that description. I soaked up everything he had to say during this course. Alan only missed two classes during the semester. They were both in October 1963 when he was out at the Millstone antenna finding the OH line. Also that fall we had a class trip to Haystack and we saw the partially completed telescope inside the radome. Little did I know how much of my life would be spent there.

Barrett started me off working on a project to measure the brightness temperature of the Moon. If you measure the lunar brightness temperature as a function of frequency and lunar phase, you can find out quite a bit about the thermal and electrical properties of the lunar surface. We made measurements with a 28-ft antenna on the roof of Lincoln Laboratory in Lexington with the 22-GHz receiver that Dave Staelin built to look for water vapor in the atmosphere of Venus. The data from this experiment went into my master's degree thesis, a requirement in the EE department. I remember fondly the trip to Ithaca, New York, to present the results of this work at the AAS meeting at Cornell in the summer of 1966. Barrett rented a car and he, Alan Rogers, Ron Allen, and I drove out there for the first AAS meeting I had ever been to. I anticipated that there would be no interest in my lunar measurements. I was wrong; preparations were being made for the manned lunar landings. Tom Gold even asked me a question about my results.

With that work done, I started what I thought was going to be a Ph.D. project looking for interstellar CH. Perhaps Alan thought he was going to repeat the OH story with CH, which has a spectrum similar to that of OH. It has a lambda doublet spectrum, and if you looked in Shklovsky's book[1] there was a frequency listed for the ground state transition of 3180 MHz, but it was very uncertain. Alan convinced a colleague of his at M.I.T., Perry Miles, to measure the frequency more accurately in his laboratory. Miles observed the transitions between an excited rotational level and the two ground state lambda doublet levels with an infrared Fabry-Perot interferometer. His result was 3380 MHz, plus or minus 40 MHz. Alan thought that was good enough for an astronomical search. He got a receiver together at NRAO; it had a traveling wave tube amplifier, and I think the system temperature was about 3000 K. We were scheduled for three days of telescope time on the 140-ft telescope, and a few days before the experiment, I remember having dinner with Alan and Perry Miles. We had a few drinks and Alan was thinking about our assigned telescope time, and he said, "What are we going to do with these three days? We might try to knock off not only CH, but SiH in the same experiment." Well, we went down to Green Bank for our three days in April 1966, and we scanned mightily around the spectrum (covering more than 100 MHz), but we never saw any molecular lines. Alan Rogers and Sandy Weinreb, who by that time was on the NRAO staff, were there, and they nursed along the flaky 100-channel autocorrelation spectrometer. We knew that the receiver worked properly because we detected the H128α recombination line in M17. Studies of recombination lines were just beginning then. About a decade later, Turner and Zuckerman[2] and Rydbeck, Ellder, and Irvine[3] finally detected the CH line. It was more than an order of magnitude below our detection limit, although we had scanned over the right frequency of 3349 MHz for the $^2\pi_{3/2}$, $J = 3/2$ $(F = 1 \rightarrow 0)$ transition. We started building a new CH receiver to use on the Millstone antenna of M.I.T. Lincoln Laboratory, but it looked as if no big block of time was going to be available there, so I went on to something else. Parts of that planned CH receiver made their way into the first medical electronics radiometer that Al used for breast cancer research.

One experiment I want to describe in some detail was the one with the Millstone–Agassiz interferometer, which was the serious beginning of my Ph.D. thesis. A lot of work was being done on OH masers. The position and limits on the sizes of OH masers had been obtained in the summer of 1966 with the Haystack–Millstone interferometer[4] and the Cal Tech interferometer[5]. The Haystack–Millstone interferometer sprang into existence, so to speak, as a result of the electronic genius of Alan Rogers, another one of Barrett's students. These instruments had baselines of about 1 kilometer and fringe spacings of about an arcminute. The angular sizes and any separation among spectral features were found to be less than 20″ and 3″, respectively. The prevailing view was that masers would never be resolved and succeeding experiments would just enable smaller limits to be placed on their angular sizes. The joke was that the papers could be written before the experiments were done and the angular size limits filled in afterwards. In retrospect, this view doesn't seem reasonable because we should have realized that the velocity spread among the maser features corresponded to a dynamical structure that could be spatially resolved. In any event, we decided to forge ahead with a longer baseline interferometer.

The idea came to link the Millstone Hill antenna with the Agassiz antenna of Harvard College Observatory. These instruments are shown in Figure 1. This experiment brought together the groups from Harvard and M.I.T., which had been competing in research on OH masers for some time. Barrett arranged the experiment with Ed Lilley, and Ed got permission to use the firetower on the Agassiz station property to hold the microwave relay link. There was a direct line of sight from this firetower to the Millstone antenna some 13 kilometers away. Alan Rogers and Hays Penfield built most of the required electronics. The interferometer was not terribly phase stable because no round-trip electrical path corrections were made. A local oscillator reference was sent from Millstone to Agassiz, and the IF signal from Agassiz was returned to Millstone over the link. The IF signals from both stations were sent to the Haystack control room over cables for

FIGURE 1. (Left) The 60-ft diameter antenna of the G. R. Agassiz station of the Harvard College Observatory in Harvard, Mass., as it appeared in 1966. (Right) The 84-ft diameter antenna of the MIT Lincoln Laboratory in Westford, Mass.

correlation with the 100-channel one-bit digital correlator. With a 13-km baseline (74,000 wavelengths), the fringe spacing was 3″. In November 1966, we turned the interferometer to look at W3(OH), our favorite source, and found that the fringe amplitudes were as expected for an unresolved source. This result set the brightness temperature up to at least 10^9 K. However, we found that the relative phases associated with maser features at different frequencies changed with hour angle. At first, we thought that the effect might be due to the polarization of the masers since both telescopes had linearly polarized feeds, the Millstone feed fixed with respect to the horizon and the Agassiz feed fixed on the sky. We quickly realized that the diurnal sinusoidal variation in phase could only be due to a position offset in the masers. An example of the relative phase variation is shown in Figure 2. I can still remember what a tremendous thrill it was to discover something

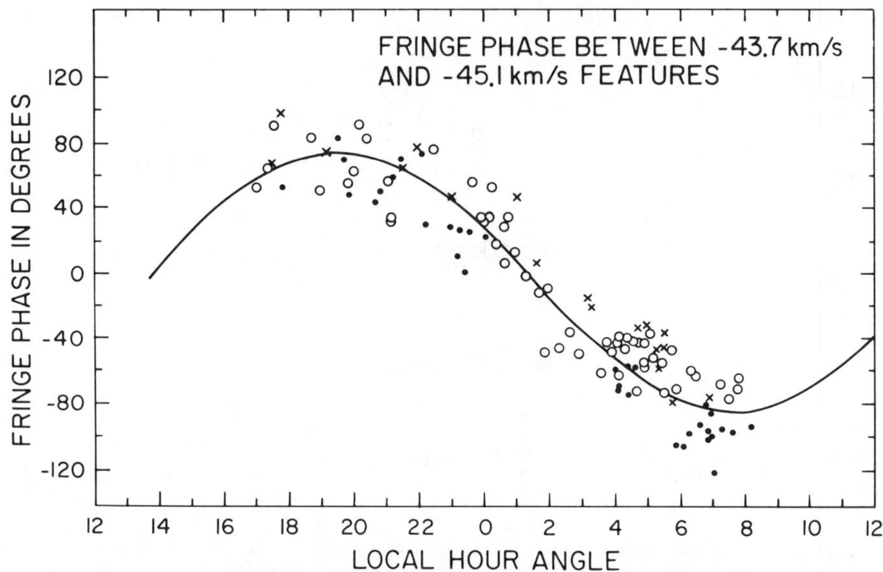

FIGURE 2. The relative fringe phase between the –43.7 km s^{-1} and –45.1 km s^{-1} features in W3(OH) versus local hour angle as observed with the Agassiz–Millstone interferometer. Each point represents a 10-minute integration. The various symbols denote measurements with different feed polarizations at Millstone. (Reproduced from ref. 6.)

new and unanticipated.

The first map of an OH maser is shown in Figure 3. This map was reported in the paper by Moran, Barrett, Rogers, Burke, Zuckerman, Penfield, and Meeks[6], which was the only paper published based on results from this interferometer. We measured the positions of three distinct maser features at –43.7, –45.1, and –46.5 km s^{-1}. The feature at –45.5 km s^{-1} was tentative and subsequently shown not to be at the position shown. One curious problem with this map is that the error bars are much too large, as is obvious from a casual inspection of the data in Figure 2. I recently reanalyzed the data and found that the 1-σ errors in position are about 0.″03. Hence, the quoted errors are in fact the 10-σ error values. I do not recall the reason for our conservativism. I remember reviewing the Haystack–Millstone data

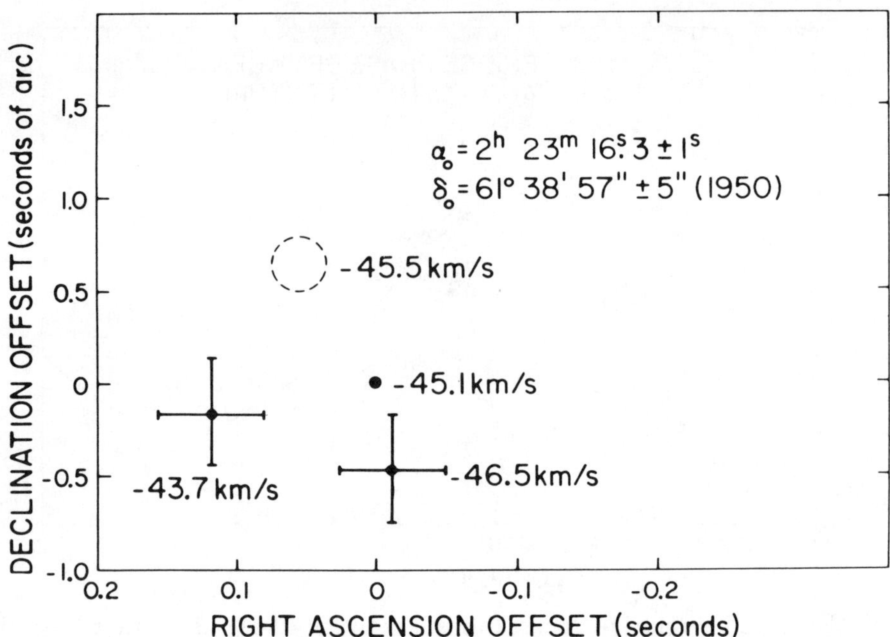

FIGURE 3. Map of the relative positions of four features in W3(OH), the first such map of an OH maser. (Reproduced from ref. 6.)

in 1967 and seeing an obvious but small systematic diurnal fluctuation of amplitude $\sim 3°$ in the relative phase data, indicative of the position offsets between features.

Since these measurements were made, many maps, of ever improving accuracy and detail, have been made of the W3(OH) maser. A recent one, constructed by Tony Garcia-Barreto[7] for his Ph.D. thesis, shows 81 distinct maser spots with their polarization parameters superimposed on a map of the background HII region. The three spots in Figure 3 can be identified among those in the top of Figure 4. The maser has not changed much over 15 years, and we did a good job of making the early measurements.

We knew that the phase stability of the Agassiz–Millstone interferometer was poor. We were in too much of a hurry to engineer it properly. However, we did measure the stability of the interferometer by setting up

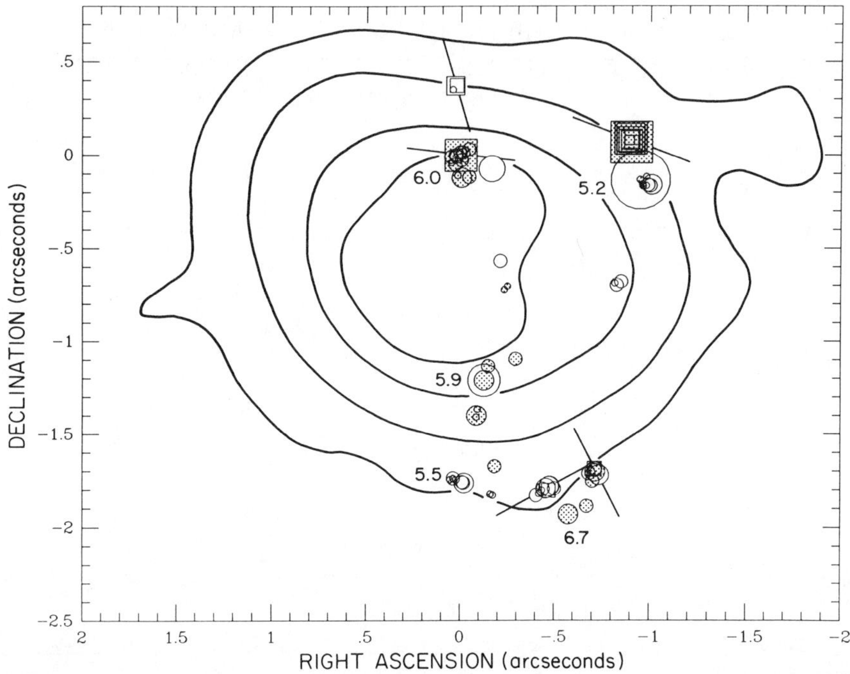

FIGURE 4. A map of the distribution of 81 OH maser features in W3(OH) superimposed on the contour map of 23-GHz continuum emission. Circles show circularly polarized features (open are LCP, stippled are RCP). Squares show elliptically polarized features. The areas of the symbols are proportional to their flux densities. The numbers are the magnetic field strengths in milliGauss derived from Zeeman pairs. The maser spots in Figure 3 are in the groups marked by 6.0 and 5.2 mG. (Reproduced from ref. 7.)

a transmitter on the firetower, which transmitted a sinewave to both telescopes. Figure 5 shows the output of the correlator as a function of time. We introduced a local oscillator offset of 0.02 Hz at one telescope such that a stable interferometer would have exhibited a sinusoidal correlator output. We then removed the local oscillator offset with this frequency (see the upper left part of the record in Figure 5). The observed drift is due to phase changes in the atmosphere, receivers, and microwave link. The coherence time was about 5 minutes. Our experience with this interferometer showed us that good phase stability was not essential for getting good results on

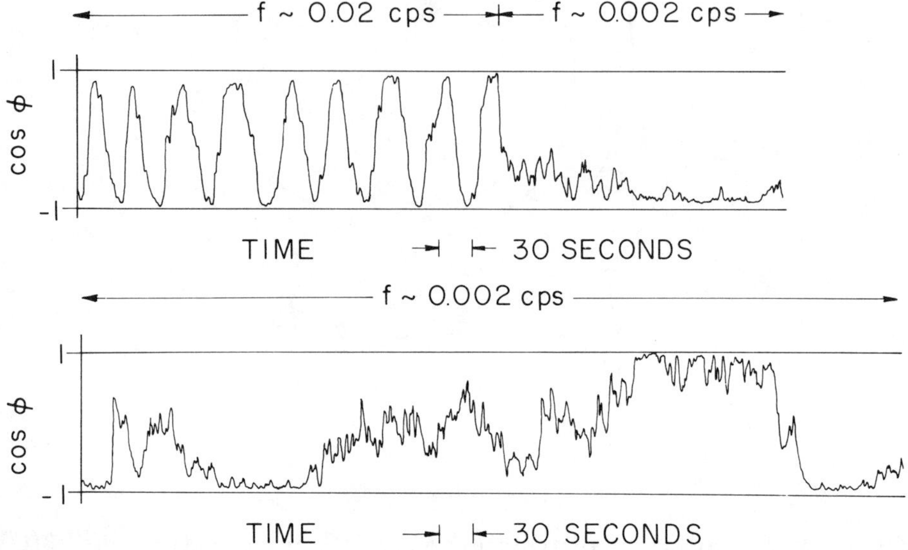

FIGURE 5. The response of the correlator of the Agassiz–Millstone interferometer to an RF sine wave test signal. In the top left section, the local oscillator of one receiver was offset by 0.02 Hz. During the rest of the record, this offset was removed (the 0.002-Hz variation is due to slow instrumental phase change caused by the atmosphere, receivers, and microwave link). (Reproduced from ref. 8.)

certain types of experiments. This insight, also appreciated by others at the time, was a key idea in the development of very long baseline interferometry (VLBI). However, no VLBI system at this wavelength had as bad a phase coherence as that of the Agassiz–Millstone interferometer!

With the exciting results of the Agassiz–Millstone interferometer in hand, we immediately began looking for a means to obtain even higher resolution. Bernie Burke learned of the NRAO–Cornell efforts to use digital tape recorders to record the IF voltages at widely spaced telescopes and correlate the data later – the technique we now know as very long baseline interferometry (VLBI). We began working with that group and built a VLBI terminal out of the available equipment and computers at Haystack. I shall not recount that development in detail; however, I would like to recall a scene that I fondly remember.

The first VLBI experiment at 18 cm was in June 1967 between Haystack and the NRAO 140-ft telescope. It was three days long, and, while the experiment was still underway, Burke arrived at Haystack with the first tapes from NRAO. I began running the cross correlation program on the CDC 3200 computer. After several false starts, the printer spewed forth the results shown in Figure 6. John Ball turned to Alan Barrett and said, "I guess Jim has his thesis wrapped up now," to which Alan replied," Well, we'll have to see what the science is."

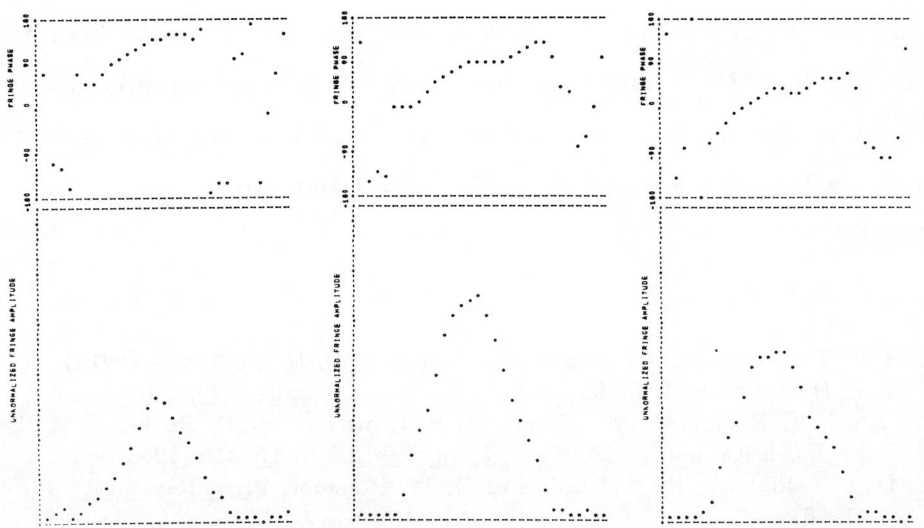

FIGURE 6. Cross power spectrum (phase, top; amplitude, bottom) for an observation of W3(OH) with the Haystack–Green Bank interferometer. The three spectra differ in the instrumental fringe frequency offset that has been removed. Each spectrum covers 5 kHz and shows only the -45.1 km s^{-1} feature. The narrow bandwidth of the maser made it relatively easy to find the proper temporal alignment of the recorded signals. The 5-kHz filters were never again used for a VLBI experiment (Reproduced from ref. 9.) Copyright 1967 by the American Association for the Advancement of Science

I would like to return now to talking about Alan Barrett as a mentor to his students. In my particular case, when I came to M.I.T., I was quite wet behind the ears; I needed a lot of guidance and support, and I got it from Alan Barrett. I am particularly grateful. I think that without his help I may not have made it through M.I.T. Another of Alan's students who felt the same way has made a rather eloquent statement about his guidance; this acknowledgment was written by Stephen Poole who did his B.S. thesis in 1971 on the medical electronics project. Poole wrote: "It is easy for me to express my deepest appreciation to Professor Alan H. Barrett for entrusting me with this, his concept, and then allowing me the freedom to make it my own dream. His enthusiasm and encouragement has kept me going, while his experience has kept me from going too far in the wrong direction. In a sense, though, I cannot thank Professor Barrett enough, for I have yet to learn how much gratitude I truly owe him. His interest and assistance has gone far beyond the bounds of this thesis. He has had more influence on my future than any other man I have met at M.I.T. I have a feeling that for years I will be richer for having made his acquaintance."

REFERENCES

1. I. S. Shklovsky, Cosmic Radio Waves (Harvard Univ. Press, Cambridge, 1960) p. 266.
2. B. E. Turner and B. Zuckerman, Astrophys. J. (Lett.), 187, L59 (1974).
3. O. E. H. Rydbeck, J. Ellder, and W. M. Irvine, Nature, 246, 466 (1973).
4. A. E. E. Rogers, J. M. Moran, P. P. Crowther, B. F. Burke, J. A. Ball, M. L. Meeks, and G. M. Hyde, Phys. Rev. Lett., 17, 450 (1966).
5. D. D. Cudaback, R. B. Read, and G. W. Rougoor, Phys. Rev. Lett., 17, 452 (1966).
6. J. M. Moran, A. H. Barrett, A. E. E. Rogers, B. F. Burke, B. Zuckerman, H. Penfield, and M. L. Meeks, Astrophys. J. (Lett.), 148, L69 (1967).
7. J.A. Garcia-Barreto, B. F. Burke, M. J. Reid, J. M. Moran, A.D. Haschick, and R. T. Schilizzi, Astrophys. J., in press (March 15, 1988).
8. J. M. Moran, Ph.D. thesis, MIT (1968).
9. J. M. Moran, P. P. Crowther, B. F. Burke, A. H. Barrett, A. E. E. Rogers, J. A. Ball, J. C. Carter, and C. Bare, Science, 157, 676 (1967).

ALAN BARRETT AND FUTURE DIRECTIONS IN ASTRONOMY

BERNARD F. BURKE
Massachusetts Institute of Technology, Cambridge, MA

We've been reminded of the great variety of things that Al is involved in; in my case I can start with the time 22 years ago when he played a most instrumental role in persuading me to come here to MIT. In fact, not only was he an important player in that event, but he also was extraordinarily helpful in easing my way into the new environment. The way sailor Joshua Slocum put it, "There's a big difference between going into the captain's cabin by climbing over the bows and by crawling in through the cabin window." In a sense I had crawled in through the cabin window since I came in as a full professor. The title sounds fine, but some of your colleagues still want to see how you earned your credits, and Al, who had come over the bows by climbing up the faculty ladder, was an enormous help when I was learning how to interact with the community and how to get something done. The instructions on how to go through that sort of tribal initiation are not written down in any book, and a knowledgable mentor, such as Al was, makes the process far smoother. Now, it would be easy to give a further contemplative review of that long history, but I thought that I would instead look at his collection of seminal papers, not to look at the past, but as a way of projecting into the future to see how Alan's work will continue to have an impact.

There are two clear themes that stand out from his collection of papers– the desire to use microwaves to study the atmospheres of the earth and other planets, and the use of spectroscopy to study the interstellar medium. We've heard how dramatic his work on interpreting the radio observations of the planet Venus turned out to be, and I can remember very clearly the reception of his paper by the general run of conventional planetary astronomers, who had a hard time believing it because it relied upon a new kind of data that some weren't ready to accept. Those are just details that are past; the principal lesson, I think, to be drawn from that work is the unexpected character of the discoveries made when you turn new equipment into new subject areas. The interstellar medium, for example, was thought to be an area of study that would be, in its spectroscopic aspect, fairly dull for the radio astronomer. Instead of having only the 21 cm line to offer, it's turned out to be instead richly varied, full of puzzles, capable of yielding deep insights, and it might even have something to say about the origins of life. We've come a long way from that original discovery, by Al and his Lincoln Lab associates, of the OH absorption lines in the spectrum of Cassiopeia A. The surprises, though, that came on the heels of that discovery, went much deeper than the discovery itself. It may in fact trace its history back to Columbia when the first maser was in Charlie Townes' laboratory, in addition to all those microwave spectrographs. Those people who had that experience were prepared with a readiness to look for situations that might perhaps be more complicated than being characterized by a density and a temperature. It has turned out that the interstellar medium is just as complex as that device in Charlie Townes' lab. The maser action has turned out to be a key tracer of where the action is (except where there are anti-masers—can you say that's where the action isn't?). The trouble at the moment is that we don't understand yet how the masers work, and it's always hard to do really deep astrophysics until you get that knowledge. Over the next twenty years, I think we'll see our physical understanding grow, and that tool that Al was so important in pioneering will become an even more incisive tool in finding out what's going on in those regions where stars are being formed. Maybe the radio measurements will in fact turn out to be the key tool, because they can penetrate the dust that shrouds the process of stellar and planetary formation.

What can we expect over the next twenty years? Here I'm going to turn a bit sober for a moment, because if you look at what is needed to carry out the programs that Al has been so instrumental in shaping, you can see that there are a variety of major capital projects that need to get started. We need large millimeter wave dishes, we need interferometric arrays, in the infrared we need telescopes that would have to be either in space or in airplanes. This means that my crystal ball is very clouded if I try to look into that part of the future. I think if progress is to come it will be contingent on a new consensus in society that I still don't detect. There has to be a joining of forces between two communities, radio and infrared, that up 'til now have been able to enjoy the role of swashbuckling adventurers. There are times, though, when you have to make marriages of convenience and I think this may be one of them. The boundary between infrared and radio astronomy is somewhat blurred, although I think one can give a formal definition: radio astronomy is where you make your detections coherently—after coherent amplification and/or mixing of the signal—whereas in the infrared you use bolometers to detect the radiation immediately. The boundary between the two has been shifting to ever shorter wavelengths and I don't think the disciplinary distinctions are going to be very important in the future. The two disciplines must work as one to help guide the nation's space program out of its present chaos, and we must encourage the National Science Foundation to step up bravely to its chartered task of encouraging forefront science.

What is farther down the line? Well, I think the next step will be—

after we understand over the next twenty years how planetary systems are formed—actually to study the planets that belong to other stars. The detection probably will be at infrared or optical wavelengths, but the methods that will be used will be the methods of radio astronomy and that, I think, is another part of Al's legacy that will appear as a theme a number of years downstream. However, these discoveries are in store, because the instruments of the future are going to be able to analyze the atmospheres of those planets—and we've found that on our sample of nine planets each planet is an individual. Each individual planet generally gives a surprising answer—preconceptions are not much help and I don't think that the nine-fold sample we have in our solar system has exhausted the possibilities. *"There are more things, in Heaven and on Earth, Horatio, than are dreamed of in your philosophy."* Now, most of those planets to be discovered in the future will show only inorganic chemistry, but there will be more than one example of an atmosphere so far from chemical equilibrium that there'll be very strong evidence for the conclusion that on such a planet there is the presence of some kind of life. Perhaps you can put it in terms of the Gaia hypothesis that Lynn Margulis has popularized: that life on a planet can be viewed as one complex organism with the atmosphere serving as the circulatory system. Thus the study of planetary atmospheres gives an insight into the nature of the living system on those planets. (Of course, I think we should say on those planets fortunate enough to bear life—we don't have to say unfortunate enough to bear life, because the planet doesn't care.) In this phase of my crystal gazing, I would guess that the time scale stretches beyond our lifetimes, but perhaps some of the younger people here will see this come to pass. There is a new age of exciting adventure waiting, and I think it'll be just as exciting as those marvelous times in the 50's and 60's when the radio astronomy world was fresh and new, like a blooming garden for those of us who were there to enjoy it.

Now, once there is life on a planet, there is a further progression that's possible: the progression from life to intelligent life, from intelligent life to life like us, or at least life that is able to think and act like us, and then develop into a technological civilization. One of the papers that Al Barrett wrote shows his spirit, his systematic, careful approach that has been emphasized in all our discussions here. Nevertheless, there's a daring side to his discussion —maybe you'd say it shows a cautiously daring approach that we can all admire very much. The following comes from the *Science* review paper on OH masers[1] in the section that handled interstellar communications. Al wrote:

> " Perhaps it has occured to some readers that the radiation....may be a precursor of interstellar communications."

Farther on, you'll find him examining the reasons why we should be a bit suspect of the OH masers:

> " It appears that the OH frequencies might be prime candidates for interstellar communications when one considers the following questions: 1) If one civilization wanted to attract the attention of another, which it suspected might be actively engaged in radio astronomy or even actually listening for signals, what better way to attract attention, than to violently upset the expected intensity ratios of the four OH lines? 2) If OH in the interstellar medium is capable of amplification, because of some form of maser inversion, even in small interstellar clouds, then wouldn't the OH frequencies be likely ones to be used for transmission of information over interstellar distances? 3) Are there, because of the coupling between OH energy levels, ways in which radio power pumped into an OH cloud at one transition frequency could render the cloud an amplifier, or a very large transmitting antenna, at another OH line frequency?"

Then of course, Al's cautious side shows up. He's been pretty daring. Then he says, wait a minute—

> " There's no evidence that the OH radiation is really interstellar signaling, and I am making no such proposal."

Well, not really. However, he is led to conclude that the observations are suggestive, even if he's carefully hedging his bets. Then, after another cautious remark:

> " The speculative character of the foregoing remark should be clearly borne in mind,"

he proceeds with

> " Interest in interstellar communication and its detection has grown markedly in recent years. Much of the effort seems to have been devoted to finding how to conduct a search at 1420 MHz. ...The possibility of an accidental discovery, which in my opinion is more likely, is rarely considered."

I think this is good advice for all young radio astronomers, to remind themselves that if they see something that they don't understand, then they certainly should leave their mind open, as Al suggests.

In our later history, NASA started an actual Search for Extraterrestrial Intelligence (SETI) program and I was inveigled to join in, I think as a responsible skeptic. Al had a chance to do so, also, and he voted with his feet—he stayed on the sidelines as a silent observer willing to look at real data with an open mind, but choosing to use his time in other ways, aimed at measuring real things. In his *Science* article he certainly articulates this point of view very well, where he advocates that every observer must have an open mind when contemplating the data and especially when puzzling over the anomalies. This attitude makes the difference between missed opportunity and serendipitious discovery. I'll avoid the temptation to make an easy prediction—that we will make interstellar contact in some time scale: 20, 50, 100, 500 years, all equally plausible, all equally foolish. The discovery will be contingent, it will not be made if one never tries, but if it is to be a serendipitous discovery, this too will never occur if the observers equip themselves with intellectual blinders. Al is certainly one who has never had intellectual blinders on as he does his work. This is where his research style really pays off. Hard work, a fine grasp of the physics, an open mind to the unexpected, a sharp eye for anomalies, a critical view that demands strong proof before acceptance, a willingness to discuss what he's working on with his colleagues, and the ability to work in a creative way with students have been Al's attributes, and he has shown that he has always enjoyed that marvelous combination of being led by the students, at the same time that he, himself, provided such fine leadership to them.

REFERENCES

1. A.H. Barrett, Science, 157, 676 (1967).

PROGRAM

Second Haystack Observatory Meeting

INTERSTELLAR MATTER

A Symposium on Interstellar Molecules and Grains
in Honor of Alan H. Barrett

Sponsors: Haystack Observatory, National Science Foundation, and Department of Physics, Massachusetts Institute of Technology

10–12 June 1987, M.I.T., Cambridge, Massachusetts
Edgerton, Germeshausen and Grier Education Center, 50 Vassar Street

Schedule of Events

All at EG & G Center unless stated otherwise

Reception 9 June, 19:30-22:00, Room 36-428.

Registration 9 June, 18:00-22:00, and 10 June, 09:00-12:00

Wednesday, 10 June

09:00	**Welcome**	J. E. Salah K. A. Smith
	Molecules as Probes of Cloud Chemistry	
09:15	Diffuse Clouds	A. Dalgarno
09:55	Dense Clouds	W. M. Irvine
10:35	Coffee Break	
10:55	Outflows and Shocks	W. J. Welch
11:35	**Poster Session**	
12:30	**Lunch**	
	Molecules as Probes of Cloud Physics	
14:00	Heating and Cooling	D. J. Hollenbach
14:40	Structure and Motions	B. G. Elmegreen
15:20	Cloud Evolution	F. H. Shu
16:00	**Cookout** at Haystack Observatory	

Busses will leave the meeting at 16:00, arriving at Haystack at about 17:00.
They will leave Haystack for Cambridge at about 21:00.
Your registration badge will serve as your bus ticket.

Thursday, 11 June

Grains and Large Molecules

08:30	Molecular Chains and Rings	G. F. Winnewisser
09:10	Grain Size, Shape, and Composition	J. S. Mathis
09:50	The "Infrared Cirrus"	M. G. Hauser
10:30	Coffee Break	
10:50	Magnetic Alignment of Grains	R. H. Hildebrand

11:30	**Poster Session**
12:30	**Lunch**

New Techniques for Studying Interstellar Matter

14:00	Millimeter and Submillimeter Telescope Arrays	D. Downes
14:40	Submillimeter and Far-Infrared Detectors	T. G. Phillips
15:20	Coffee Break	
15:40	Large Millimeter and Submillimeter Telescopes	R. E. Hills
16:00	Poster Review	P. T. P. Ho
		M. J. Reid

Banquet at Royal Sonesta Hotel, 5 Cambridge Parkway, Grand Ballroom B

18:30	Reception – cash bar	
19:30	Dinner	
	After Dinner Talk	C. H. Townes

Friday, 12 June

Alan H. Barrett: Scientist, Colleague, and Teacher

08:45	**Opening Remarks**	J. Allen J. I. Friedman
	Scientific Contributions	
09:00	Medical Radiometry	N. L. Sadowsky
09:30	Planetary Atmospheres	D. H. Staelin
10:00	Interstellar Molecules	P. Thaddeus
10:30	Circumstellar Molecules	L. E. Snyder
11:00	**Poster Session**	
12:00	**Lunch**	
	Career Recollections	
13:30	Naval Research Laboratory	C. H. Mayer
	University of Michigan	W. E. Howard III
	M.I.T.	J. B. Wiesner H. J. Zimmermann M. L. Meeks R. J. Allen B. F. Burke J. M. Moran
	Closing Ceremony	P. C. Myers
16:00	**Reception** at Center for Space Research, Marlar Lounge (Room 252)	

Saturday, 13 June

Outing hosted by the Barretts, Lake Winnipesaukee, New Hampshire (transportation provided)

Speakers will have 35 mm slide projector and overhead transparency projector available. Poster presenters will have an area about 3 feet wide and 4 feet high available for the duration of the meeting. Posters should be in place by 09:00 Wednesday June 10, and should be removed by 17:00 Friday June 12.

Author Index
(indexed by first page of article)

Source Index

(indexed by first page of article)

Subject Index
(indexed by first page of article)